Advances in Intelligent Systems and Computing

Volume 971

The series "Advances in Intelligent Systems and Computing" contains publications on theory, applications, and design methods of Intelligent Systems and Intelligent Computing. Virtually all disciplines such as engineering, natural sciences, computer and information science, ICT, economics, business, e-commerce, environment, healthcare, life science are covered. The list of topics spans all the areas of modern intelligent systems and computing such as: computational intelligence, soft computing including neural networks, fuzzy systems, evolutionary computing and the fusion of these paradigms, social intelligence, ambient intelligence, computational neuroscience, artificial life, virtual worlds and society, cognitive science and systems, Perception and Vision, DNA and immune based systems, self-organizing and adaptive systems, e-Learning and teaching, human-centered and human-centric computing, recommender systems, intelligent control, robotics and mechatronics including human-machine teaming, knowledge-based paradigms, learning paradigms, machine ethics, intelligent data analysis, knowledge management, intelligent agents, intelligent decision making and support, intelligent network security, trust management, interactive entertainment, Web intelligence and multimedia.

The publications within "Advances in Intelligent Systems and Computing" are primarily proceedings of important conferences, symposia and congresses. They cover significant recent developments in the field, both of a foundational and applicable character. An important characteristic feature of the series is the short publication time and world-wide distribution. This permits a rapid and broad dissemination of research results.

** Indexing: The books of this series are submitted to ISI Proceedings, EI-Compendex, DBLP, SCOPUS, Google Scholar and Springerlink **

More information about this series at http://www.springer.com/series/11156

Waldemar Karwowski ·
Stefan Trzcielinski · Beata Mrugalska
Editors

Advances in Manufacturing, Production Management and Process Control

Proceedings of the AHFE 2019 International Conference on Human Aspects of Advanced Manufacturing, and the AHFE International Conference on Advanced Production Management and Process Control, July 24–28, 2019, Washington D.C., USA

 Springer

Editors
Waldemar Karwowski
Department of Industrial Engineering
and Management System
University of Central Florida
Orlando, FL, USA

Stefan Trzcielinski
Poznan University of Technology
Poznan, Poland

Beata Mrugalska
Poznan University of Technology
Poznan, Poland

ISSN 2194-5357   ISSN 2194-5365   (electronic)
Advances in Intelligent Systems and Computing
ISBN 978-3-030-20493-8   ISBN 978-3-030-20494-5   (eBook)
https://doi.org/10.1007/978-3-030-20494-5

This Springer imprint is published by the registered company Springer Nature Switzerland AG
The registered company address is: Gewerbestrasse 11, 6330 Cham, Switzerland

Advances in Human Factors and Ergonomics 2019

AHFE 2019 Series Editors

Tareq Ahram, Florida, USA
Waldemar Karwowski, Florida, USA

10th International Conference on Applied Human Factors and Ergonomics and the Affiliated Conferences

Proceedings of the AHFE 2019 International Conference on Human Aspects of Advanced Manufacturing, and the AHFE International Conference on Advanced Production Management and Process Control, held on July 24–28, 2019, in Washington D.C., USA

Advances in Affective and Pleasurable Design	Shuichi Fukuda
Advances in Neuroergonomics and Cognitive Engineering	Hasan Ayaz
Advances in Design for Inclusion	Giuseppe Di Bucchianico
Advances in Ergonomics in Design	Francisco Rebelo and Marcelo M. Soares
Advances in Human Error, Reliability, Resilience, and Performance	Ronald L. Boring
Advances in Human Factors and Ergonomics in Healthcare and Medical Devices	Nancy J. Lightner and Jay Kalra
Advances in Human Factors and Simulation	Daniel N. Cassenti
Advances in Human Factors and Systems Interaction	Isabel L. Nunes
Advances in Human Factors in Cybersecurity	Tareq Ahram and Waldemar Karwowski
Advances in Human Factors, Business Management and Leadership	Jussi Ilari Kantola and Salman Nazir
Advances in Human Factors in Robots and Unmanned Systems	Jessie Chen
Advances in Human Factors in Training, Education, and Learning Sciences	Waldemar Karwowski, Tareq Ahram and Salman Nazir
Advances in Human Factors of Transportation	Neville Stanton

(continued)

(continued)

Advances in Artificial Intelligence, Software and Systems Engineering	Tareq Ahram
Advances in Human Factors in Architecture, Sustainable Urban Planning and Infrastructure	Jerzy Charytonowicz and Christianne Falcão
Advances in Physical Ergonomics and Human Factors	Ravindra S. Goonetilleke and Waldemar Karwowski
Advances in Interdisciplinary Practice in Industrial Design	Cliff Sungsoo Shin
Advances in Safety Management and Human Factors	Pedro M. Arezes
Advances in Social and Occupational Ergonomics	Richard H. M. Goossens and Atsuo Murata
Advances in Manufacturing, Production Management and Process Control	Waldemar Karwowski, Stefan Trzcielinski and Beata Mrugalska
Advances in Usability and User Experience	Tareq Ahram and Christianne Falcão
Advances in Human Factors in Wearable Technologies and Game Design	Tareq Ahram
Advances in Human Factors in Communication of Design	Amic G. Ho
Advances in Additive Manufacturing, Modeling Systems and 3D Prototyping	Massimo Di Nicolantonio, Emilio Rossi and Thomas Alexander

Preface

Contemporary manufacturing enterprises aim to deliver a great number of consumer products and systems through friendly and satisfying working environments for people who are involved in manufacturing services. Human-centered design factors, which strongly affect manufacturing processes, as well as the potential end-users, are crucial for achieving continuous progress in this respect. Researchers around the world attempt to improve the quality of consumer products and working environments. The AHFE International Conference on Advanced Production Management and Process Control (APMPC) promotes the exchange of ideas and developments in production, sustainability, life cycle, innovation, development, fault diagnostics, and control systems. It addresses a spectrum of theoretical and practical topics. It provides an excellent forum of exploring frontiers between researchers and practitioners from academia and industry. It offers the possibility of discussing research results, innovative applications, and future directions.

We believe that such findings can either inspire or support others in the field of manufacturing and process control to advance their designs and implement them into practice. Therefore, this book is addressed to both researchers and practitioners.

The papers presented in this book have been arranged into nine sections. The first three sections focus mainly on topics in advanced manufacturing, while the remaining six sections focus on topics related to advanced production management and process control.

Section 1   Ergonomics in Manufacturing
Section 2   Agile Manufacturing
Section 3   Competencies in Work Environment
Section 4   Human Aspects in Industrial and Work Environment
Section 5   Strategic Decision-Making Models in Manufacturing and Service Systems
Section 6   Manufacturing Aspects of Work Improvement

The contents of this book required the dedicated effort of many people. We would like to thank the authors, whose research and development efforts are published here. Finally, we also wish to thank the following editorial board members for their diligence and expertise in selecting and reviewing the presented papers:

Advanced Manufacturing

Madalena Araujo, Portugal
Dominique Besson, France
Lucia Botti, Italy
Alan Chan, China
Keyur Darji, India
Enda Fallon, Ireland
Sarah Fletcher, UK
Weimin Ge, China
H. Hamada, Japan
Irena Hejduk, Poland
Joanna Kalkowska, Poland
Aleksandr Kozlov, Russia
Guangwen Luo, China
Preeti Nair, India
Edmund Pawlowski, Poland
Aleksandra Polak-Sopinska, Poland
Vesa Salminen, Finland
Antonio Lucas Soares, Portugal
Lukasz Sulkowski, Poland
Gyula Szabó, Hungary
Yingchun Wang, China
Marc-Andre Weber, Germany
Hanna Wlodarkiewicz-Klimek, Poland

Production Management and Process Control

Salvador Ávila Filho, Brazil
Mihai Dragomir, Romania
Murray Gibson, USA
Akihiko Goto, Japan
Adam Hamrol, Poland
Aidé Aracely Maldonado Macias, Mexico
Jörg Niemann, Germany
Tomoko Ota, Japan
Silvio Simani, Italy
Yusuf Tansel İç, Turkcy
Magdalena Wyrwicka, Poland

July 2019

Waldemar Karwowski
Stefan Trzcielinski
Beata Mrugalska

Contents

Production, Quality and Maintenance Management

Ergonomics in Manufacturing

Development of an Intelligent Robotic Additive Manufacturing Cell for the Nuclear Industry

Richard French$^{(\boxtimes)}$, Hector Marin-Reyes, Gabriel Kapellmann-Zafra, and Samantha Abrego-Hernandez

Physics and Astronomy, The University of Sheffield,
Hicks Building, Hounsfield Road, Sheffield S3 7RH, UK
{R.S.French, H.Marin-Reyes, G.Kapellmann,
S.Abrego}@sheffield.ac.uk

Abstract. Applications of Advanced manufacturing methods in the nuclear industry to ensure quality, security, process codes and standardisation are increasingly needed to ease adoption of new technologies. Many assemblies and decommissioning tasks are still heavily dependent on experienced human engineers and practitioners. Human error in production plays a large part in the development of standardisation to avoid defects and increase productivity. Risks to humans, previously considered as "part of the job" are no longer acceptable. Within European manufacturing, a greater problem exists; a dwindling skilled workforce capable of delivering high precision manufactured products. To address these issues this paper describes the motivation, design and implementation phases of the SERFOW (Smart Enabling Robotics driving Free Form Welding) project, which is an automated fusion-welding cell, linking future nuclear industry manufacturing requirements by mimicking human skill and technical experience combined with academic knowledge and UK based innovation. Development of key machine vision systems combined with novel robotic grasping technology and experienced welding engineers has made possible the construction of a potentially disruptive robotic manufacturing platform.

Keywords: Robot · Grasper · Ergonomic·3D · Vision system · TIG welding · Additive manufacturing

1 Introduction

Advances in novel manufacturing processes demand the application of Advanced Manufacturing Technologies (AMT), to obtain optimum results regarding quality, productivity, efficiency and responsiveness to the marketplace changes. This technology-centred approach is the design vehicle in which the majority of these manufacturing processes follow. Even though AMT can help to reduce direct labour, set-up time, inventory and manufacturing lead times [1], humans still playing an important role in delivering performance, but attributing cost within the organisations and manufacturing process. The importance of working towards the consideration of a combined human-design approach will drive improvements in ergonomics and safety

© Springer Nature Switzerland AG 2020
W. Karwowski et al. (Eds.): AHFE 2019, AISC 971, pp. 3–13, 2020.
https://doi.org/10.1007/978-3-030-20494-5_1

whilst reducing security risks. Failure to adopt will affect the performance and future capabilities of manufacturing companies [2]. Innovation in the nuclear manufacturing industry is correctly mitigated by risk; therefore, the industry is slow to adopt new technology capable of supporting the development of components used on-site for low carbon power generation. Within the UK, investment is steadily ramping up to ensure an augmented clean energy supply can fill the energy production void that fossil fuels would otherwise occupy should deployment of renewable energy not meet the anticipated power demand. By 2030 most existing (AGR) Advanced Gas Cooled reactors within the UK will be retired [3] and subsequently decommissioned until 2097 [4]. Preparations to adapt and develop new technologies for deployment in the next generation of advanced reactors, which could potentially be operational by 2030, is underway. However, who is actually going to make them? [5], Fig. 1.

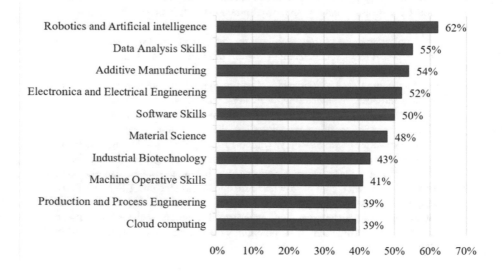

Fig. 1. Top 10 UK manufacturing technical skills shortages extracted from [5].

The backbone of UK manufacturing is supported by small manufacturing companies (SME) who work across all sectors, somewhat similar to Germany's "Mittlestand" SME Sector [6]. However, to be effective, they must invest in technology to augment and develop human skills to remain competitive against low-wage emerging economies.

The joining of metallic materials permanently by applying heat (fusion) or pressure (solid state) [7, 8] is effectively known as welding. Different types of fusion welding such as spot welding, metal inert gas (MIG), and tungsten inert gas (TIG or GTAW), to mention a few, are widely used by industry [9]. The welding process can build-up components through material deposition layer-by-layer which is well known as Additive Manufacturing, AM [10]. The majority of welding procedures are performed by highly skilled welding engineers because of the complex geometries of the products as the weld form and layer build up. These manual welding procedures involve risks such as heavy equipment handling, burns, toxic gas inhalation, ultraviolet and electromagnetic radiation which adversely affect the health of the human welders [7, 8]. Industrial accidents due to the risks mentioned above could represent costs related to indemnification, medical cost, re-staffing and damage to the equipment [9]. Critical human factors such as, the ageing workforce seeing retirement of these highly skilled welders when combined with the length of recruitment and the long term learning timescale required to produce new, skilled welding engineers within a declining workforce must be taking into account as part of the reason for the development of automated processes [10]. The ambition and vision of future high value manufacturing industries, such as nuclear, involves the implementation of Industry 4.0 technologies which ensures short development periods, producing customisable components on demand. In this case it requires adaptation to human needs, security, sustainability and resource-efficiency [11]. To mimic and automate the welding and AM process, robotic and computerised vision systems must be integrated to generate data and provide feedback to control the welding or AM platform.

This project aims to produce a low-cost prototype that is capable of providing welding and AM. The Smart Enabling Robotics driving Free Form Welding (SERFOW) concept is an automated welding cell which takes into account industrial requirements and the human-centred approach design to eliminate risks and improve the ergonomics of the process. This collaborative research project has been carried out by i3D Robotics LTD (UK), the Shadow Robot Company Limited (UK) and The University of Sheffield (UK) associating the nuclear industry demand with academic knowledge and innovation. Development of new autonomous AM technologies, as depicted here, adds flexibility with the rapid dynamic adaption of production tasks, reducing the high financial penalties of more complex traditional systems.

2 Methodology

The work presented in this paper was focused on the development of automated welding and AM cell designed from a technology-human perspective to improve quality, efficiency, ergonomics, safety and security in the nuclear industry. This prototype aims to mimic highly skilled fusion welding engineers by the integration of a 3D vision system and a robotic manipulator, which can perform layer-by-layer AM by means of automated fusion welding. The SERFOW cell is comprised of the following components: the main structure, the robot arm and smart gripper, the welding power source, 3D stereo vision and security system.

The main structure was made adaptable through the use of extruded aluminium profile sections with T-slot channels to allow the flexible attachment of cameras, sensors and other devices. Essential design points under consideration were to minimum size, reduce floor space, providing flexibility for design adaptation in future use, ergonomics and security. The Universal Robots UR-10 robot [12] arm and the Shadow Robot Smart Grasping System SGS [13] were used to pick and place the work pieces from the loading tray to the automated turn-table.

The TIG welding system (wire feeder and torch) was automated by two linear positioners in the X and Z axis. The safety system of the cell was designed according to ISO 10218-2 required for industrial collaborative robots. It consisted of two start buttons outside the welding and robot movements to ensure the security of the operators. Additionally, two of emergency or E-Stop buttons were installed.

Integration and assembly of the above-mentioned components with the innovative 3D stereo vision system developed by I3D robotics [14] is illustrated in Fig. 2.

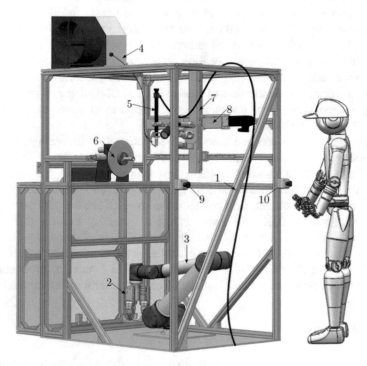

Fig. 2. SERFOW 3D CAD rendering: (1) aluminium structure, (2) robotic arm, (3) smart grasper, (4) wire feeder, (5) torch, (6) turn-table, (7) "Z" and (8) "X" linear motor drives, (9) left-hand, and (10) right-hand safety buttons.

The assembly, fusion welding, and AM layer build-up trials of this prototype were performed in the Enabling Sciences for Intelligent Manufacturing Laboratory at the University of Sheffield. The samples used for the welding and AM were made of 316L Stainless Steel which represent elements found in the design of nuclear heat exchanger manifolds, Fig. 3. A detailed explanation of the robotic, 3D stereo vision, welding and security system are presented in the following sections of this paper.

Fig. 3. 3D CAD rendering of the representative heat exchanger manifold components.

3 Smart Robot Arm and Grasper System

The six degrees-of-freedom UR10 robotic arm and the Shadow Robotics Smart Grasper, SGS, were required to pick and place the representative work pieces from the jig in the loading tray and deliver accurately to the automated rotary table.

The UR10 robotic arm is a collaborative robot that can withstand a payload mass of 10 kg; it can reach up to 1.3 m with a repeatable accuracy of 0.1 mm.

The Shadow Robotics Smart Grasper [15] system is a nine degrees-of-freedom gripper that is modular, flexible, robust, cost-effective and can tolerate working in radioactive environments [16, 17]. Its flexibility allows for the grasping of small and large complex shapes and objects up to 2 kg weight. The modularity of the SGS system offers a variety of configurations from two to three fingertips. In this development, a two-fingertip configuration was used to grab the working piece with good dexterity as Fig. 4 shows. The third fingertip was later used to stabilise the components during AM operations. The UR10 robot arm and Shadow's Smart Grasper are interfaced and controlled by ROS-Industrial software [18]. The use of ROS-Industrial has allowed the customisation of the inverse kinematics for manipulators. The coordinates acquired by the I3D robotics 3D stereo vision system when scanning robot position, dictate the location and movements meaning that manual programming was no longer needed.

4 Automatic Welding System

The welding system used is comprised of what could be considered an industry standard Miller Dynasty 350 TIG power source. This is effectively a reliable and proven generic power unit that can provide GTAW (gas tungsten arc welding) capabilities to the AM cell. Conducting the arc from the power source was via an umbilical cable to a machine type TIG Torch and the additive wire was delivered by means of a TIGFEED 40 automatic wire feeder. In this application a stainless steel filler wire with a high silicon content was selected for its weldability, wetting effect in the weld pool and ability to deliver layered weld build up or AM. For later production applications this filler wire would not be suitable due to having poorer metallurgical performance when compared with more advanced filler wires or additive materials.

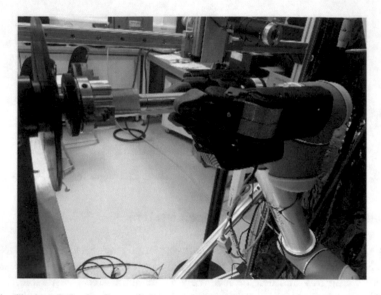

Fig. 4. Shadow Robotics Smart Grasper system holding a work piece with three fingers.

4.1 Linear Motor Driver System and Automatic Turn-Table

Control and automation of the welding torch were by means of an adapted mounting onto two linear stages produced by Parker Hannifin. Motorisation of these linear stages was by means of advanced servo motors from Kollmorgan enabling accurate, automated movements in the X and Z axis of the cell. This linear positioning system was covered with aluminium sheet to protect from internal contamination and trapping of fingers or clothing and electrical limit switches were added to avoid physical collisions in fault conditions.

The rotary turntable was a modified, low-cost automatic unit with a modified Maxon DC motor coupled to a simple speed controller with on/off switches. Communicating via the RS-485 asynchronous serial protocol, the system can either run at a set speed and then measure and match the absolute encoder, or the motor control systems may be changed to have better control of the speed. Isolation of the electronics components and motor were implemented to avoid short-circuiting, back EMF and HF conductance risks associated with welding.

5 Smart 3D Stereo Vision System for Welding Process

Vision systems for industrial inspection and assembly are widely used in automated manufacturing processes by identifying and processing images with a similar mechanism as to what a human eye would use. The vision systems obtain the images and then conduct both the processing and analysis of the image in order to provide accurate feedback to the robotic system. This feedback will allow control over the required AM and robotic path planning and trajectory parameters in order to improve the quality or security of the process. Additionally, the stereo vision obtains the 3D information from the acquired images and matches the features between these images. Algorithms such as [19] have been developed to overcome the commonly found "correspondence problem" [20] by making a correlation between points in the image pairs.

Stereo images are usually rectified, such that corresponding features lie on the same pixel line in each image [21]. Stereo vision utilises the fact that 3D points in space project to distinct 2D pixel locations in images of the scene when acquired from two different locations. The differences in pixel coordinates in the images allow reconstruction of the 3D coordinates from the images. I3D Robotics (I3DR) proposed and implemented an advanced 3D stereo vision camera system with auto-focus capabilities to be used in this prototype. This system was focused on identification of the stainless steel work-pieces for grasping, confirmation of grasp and part/torch location prior to weld and welding feedback. This vision system was controlled by the Phobos and PosCam systems [14].

5.1 Phobos Camera System

Phobos is a high-resolution stereo camera developed by I3D Robotics that recognises and obtains objects dimensional characteristics such as profile, shape and depth, shown in Fig. 5 using sub-routines constructed in HALCON software [22] as previously developed in iView project [23].

Fig. 5. Image of blocks with different shapes, sizes and orientations obtained by the Phobos camera, where the colour gradient indicates different height variations of the blocks.

To obtain the positional coordinates of the components either to be joined by fusion welding or to undergo the AM process, the SERFOW cell's Phobos camera was positioned above the loading area as illustrated in Fig. 6a. This task was essential for the UR-10 robot arm and SGS to record the exact location of the pieces allowing them to transport the components accurately to, and install in the automated rotary turntable.

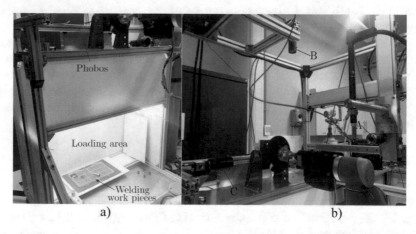

a)   b)

Fig. 6. (a) Phobos camera position within the SERFOW cell, (b) PosCam system comprised of cameras A, B and C.

White form board walls enclose the component loading area and backlighting in different angles was added to effectively control the highly reflective nature of the metallic work-pieces or components.

Following the successful implementation of the lightbox for the loading area, the Phobos system could reliably acquire the work-piece positional coordinate data. A considerable amount of refinement to achieve the appropriate set-up and accurate calibration was required.

5.2   PosCam System

The PosCam system consists of three cameras model DMK23UM021 located in the welding area as shown in Fig. 6b. The horizontal camera A was attached and centrally aligned to the welding torch tip. This allowed the precise positioning of the welding torch tip to the work piece using the linear motor drives X and Z. This position was absolutely critical to the success of fusion welding and AM. The precision is deter- mined by the lens choice providing a field of view and a working distance of 150 mm. The vertical camera B and horizontal camera C were mounted on the welding structures chassis. Both cameras were needed to monitor the work piece position. The main purpose of these cameras was to inspect and monitor the welding or AM process from different angles. 3D models of the conducted weld were produced by camera A as shown in Fig. 7 in order to give feedback and control the voltage, linear welding speed, distance between arc tip and AM filler wire delivery speed.

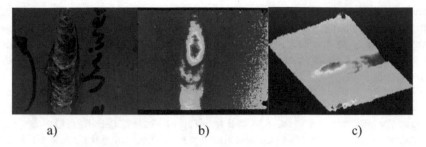

a)   b)   c)

Fig. 7. (a) 2D actual welding image, (b) 2D processed welding image and (c) 3D representation of the welding image of the weld.

6 Implementation and Welding Trials

The integration and assembly of the subsystems described previously was successfully performed. Welding trials on a 316L Stainless Steel samples were conducted as shown in Fig. 8. This was to establish the correct fusion of additive filler with the work piece and then later AM layering of material.

Fig. 8. Fusion AM trials performed in a 20 mm 316L Stainless Steel tube.

7 Conclusions

The welding process in high-value manufacturing industries such as those who produce components for nuclear power plants, is largely dependent on human, manual welding engineers because of the complexity of the components and design standardisation.

Current barriers of adoption for automation are attributed to both the material cost, but also the high labour cost of the human skilled workforce. The human–robot participatory design approach to adopt technology capable of automation and minimising human exposure to risks within the welding process has been fully accomplished in this project.

During the build-up of weld filler layers or additive manufacturing, constant monitoring by skilled welding engineers is needed to inspect the multiple arrays of changing parameters during the AM and welding processes. The SERFOW design and integration of the SGS end effector and six degrees of freedom collaborative robot with the automatic welding and AM system, required the smart 3D stereo positional data to function correctly. System performance evaluation and further analysis of the replication of highly skilled human welding engineers required a series of welding trials to be conducted. The results of the welding trials have provided an accurate initial assessment to demonstrate initial implementation and function of the SERFOW cell. The data captured during the later trials have proved that the cell performed the automatic welding and AM processes successfully, however subsequent improvement of metallurgical performance in the added layers is required. The project has responded partially to the desire for a lights out, Industry 4.0 factory of the future for high value manufacturing, but has shown an easy to adopt methodology to include human input without detriment to product output via collaborative automation and robotics.

Acknowledgments. The authors would like to thank Ben Crutchley from Industrial 3D Robotics, Fotios Papadopoulos from Shadow Robotics company, Ben Kitchener, Kieren Howarth and Samuel Edwards from the Enabling Sciences for Intelligent Manufacturing Laboratory at the University of Sheffield for their help and their invaluable support.

References

1. Co, H.C., Patuwo, B.E., Hu, M.Y.: The human factor in advanced manufacturing technology adoption: an empirical analysis (2011)
2. Slatter, R.R., Husband, T.M., Besant, C.B., Ristic, M.R.: A human-centred approach to the design of advanced manufacturing systems. CIRP Ann. - Manuf. Technol. **38**, 461–464 (1989)
3. Bryers, J., Ashmead, S.: Preparation for future defuelling and decommissioning works on EDF Energy's UK fleet of Advanced Gas Cooled Reactors (2016)
4. Authority, N.D.: Nuclear Provision: Explaining the Cost of Cleaning Up Britainâ A Zs Nuclear Legacy. Corp, Rep (2015)
5. UK Manufacturing Skills Shortages, Leadership and Investment White Paper by Cranfield University, https://www.cranfield.ac.uk/~/media/files/events/nmd/nmd-2017/nmd-2017-white-paper.ashx
6. Hancké, B., Coulter, S.: The German manufacturing sector unpacked: institutions, policies and future trajectories. (2013)
7. Lancaster, J.F.: The physics of welding. Phys. Technol. **15**, 73 (1984)
8. Norrish, J.: Advanced Welding Processes. Elsevier (2006)
9. Jeffus, L.: Welding: Principles and Applications. Nelson Education (2011)

10. Silva, R.J., Barbosa, G.F., Carvalho, J.: Additive manufacturing of metal parts by welding. IFAC-PapersOnLine **48**, 2318–2322 (2015)
11. Lasi, H., Fettke, P., Kemper, H.-G., Feld, T., Hoffmann, M.: Industry 4.0. Bus. Inf. Syst. Eng. **6**, 239–242 (2014)
12. Universal robots: introducing the UR10 collaborative industrial robot. https://www.universal-robots.com/products/ur10-robot/?gclid=EAIaIQobChMIy-XPmuq24AIVDuR3Ch1fswa8EAAYASAAEgKsnPD_BwE
13. Shadow Robotics Company Ltd: introducing the new modular grasper. https://www.shadowrobot.com/products/modular-grasper/
14. Industrial 3D Robotics: Phobos. http://i3drobotics.com/phobos
15. Shadow Robotics Company Ltd: modular grasper documentation. https://modular-grasper.readthedocs.io/en/latest/
16. French, R., Cryer, A., Kapellmann-Zafra, G., Marin-Reyes, H.: Evaluating the radiation tolerance of a robotic finger. In: Annual Conference Towards Autonomous Robotic Systems, pp. 103–111. Springer (2018)
17. French, R., Marin-Reyes, H., Kourlitis, E.: Usability study to qualify a dexterous robotic manipulator for high radiation environments. In: 2016 IEEE 21st International Conference on Emerging Technologies and Factory Automation (ETFA), pp. 1–6. IEEE (2016)
18. ROS-industrial: ROS for LabVIEW by Tufts University. www.clearpathrobotics.com/assets/guides/ros/ROSforLabVIEW.html
19. Bogue, R.: Robots in the nuclear industry: a review of technologies and applications. Ind. Robot. Int. J. **38**, 113–118 (2011)
20. Tippetts, B., Lee, D.J., Lillywhite, K., Archibald, J.: Review of stereo vision algorithms and their suitability for resource-limited systems. J. R.-Time Image Process. **11**, 5–25 (2016)
21. Kawatsuma, S., Fukushima, M., Okada, T.: Emergency response by robots to Fukushima-Daiichi accident: summary and lessons learned. Ind. Robot. Int. J. **39**, 428–435 (2012)
22. MVTec Software GmbH: HALCON – the power of machine vision. https://www.mvtec.com/products/halcon/
23. French, R., Marin-Reyes, H.: Underpinning UK high-value manufacturing: development of a robotic re-manufacturing system. In: 2016 IEEE 21st International Conference on Emerging Technologies and Factory Automation (ETFA), pp. 1–8. IEEE (2016)

Manageable and Scalable Manufacturing IT Through an App Based Approach

Christian Knecht[1]([⊠]), Andreas Schuller[1], and Andrei Miclaus[2]

[1] Fraunhofer Institute for Industrial Engineering IAO,
Nobelstraße 12, Stuttgart 70569, Germany
{christian.knecht,andreas.schuller}@iao.fraunhofer.de
[2] Karlsruhe Institute for Technology KIT,
Vincenz-Prießnitz-Str. 1, Karlsruhe 76131, Germany
miclaus@teco.edu

Abstract. Software is playing an ever-increasing role in providing the flexibility and efficiency required for handling complex processes. However, the IT capabilities of manufacturers are mostly limited to legacy systems or expensive, slow and inflexible development of new software. Therefore, new paradigms are needed for managing and scaling modern manufacturing processes with the help of software systems. This paper presents the organizational concepts of a software eco-system for the shop floor based on independent software modules called Apps. Additionally, we present the methodology used in the creation of industrial Apps. These concepts were successfully implemented in real world manufacturing scenarios. We focus on describing the specific challenges and requirements for industrial Apps, the technical architecture of the ScaleIT platform and a step by step process model to identify App ideas. An evaluation in the form of a questionnaire describes the assessment of the App-based approach by an industrial consortium.

Keywords: Industry 4.0 · IIoT-Platform · Micro Services, Micro Frontends · Edge Computing · Digitization · Shadow IT

1 Introduction

Staying competitive in today's globalized industrial markets is a complex cross-functional endeavor. Manufacturing enterprises need to find the right balance between optimum quality standards, quickly delivering products, customer orientation, and on-demand flexibility through customization of products, systems, and services [1, 2].

Continuously optimizing the production process is imperative for reaching these goals. Digitization and a flexible and agile IT are key factors in this regard [3]. The IT capabilities of manufacturing enterprises must become faster, highly scalable, and easily manageable for efficient accomplishment of the workers' tasks. Value creating software on the shop floor needs efficient and effective requirements engineering, implementation and deployment in order to adapt to the fast pace and changing business needs. Additionally, the costs for maintaining the production IT are a significant factor and must be continuously lowered to maintain acceptable margins [4].

© Springer Nature Switzerland AG 2020
W. Karwowski et al. (Eds.): AHFE 2019, AISC 971, pp. 14–26, 2020.
https://doi.org/10.1007/978-3-030-20494-5_2

In contrast, today's IT landscape in manufacturing is often rigid, siloed, and comprises mainly of isolated non-reusable custom-built applications (often based on spreadsheets) and monolithic single vendor solutions, e.g. Enterprise Resource Planning (ERP) or Manufacturing Execution Systems (MES). These are mostly used to control day-to-day operations [5]. While such systems offer a high degree of stability and extensive customization capabilities, companies are often dependent and have no control over further developments and updates of the software. Investing significant resources for company specific customizations is commonplace. In addition, strict and rigid licensing policies (single terminal or single user restriction) and non-flexible price policy adjustments entail high costs. Manufacturing companies are often unable to migrate due to proprietary data formats and closed interfaces. Further, integrating or upgrading to other technologies, e.g. cyber-physical and cloud systems, requires substantial IT engineering effort which is bottlenecked by the limited resources of the manufacturer. Therefore, digital tool creation often fails early on, and operators are left without adequate software support.

As outlined in Miclaus et al. [6], the research project ScaleIT proposes a concept for a technological platform based on an App paradigm that allows for rapid introduction of new IT especially in small to medium-sized industrial enterprises. It turned out that App technology is a suitable approach for the diverse shop floor requirements. By modularizing the software landscape, it becomes possible to change the IT governance to attribute responsibilities and functionalities to different stakeholders. Additionally, the flexible development of Apps, using state of the art web technology, allows for user-level customization based on roles, localization or the operating context.

We base our methodology for creating industrial software (Apps) on the technological capabilities of the ScaleIT platform.

2 Apps for the Industrial Shop Floor

The term *App* was originally used in the context of mobile platforms such as iOS and Android. However, there is yet no established common definition [7]. An App is essentially a software application or web application with a limited range of functions. Apps have become a major game changer for consumers, given ubiquitous technological platforms, an ecosystem of developers and leveraging modern UX techniques. Apps opened the possibility of creating solutions for very specific problems. Following the motto *There's an App for that*, iOS and Android platforms offer digital solutions for any task a user might want to undertake.

Apps must be intuitive, self-descriptive, and conform to user expectations, in order to be efficient tools in manufacturing enterprises. In addition, the installation of Apps needs to be kept as simple as possible, e.g. as downloadable components via an App store. The App should be ideally preconfigured and immediately usable after installation. Payment needs to be carried out comfortably and the costs can be kept low due to lower development effort and a higher sales volume. Additionally, for most use cases, there are a variety of Apps that offer similar functionality - allowing the user to compare alternatives and choose the best fit [8].

In industrial environments, Apps may serve as single purpose tools that leverage digital technologies in order to assist in or eliminate manual tasks needed in manufacturing processes [5]. Different platforms using the App concept can be seen in Table 1.

Table 1. Platforms for the Industrial Shop Floor using the App concept (R = research, A = association, C = commercial)

Name	Type	Term	Characteristics
ScaleIT [6]	R/C	Industry 4.0 Apps	Microservices, Micro Frontends, role-based approach, Shadow IT, open source
Apps4aME [3]	R	eApps	Adaptable, context-aware, cross-linkable
Industrial Data Space [9]	A	Smart services	Data sovereignty in business ecosystems
Virtual Fort Knox [10]	C	Production-related software solutions	Integrated security across all components, cloud-based
AXOOM	C	Industry 4.0 Apps	Digital production platform for the industry
ThingWorx [11]	C	Things	Model-based application development, digital twin characterized by properties, services, events and subscriptions
SAP Fiori [12]	C	ERP/PLM Apps	Role-based approach (specific views for each role), rapid deployment solution, extendable
OpenERP/Odoo [13]	C	ERP/PLM Apps	User friendly, easy integration, open source
Adamos	C	Apps	End-to-end transparency, scalability, on-premise integration
Cybus	C	Data based industry services	Modular, plug-and-play-connectivity, extendable
Predix	C	Industrial applications	scalable, asset-centric data foundation and digital twins, cloud-based
Cumulocity IoT	C	IoT solutions	Cloud-based, plug-and-play-connectivity

The requirements for the industrial shop floor are usually different than those for smartphones or tablets of a private end user (compare full list of requirements in Table 2). End users, for example, often buy new mobile devices and may even switch the respective App platform after a few years. However, machines and the associated software systems in manufacturing enterprises usually have a lifetime of many decades. Changes in ongoing processes such as updates or addition of novel components is a high-risk change. Only small timeslots between shifts or on weekends can be used for maintenance or installation of new software. The requirements for robustness and security of applications are quite high in the industrial context. Additionally, Apps require a higher degree of interconnection in order to generate the necessary business value [6].

Unlike the smartphone App, the industrial user (operators, shift managers or process owners) does not decide to install or update software on the factory floor, but the manufacturing IT. Therefore, the IT plays the role of gatekeeper in current manufacturing organizations. However, only the technological stack, safety and security should be managed by the company IT. Decisions on functionality, usability and overall lifecycle of software applications should be managed by the process engineers and managers on the shop floor. They have far better knowledge of the requirements of the day-to-day business of the shop floor. Industry 4.0 business needs and the corresponding business models are becoming more alike their software counterparts (Freemium, Pay-Per-Use, Subscription) [14]. Therefore, we argue that software needs to be flexible and to act according to the multitude of circumstances required by the enterprise. This is also a reason why current App models that focus on the consumer side cannot work for industrial applications. Additionally, a different degree of flexibility is needed in managing license models compared to the monolithic purchases of the past or simple purchases on commercial App stores.

Table 2. App requirements for the industrial shop floor (CR = Customer Requirement, IR = Industrial Requirement)

#	Requirements	Category	Type
(1)	Clear purpose and limited range of functions	Usability	CR
(2)	Intuitive, self-descriptive, conforming to user expectations	Usability	CR
(3)	Usable immediately after installation	Usability	CR
(4)	Comfortable payment, reasonable prices and flexible policies	Platform	CR
(5)	Simple access, findable and comparable	Platform	CR
(6)	Test/Trial version available	Functionality	CR
(7)	Secure, robust and consistent (fail-proof and safe)	Functionality	CR
(8)	High degree of connectivity and interconnection	Functionality	IR
(9)	Context aware, possible to use emerging technologies like NFC, real-time localization or embedded sensors	Functionality	IR
(10)	Networking and linking of information from the physical production and the digital world	Functionality	IR
(11)	Easy installation/deployment/distribution and updating, also possible without internet connection (Edge Computing)	Platform	IR
(12)	Long term support (LTS)	Support	IR
(13)	Centrally manageable, configurable and adaptable	Management	IR
(14)	Role based composition/allocation, views and permissions	Management	IR
(15)	Hardware and Operating System independent	Interoperability	IR
(16)	Support for cultural changes in the transition to a more software-oriented enterprise	Culture	IR
(17)	Engineering tools and documentation supporting developers	Development	IR
(18)	Based on standardized and open data models and interfaces	Standardization	IR
(19)	Traceability of actions and information	Audit	IR

An important aspect, often overlooked, is the necessity to support the cultural shift of an enterprise from classical manufacturing to a more software-oriented enterprise. Agility and lean thinking must be adopted in regard to the software development efforts. Documentation and guidelines on how to create, structure, distribute, deploy, license and maintain Apps on the platform, are necessary to avoid common pitfalls and accelerate the ramp up of value generating software systems.

3   Manageable and Scalable Industrial App Ecosystem

A modern manufacturing ecosystem comprises both humans and machines (including software). Structure, rules and guidelines foster an efficient and effective cooperation. The ScaleIT Ecosystem allows for maximal freedom within the boundaries of an App while allowing stakeholders to exercise their governance on the aspects that are directly linked to their responsibilities in the enterprise. Additionally, it provides a platform for building and deploying Industry 4.0 Apps such that the value chain can be digitized systematically step-by-step.

3.1   Role Based Organization

Within the ScaleIT platform ecosystem there are several relevant roles and interactions between these roles (compare Fig. 1). For each role, ScaleIT offers its own views, tools and support in order to better fit the roles' skills and requirements.

The roles of *Platform Operators* and *App Developers* are occupied by either the manufacturing enterprise using ScaleIT technologies or by contracting external vendors. Operators, engineers, quality assurance or other workers are considered *Shop Floor Personnel*. Employees holding this role use the Apps and profit the most from the improved software toolbox while dealing with their daily tasks and workflows.

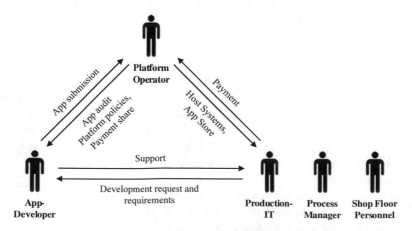

Fig. 1. Roles in the ScaleIT platform ecosystem and the corresponding interactions between them.

Platform Operators are in charge of the App store – the main entry point of Apps onto the manufacturing enterprise. They regulate which Apps gain access, payment, customer billing, receive a share from the sale of the App and transfer a part to the App developers. *Platform Operators* are also responsible for setting up host systems. ScaleIT provides them open source utilities and guidelines for setting up the platform.

The *Production-IT* is responsible for the installation of services and Apps, the configuration of the most important parameters, software lifecycle management, monitoring the status, starting/stopping, upgrading, and uninstalling Apps.

Process Managers are responsible for providing each person on the shop floor with access to appropriate tools that only show relevant information and interaction possibilities for the job at hand. They use the Apps available in the App store or commission the creation of new ones from the *App Developers*. *Process Managers* decide when it is time to retire an App because it is no longer useful. The *Shop Floor Personnel* has no task in setting up the platform, they are only beneficiaries.

Finally, *App Developers* create Apps for the App Store and therefore need to accept the terms of the *Platform Operator*. They are also responsible for providing support for their Apps. Ideally, they get requirements from the manufacturing companies. Developers profit from the ScaleIT platform because they reach many potential customers through the distribution platform. Additionally, deployment and license fees are simplified such that they can focus on implementing the business logic of the Apps. If customers are satisfied with previous Apps, they might order more developments from the same developers.

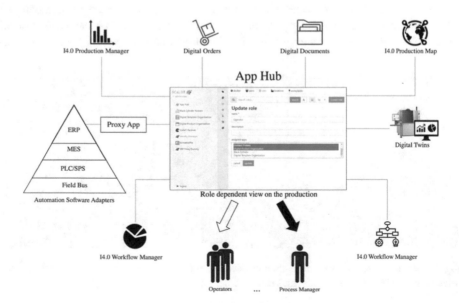

Fig. 2. Flexible assignment of Apps to specific roles via an administrative view. The Apps cover all aspects of the shop floor including workflows and control and legacy systems.

Therefore, ScaleIT Apps can be flexibly combined to form an overall system for any given user role. An administrative view of the Identity Management allows Process Managers to assign Apps to different roles easily (compare Fig. 2). For example, it can be defined that all shift supervisors should have a job-scheduling App and an App for condition monitoring of the entire plant. Employees in production, however, have an App for viewing work instructions, an App for error detection, and an App for monitoring machine conditions.

3.2   Manageable and Scalable IT Architecture

Apps need a runtime with an appropriate design and structure to execute in. In ScaleIT, Apps are *containerized* using the Docker technology. In most scenarios, many Docker containers run inside an App (web-UI, database etc.). However, the App is a self-contained entity, where neither process managers nor users can see the internal structure. This provides both isolation and simplicity, while allowing the developers to use any technology stack they prefer.

An important factor is the addition of platform specific functionality to the App via separate Docker containers (called sidecars). The sidecars allow the App code to be independent of the operator and the specific platform implementation. The developer imports the necessary sidecars from the Platform Operator and is thus integrated into the ecosystem. A further feature is the availability of web-based user interfaces for every App (as mandated by the guidelines). This allows for more complex role-based customizations as presented above.

Figure 3 Illustrates the operational layers and the corresponding organizational roles. Different types of Apps reside on each layer thus having a clear separation of concerns. Apps in the layers above use functionality from the Apps or systems below.

App ecosystem	**Domain-Apps** OEE-App, Condition Monitoring App, Job-Scheduling App...		Process manager
	Business-Essentials App Hub, Device Management...		Production IT
App Infrastructure	**Platform-Essentials** App Store, Identity Management, Message Broker...		Production IT
	Container Management Docker, Rancher, Git (Catalogue-Repository)		Platform Operator
OS / Cloud infrastructure	Host System	Host System	Host System	Platform Operator

Fig. 3. ScaleIT architecture differentiating between different operational layers (left) and the roles responsible for maintenance and usage of the corresponding Apps (right).

Apps deployed on the above architecture provide solutions for challenges described in Table 2 in the following manner:

- **Manageable IT:** ScaleIT provides a capable IT architecture that caters to the human and organizational needs using modularized software blocks called Apps. The *platform as a sidecar* as architectural concept allows for mixing and matching vendor and customer specific functionalities. For example, companywide single sign-on mechanisms can be implemented this way without modifying the core application code. *Requirements targeted: 1, 13, 14, 16, 19.*
- **Installation/deployment and updates:** The App Store makes it quite straightforward to install, update and monitor Apps. The containerized microservice approach brings software development and administration closer together (Agility and DevOps capabilities). Engineering tools and ready to use App skeletons (software stencils) provide a quick start for both experienced and beginner developers alike. *Req.: 1, 2, 3, 4, 5, 6, 11, 15, 17.*
- **Usability/User-centeredness:** Web technologies are used for platform-independent, responsive and easy-to-use UIs on all devices or operating systems that have a modern browser, including mobile devices such as tablets. Micro-front-ends (web components) display all Apps under a common user interface and mechanisms that allow for cross-app interaction. *Req.: 1, 2, 3.*
- **Business models:** Through the high modularity and versatility of the system, many modern business models can be implemented and combined within one ecosystem. *Req.: 12, 18.*
- **Scalability and Security:** Performance scalability and security mechanisms used in the web are employed in ScaleIT. This ensures that companies benefit from state-of-the-art technologies providing security and safety mechanisms for personnel and data. *Req.: 7.*
- **Legacy systems and connectivity:** Existing IT systems (legacy or otherwise) can be connected via abstracted intermediate layers implemented as Proxy Apps (ERP, MES proxies). The architecture accepts any web communication protocol. However, we found that a central MQTT message broker and standardized JSON data formats can be used to easily connect Apps (inter App communication) or realize digital retrofitting (Digital Twins). *Req.: 8, 10.*
- **Licensing:** Platform Operators can enforce contractual licenses using licensing sidecars and a centralized license management App. Developers thus have the freedom to offer multiple licensing schemes and customers are able to choose. *Req.: 4.*
- **Open for innovation:** New Apps can coexist with old ones through virtualization and the modular design. This opens the door for innovative applications and the modern Edge Computing paradigm. *Req.: 7, 9.*

4 Methodological Approach to Identifying App Ideas

4.1 Workshops with Stakeholders

The methodological support from the initial idea up to the finished product (App) was an important concern in the ScaleIT project. Therefore, a multi-stage process was developed in which the following principles were applied:

- Start from the problem and with the users in mind (following User Centered Design), rather than a technologically driven approach.
- Iterative design: quickly sketch a concept idea and test it with users instead of waiting for complete implementations.
- Lean UX: small work packages, interdisciplinary teams, minimal viable products (MVP), gaining experience with real users quickly.

The chosen procedure corresponds to a context-of-use analysis [15] with the goal of identifying company use cases that would benefit the most from being implemented as an Industry 4.0 App. Before developing the Industry 4.0 App, it should be clarified, what exact goals should be achieved and how they can be quantified. These goals serve later as a success evaluation of the new development. Potential benefits stemming from the development of a new App solution are diverse. Advantages are for e.g., saving time in production steps, significantly reducing the workers' workload, avoiding unnecessary work steps, improved guidance and support for new employees, simplification of communication between employees and systems, possible savings in operational resources, decentralized and rapid information transfer.

The central idea of the approach is that the users, their opinions, wishes, and their everyday professional challenges (i.e. pain points) are possible starting points for new concepts and ideas. This favors a pragmatic, application-oriented and goal-oriented approach, which is opposite to a purely technically driven approach.

The course of action consists of four steps, which are carried out on-site as a Site Visit in a company that wants to implement a new application. The approach follows the entire horizontal process chain in smaller companies and can be divided up in sub-process level in larger companies.

1. Analysis of people and roles in process steps: typical tasks, creation of profiles.
2. Information and data: Protocol relevant data, documents, software used in each process step.
3. Working context: e.g. physical arrangement workstation, volume levels, hardware used (e.g. terminals or tablets).
4. Processes and relationships: Understanding steps and relationships, creating a common image (e.g. as a user story map, compare Fig. 4).

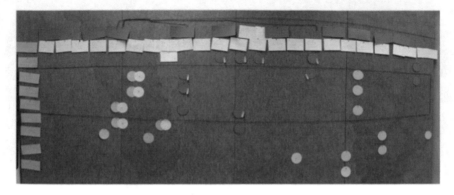

Fig. 4. Documentation and formalization of the process chain as a user story map for the identification of media breaks and optimization points.

The goal of the workshop is to reach a common understanding and prioritization of the important digitization potentials in the company among stakeholders. Therefore, the potential analysis is a good basis for further steps towards the conception and prototyping of interactive solutions. Although the operative employees spend their time involved in the discussion, the procedure generated a multitude of ideas and suggestions for improvement within the ScaleIT project. Using this approach, significant digitization and improvement potential can be identified in just a few days, provided that the key stakeholders in the company invest the necessary time and effort to provide the relevant information. The human being is therefore, always the starting point of the development and all App ideas have their origin in relevant operational practice.

4.2 Paper Prototyping with Focus Groups

An additional method that was employed during the project consisted of paper prototyping with selected focus groups. It was used as an iterative design process in a workshop format. The manufacturing employees that participated were able to ideate between 1 and 3 App ideas per session together with a corresponding use case that can help with their day-to-day tasks. This shows that App design is doable by non-developers, has good results and that there is vast potential for digital improvement in manufacturing enterprises.

The initial results of the workshop format have been promising because they quickly led to the creation of further related ideas as well as potential Apps across domains. The paper prototyping kit (see Fig. 5) has been released as a reusable workshop kit under an open source license[1].

---

[1] https://github.com/ScaleIT-Org/workshop-app-prototyping.

Fig. 5. The ScaleIT App Prototyping Kit used in workshops has been well received by participants and yielded good use cases and prototypes (right) in just a few hours of engagement.

5   Evaluation

In order to evaluate the App-based approach, a questionnaire was used to target project partners and interested external companies who wanted to employ the ScaleIT technology and methodology. The questionnaire covered several aspects besides demographic data: inquiries into the current situation of the manufacturing-IT, the App-based approach, the assessment of concrete App ideas, the App development process, virtualization technologies and open source activities.

Mainly executive managers participated, as well as process managers and IT-personnel from the following fields: Mechanical Engineering, Electrical Engineering, Optics & Optoelectronics, Measuring Technology, Automotive and Semiconductor Technology. In sum we have received 30 answers of which 17 were complete.

Concerning the current situation of the company IT all asked companies have WLAN on their industrial shop floor and 75% have access to the internet. 91% use or think about using mobile devices like tablets, smart phones, smart watches or glasses.

Following important functions or processes were listed as currently insufficiently covered by existing manufacturing IT: *transparency, real-time availability of production data, flexible IT policies related to new developments* and *order processing*.

Overall, the ScaleIT value propositions as described in 3.2 were rated positively. The most important value proposition was that the shop floor IT becomes as easy to use as Apps on the smartphone. The assessment of App ideas was performed with a 5-point Likert scale (from strongly disagree to strongly agree) including the assessment of complexity, time saving potential, usefulness, ease of use, learnability, understandability and possibility of usage with respect to work rules and regulations. Most of the participants stated that they would potentially use each of the evaluated Apps, and they see potential for saving time in production. Most participants would prefer one time App payments or a yearly payment compared to monthly subscriptions. The participants added that they already have new App ideas for novel use cases.

Regarding the App development half of the participants stated *less than one month* as an adequate time for the development of an App, 17% would prefer *less than 3 months* and 33% *less than one year*. As iterative cycles for meetings with developers, half of the participants would prefer *two times a month*, 33% *every two months* and 17% *every 3 months*. Interestingly, 83% of the companies would try out Apps that are still in development and not yet ready for production.

Positive statements about the container-technology that is the base for the App isolation in ScaleIT were *scalability, isolation,* and improvement of the process through *continuous integration* and *continuous delivery*. Negative aspects have been *the solution complexity* and *safety of Docker containers that requires further know-how*.

6 Conclusion and Outlook

This paper presents the organizational concepts of a software ecosystem for the shop floor based on independent software modules called Apps. We used the ScaleIT technological platform based on the App paradigm that allows the rapid introduction of new IT particularly in small to medium industrial enterprises. By modularizing the software landscape, control can be given to stakeholders with different roles. We presented a methodology for creating industrial Apps based on user centered iterative design, paper prototyping and lean UX techniques. This is necessary as the majority of manufacturing enterprises lack a flexible IT. In our experience, the introduction efforts are significantly lower than with current large-scale software engineering solutions commonplace in the industry. The feedback from the questionnaire and the workshop participants has been positive showing clear potential that can be harnessed.

The methods presented here lay the ground work for more flexibility and speed regarding manufacturing IT. In order to get quantitative results, a large-scale study is necessary to assess the impact of these methods on important manufacturing KPIs.

Expanding further upon the ideas presented here, new (smart) service concepts can be set up, which scale to connected cross-site and cross-company production chains. The building block-like approach based on interacting Apps is the ideal basis for a new generation of distributed, heterogeneous service ecosystem for medium-sized businesses that allows services beyond company boundaries.

Acknowledgements. This work was funded by the German Federal Ministry of Education and Research (BMBF) in the program "Innovations for Production, Service and the Work of Tomorrow" (02P14B180ff) and managed by the lead partner Karlsruhe PTKA[2]. Responsibility for the content of this publication lies with the authors.

---

[2] ScaleIT Industry 4.0 project page, https://scale-it.org.

References

1. Bauer, W., Hämmerle, M., Schlund, S., Vocke, C.: Transforming to a hyper-connected society and economy – towards an "Industry 4.0." Procedia Manuf. **3**, 417–424 (2015)
2. Geissbauer, R., Vedso, J., Schrauf, S.: Global Industry 4.0 Survey (2016). https://www.pwc.com/gx/en/industries/industries-4.0/landing-page/industry-4.0-building-your-digital-enterprise-april-2016.pdf
3. Volkmann, J.W., Landherr, M., Lucke, D., Sacco, M., Lickefett, M., Westkämper, E.: Engineering apps for advanced industrial engineering. Procedia CIRP. **41**, 632–637 (2016)
4. Waibel, M.W., Oosthuizen, G.A., du Toit, D.W.: Investigating current smart production innovations in the machine building industry on sustainability aspects. Procedia Manuf. **21**, 774–781 (2018)
5. Kulvatunyou, B., Ivezic, N., Morris, K.C., Frechette, S.: Drilling down on smart manufacturing – enabling composable apps. Manuf. Lett. **10**, 14–17 (2016)
6. Miclaus, A., Clauss, W., Schwert, E., Neumann, M.A., Mütsch, F., Riedel, T., Schmidt, F., Beigl, M.: Towards the Shop Floor App ecosystem: using the semantic web for gluing together apps into Mashups. In: Proceedings of the Seventh International Workshop on the Web of Things – WoT '16, pp. 17–21. ACM Press, Stuttgart, Germany (2016)
7. Gröger, C., Silcher, S., Westkämper, E., Mitschang, B.: Leveraging apps in manufacturing. A framework for app technology in the enterprise. Procedia CIRP. **7**, 664–669 (2013)
8. Wilhelm, T.: Was zeichnet eine gute App aus? (2011). https://www.usabilityblog.de/was-zeichnet-eine-gute-app-aus/
9. Otto, B., ten Hompel, M., Wrobel, S.: Industrial data space. In: Neugebauer, R. (ed.) Digitalisierung, pp. 113–133. Springer Berlin Heidelberg, Berlin, Heidelberg (2018)
10. Holtewert, P., Wutzke, R., Seidelmann, J., Bauernhansl, T.: Virtual fort knox federative, secure and cloud-based platform for manufacturing. Procedia CIRP. **7**, 527–532 (2013)
11. Gallasch, A.: ThingWorx–Plattform zur Integration herausfordernder Anforderungen auf dem Shopfloor. In: Produktions-und Verfügbarkeits-optimierung mit Smart Data Ansätzen, pp. 83–92 (2018)
12. Mathew, B.: Introduction to SAP Fiori. In: Beginning SAP Fiori, pp. 1–23. Springer (2015)
13. Ganesh, A., Shanil, K.N., Sunitha, C., Midhundas, A.M.: OpenERP/Odoo - an open source concept to ERP Solution. In: 2016 IEEE 6th International Conference on Advanced Computing (IACC), pp. 112–116. IEEE, Bhimavaram, India (2016)
14. Gausemeier, J., Wieseke, J., Echterhoff, B., Isenberg, L., Koldewey, C., Mittag, T., Schneider, M.: Mit Industrie 4.0 zum Unternehmenserfolg–Integrative Planung von Geschäftsmodellen und Wertschöpfungssystemen. Paderborn (2017)
15. Richter, M., Flückiger, M.D.: Usability Engineering kompakt benutzbare Produkte gezielt entwickeln. Springer Vieweg, Berlin (2013)

Ergonomics Principles for the Design of an Assembly Workstation for Left-Handed and Right-Handed Users

Lucia Botti[1]([✉]), Alice Caporale[2], Maddalena Coccagna[2], and Cristina Mora[1]

[1] Department of Industrial Engineering, University of Bologna, Viale Risorgimento 2, 40136 Bologna, Italy
{lucia.botti5, cristina.mora}@unibo.it
[2] Department of Architecture, University of Ferrara, Via Della Ghiara 36, 44121 Ferrara, Italy
alice.caporale@student.unife.it, cnm@unife.it

Abstract. Pain, fatigue and musculoskeletal disorders may be the consequence of awkward and inadequate postures caused by the improper design of work activities and poorly designed workstations. The good design of workstations following the ergonomics principles is necessary to avoid these adverse effects. This study aimed to define a set of ergonomics principles for the design of assembly workstations for both left-handed and right-handed users. An experimental study was performed at the university laboratory involving both left-handed and right-handed users for the assembly of a centrifugal electric pump. The experimental study proved that left-handed and right-handed users differently interact with the objects in the space. Four fundamental ergonomics principles for the design of assembly workstations for both left-handed and right-handed users have been defined, based on the results of the laboratory test. The proposed principles help designers and practitioners during the design of assembly workstations, proving the importance of the ambidextrous design of workstations.

Keywords: Human factors · Ambidextrous design · Ergonomics design · Assembly workstation

1 Introduction

Pain, fatigue and musculoskeletal disorders may be the consequence of awkward and inadequate postures caused by the improper design of work activities and poorly designed workstations. Such disorders may themselves affect the control of the working posture, increasing the risk of errors and reducing the productivity and the quality of the work. The good design of workstations following the ergonomics principles is necessary to avoid these adverse effects [1, 2]. Ergonomics and the study of human factors aim to understand the interactions among humans and other elements of a system, in order to optimize human well-being and overall system performance [3]. The ergonomics principles should be considered in the design of workstations, staff

© Springer Nature Switzerland AG 2020
W. Karwowski et al. (Eds.): AHFE 2019, AISC 971, pp. 27–39, 2020.
https://doi.org/10.1007/978-3-030-20494-5_3

facilities, tools and the overall work environment, including working and non-working areas. A good workplace design can enhance employees' satisfaction and well-being, while preventing the negative effects of poor workplace design, e.g. reduced productivity, quality and reliability [4]. Both the employers and employees should participate in the workplace design and the re-design and provide their suggestions for improving the occupational environment. The design of a workplace should consider the different physical and psychological needs and characteristics of the user population, i.e. the workers, including left-handed people, users with disabilities, and anthropometric extremes, such as pregnant women, obese people or any individual outside the 5th to 95th percentile range [2]. The International Standards of the International Standard Organization (ISO), as the ISO 14738 [5] and the ISO 11226 [1], provide the principles for the ergonomic design and re-design of workplaces and products, describing the value of ergonomics and the importance of assuming the proper working postures at work. Furthermore, such standards define recommended limits for static working postures without any or only with minimal external force exertion, considering other parameters as the body angles and the holding time, i.e. the duration that a static working posture is maintained. The ergonomics standards and guidelines for the design of workstations do not cover the user laterality and differences between left-handed and right-handed users. Laterality is the human preference for one side of the body over the other. Lateral preferences cause differences in task behavior between left- and right-handers individuals, e.g. left-handers take longer to begin a task and right-eared individuals have a more disinhibited approach [6]. Both genetic factors and cultural/environmental factors have been suggested as the causes for this difference. Handedness is generally believed to be determined by genetics; however, the environment affects the human preferences and provides the context and opportunity to develop a manual asymmetry [7, 8]. The individual lateral preference may be forced by circumstances or situations that happened after an unexpected event. Unfortunate occurrences as accidents and injuries may result in temporary or permanent disabilities, such as the impossibility to use the dominant hand for performing manual tasks or the temporary inability to reach far, bend or twist the dominant hand without discomfort or pain.

This study aimed to define a set of ergonomics principles for the design of assembly workstations for both left-handed and right-handed users. An experimental study was performed at the university laboratory involving both left-handed and right-handed users for the assembly of a centrifugal electric pump. The 60% of the sample were right-handed users, while the 40% were left-handed. Each user was asked to perform the operations for the electric pump assembly. The assembly activity was simulated at a self-adaptive assembly workstation aiming to assess the preferred position of the assembly accessories and the pump components for the assembly task. The laterality of eye, hand and foot has been assessed during the study. The results show that left-handed and right-handed users act differently during the assembly operations. The experimental study proved that left-handed and right-handed users differently interact with the objects in the space. Specifically, the lateral preference affects the way in which the user interacts with the workspace. Four fundamental ergonomics principles for the design of assembly workstations for both left-handed and right-handed users have been defined, based on the results of the laboratory test. The proposed principles help designers and practitioners during the design of assembly workstations, proving the importance of the ambidextrous design of workstations.

2 Literature Review

Left-handed people have been the 10% of the worldwide population for about 5000 years [9]. The prevalence rate of left-handed varies with different factors, as gender, age and culture. The 4–6% of the overall population actually prefers to use the left upper limbs and feels more comfortable to perform manual tasks with the left hand. Instead, 75 in 100 people prefer to use the right hand and 15–20 in 100 people do not show a clear preference for one hand [10]. Laterality preferences are expressed by the function of the hands and other organs, e.g. feet, ears and eyes, as an expression of the brain laterality in terms of handedness, footedness, earedness, and eyedness [11]. The design of products and processes should consider the laterality preferences of feet, ears and eyes, together with the handedness of the users. The 80% of the population have a homogeneous preference for hand and foot, e.g. right hand and right foot, or left hand and left foot, or no preference for hand and foot. About 55 in 100 people have a homogeneous preference for hand and eye, while less than 40 in 100 have a homogeneous preference for hand, eye and foot [10]. Furthermore, left-handed people are more confident in performing manual tasks with the right arm, compared with right-handed people, who show less ability with their non-dominant hand [12]. For this purpose, the design of products and processes for ambidextrous users is useful for both left-handed people and right-handed people who may lose their dominant hand. However, the constant and exclusive use of the non-dominant hand may allow the user to reach speed and precision performances comparable with the dominant hand [13]. Prevalence of left-handedness is generally assessed by questionnaires asking which hand is preferred for a set of manual tasks [14]. In this study, a questionnaire was given to verify the user dominance of hand, eye and foot. The questionnaire was structured following the Edinburgh Handedness Inventory [15] and the Lateral Preference Inventory [16] methodologies for the measurement of handedness, footedness and eyedness. After filling in the questionnaire, each user was asked to perform an assembly task at a laboratory assembly workstation. A camera was positioned close to the workstation to videotape the assembly task during the tests. This observation method helps to assess worker muscular behavior and the physical exposure to the risk factors for developing work-related musculoskeletal disorders [17–19].

3 Materials and Method

This study aimed to define a set of ergonomics principles for the design of assembly workstations for both left-handed and right-handed users. The experimental study was performed at the laboratory of the Department of Industrial Engineering of the University of Bologna, Italy. 30 testers were involved, including both left-handed and right-handed users, for the assembly of a centrifugal electric pump at an assembly workstation. The testers were under-graduate and graduate students, researchers, PhD candidates, full and associate professors at the Department of Industrial Engineering of the University of Bologna. Before starting the assembly task, each participant was required to complete a questionnaire, aiming to retrieve information about the participants laterality of hand, foot and eye. Each test was filmed with a Canon 77D + Tamron 18-200.

Humans' laterality is commonly assessed via questionnaires, e.g. the Annett [20], Edinburgh [15], Healey [21] and Waterloo [22] tools. A simple laterality questionnaire was constructed to investigate the participants laterality of hand, foot and eye. Guidelines for filling in the questionnaire were given to participants at the beginning of each test. The participants were asked to complete the questionnaire before starting the assembly task. The questionnaire included a set of personal information, e.g. name, surname, age, gender and height, and a focus on the laterality preferences for daily activities. Specifically, participants were asked which hand they use to write (hand laterality), which foot they would use to jump on one foot (foot laterality) and which eye remains open while winking (eye laterality). Four possible answers were proposed for each of questions on the laterality preferences: 'Left', 'Right', 'Left or Right indifferently' and 'No reply' (if the user could not answer the question) [23].

3.1 The Assembly Workstation and the Assembly Task

An assembly workstation was arranged in the laboratory of the Department of Industrial Engineering of the University of Bologna. The workstation consisted of three roller conveyors (one in the central position and two lateral) and a metallic frame. The roller conveyor supported the piece during the assembly process, while the metallic frame consisted of two adjustable modules and a supporting structure. The two modules contained the boxes with the parts and components for the assembly task. The user performed the assembly operations on the main roller conveyor (Fig. 1). Each participant assembled a horizontal multistage centrifugal electric pump (Fig. 2). Specifically, the reference assembly tasks consisted in assembling the components of the pump: a crankcase, a seal housing disk, a lantern and four screws.

Fig. 1. Assembly workstation used in this study. The workstation consists of three roller conveyors (one in the central position and two lateral) and a metallic frame. Two front modules contain the boxes with the parts and components for the assembly task.

Fig. 2. Parts of the horizontal multistage centrifugal electric pump. From the left: seal housing disk, pump crankcase, lantern.

The assembly process started with the crankcase placed on the central roller conveyor and the other components in the boxes on the two front modules. The position of lanterns, disks and screws varied during each test (Fig. 3).

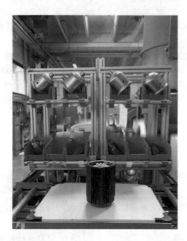

Fig. 3. Components of the horizontal multistage centrifugal electric pump in the assembly workstation.

Each participant placed the seal housing disk on the pump crankcase, then positioned the lantern on the disk and placed the screws to seal the components. After screwing the screws, the user finished the assembly cycle pressing the stopping pedal on the floor with the foot. Three pedals were available under the central conveyor of the workstation: one pedal was in central position and two were side pedals. The choice of pedal and boxes to retrieve the components was free and for the participants to determine.

The detailed assembly tasks are described in the experimental protocol in the following Sect. 3.2.

3.2   The Experimental Protocol

The experiment consisted in four consecutive assembly operations O1, O2, O3 and O4. Each assembly operation was the succession of four to five assembly tasks, from T1 to T5. The initial position of the pump crankcase was different for each assembly operation, i.e. the crankcase was at right side of the user in O1 and O3, while in O2 and O4 the component was at the left side. The position of the components on the front modules varied for each assembly operation, i.e. the seal housing disks were positioned in the four lower boxes in O1 and O2, and the lantern were positioned in the four upper boxes. The position of the components was inverted for O3 and O4. The screws were in the lower boxes throughout the test. The following Table 1 shows the detailed assembly operations and tasks performed in each test.

Table 1. Assembly operations and tasks performed in each test.

Assembly operation	Initial position of the components	Assembly tasks
O1	Pump crankcase: right. Lanterns: upper boxes. Seal housing disks: lower boxes Screws: lower boxes	T1. Move the pump crankcase in central position T2. Choose a seal housing disk and position it on the crankcase T3. Choose a lantern and position it on the disk T4. Take four screws and screw the lantern and the disk T5. Disassemble the components
O2	Pump crankcase: left Lanterns: upper boxes Seal housing disks: lower boxes Screws: lower boxes	T1. Move the pump crankcase in central position T2. Choose a seal housing disk and position it on the crankcase T3. Choose a lantern and position it on the disk T4. Disassemble the components
O3	Pump crankcase: right Lanterns: lower boxes Seal housing disks: upper boxes Screws: lower boxes	T1. Move the pump crankcase in central position T2. Choose a seal housing disk and position it on the crankcase T3. Choose a lantern and position it on the disk T4. Take four screws and screw the lantern and the disk T5. Disassemble the components
O4	Pump crankcase: left Lanterns: lower boxes Seal housing disks: upper boxes Screws: lower boxes	T1. Move the pump crankcase in central position T2. Choose a seal housing disk and position it on the crankcase T3. Choose a lantern and position it on the disk T4. Disassemble the components T5. Press one stopping pedal on the floor with the foot

Each participants performed the assembly operations and tasks in Table 1. The choice of the boxes and pedal was free. The aim was to investigate the preferences of the participants and to determine which components and pedals were mostly adopted by left-handed and right-handed users.

4   Results

This section introduces the results of the experimental study. 30 users participated in this study. 18 participants were right-handed users while the remaining 12 were left-handed users. The following Figs. 4 and 5 show the synthesis of the choices of the users. Specifically, the eight squares in each figure represent the boxes containing the pump components in the assembly workstation.

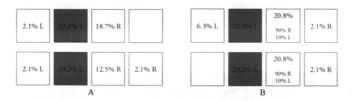

Fig. 4. Boxes chosen by right-handed users when the initial position of the pump crankcase is on the right (A) and on the left (B) side of the roller conveyor.

Fig. 5. Boxes chosen by left-handed users when the initial position of the pump crankcase is on the right (A) and on the left (B) side of the roller conveyor.

18 right-handed users performed a total number of 72 manual tasks to retrieve the pump components, i.e. the seal housing disk and the lantern, from the boxes on the assembly workstation. Figure 4A shows that when the pump crankcase initial position is on the right (OA1 and OA3), almost one third of right-handed users prefer to retrieve the pump components, i.e. the seal housing disk and the lantern, with their right hand from the central right boxes (27.8% for the upper boxes and 29.2% for the lower boxes). Similarly, Fig. 4B shows that when the pump crankcase initial position is on the left (OA2 and OA4), a significant percentage of right-handed users prefer to retrieve the pump components from the central boxes (19.4% for the upper boxes and 23.7% for the lower boxes). However, the preferred hand to retrieve the pump components in the upper boxes is the right (91.9%), while users prefer to use the left hand to retrieve the components from the lower boxes (88.2%). Figure 5 shows the boxes chosen by left-handed users and the preferred hand used to retrieve the pump components. 12 left-handed users performed a total number of 48 manual assembly tasks which involved the retrieval of the seal housing disk and the lantern from the boxes on the assembly workstation. Figure 5A shows that when the pump crankcase initial position is on the right, almost one third of the left-handed users prefer to retrieve the pump components (the seal housing disk and the lantern) with their left hand from the central left boxes (29.2% for the upper boxes and 33.3% for the lower boxes). Similarly, Fig. 5B shows that when the pump crankcase initial position is on the left, the most of the users prefer to use their left hand to retrieve the components from the central left boxes (20.8% for the upper boxes and 29.2% for the lower boxes). However, a significant percentage of the left-handed users uses the central right boxes (20.8% for the upper boxes and 20.8% for the lower boxes). In such cases, the preferred hand to retrieve the pump components is the right for both the upper and the lower boxes (90.0%). Such results suggest that if the user laterality is not indulged by the initial position of the first component to

retrieve (i.e. the pump crankcase), the user prefers to use the non-dominant hand. The following Fig. 6 shows the boxes chosen by left-handed users (A) and right-handed users (B), regardless the initial position of the pump crankcase.

Fig. 6. Boxes chosen by left-handed users (A) and right-handed users (B), regardless the initial position of the pump crankcase.

Specifically, right-handed users preferred to retrieve the pump components from the central right boxes (23.6% for the upper boxes and 24.3% for the lower boxes). The dominant hand was preferred by the most part of the users (97.1% for the upper boxes and 100% for the lower boxes). Left-handed users preferred to retrieve the components with their dominant hand from the central left boxes (25.0% for the upper boxes and 31.3% for the lower boxes). The total number of the choices of the users, regardless their handiness, is in Fig. 7.

Fig. 7. Boxes chosen by the participants in this study, regardless their handiness.

Figure 7 shows the use frequency of the boxes and the preferred hand for the 30 participants in this study. The four central boxes show similar use frequency and the difference with the lateral boxes is evident, i.e. both left-handed and right-handed users prefer to retrieve the pump components from the central boxes.

Figure 8A shows the boxes chosen by left-handed users to retrieve the screws. Specifically, two boxes containing the screws were available in the assembly workstations. One box was in the left side of the assembly workstation and the second box was in the opposite side. The participants were asked to retrieve the screws twice during the text, i.e. in OA1 and OA3, i.e. each participant performed 8 screwing actions. The two squares in Fig. 8A show the use frequency of the boxes containing the screws (50.0%). The four circles in Fig. 8A show the screws on the pump. Left-handed users preferred to screw the left screws with their dominant hand (91.7%L and 87.5%L in Fig. 8A). The right hand was preferred to screw the right screws (70.8%R and 70.8% R in Fig. 8A). Left handed users performed a total number of 96 screwing actions, of which 59.4% were performed with the dominant hand. The remaining 40.6% of the screwing actions were performed with the non-dominant hand. Similarly, Fig. 8B shows the preferences of right-handed users. The two right squares show the use frequency of the boxes containing the screws (17.0% for the left box and 83.0% for the right box). The four circles in Fig. 8B show the screws on the pump. Right-handed users preferred to screw with their dominant hand 3 out of 4 screws (94.4%R, 66.7%R and 97.2%R in Fig. 8B). Finally, right handed users performed a total number of 144 screwing actions, of which 75.0% were performed with the dominant hand. The remaining 25.0% of the screwing actions were performed with the non-dominant hand. These results suggest that left-handed users use the non-dominant hand more interchangeably compared with right-handed users. This phenomenon is the consequence to the developed capacity to adjust to use objects realized for right handers. Figure 9 shows the use frequency of the three pedals: 17 participants revealed in the questionnaire that their dominant foot is right; 12 participants stated that their dominant foot is left; one participant revealed no foot dominance. The users with right footedness preferred to push the right stopping pedal with their right foot (Fig. 9A).

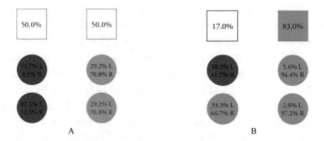

Fig. 8. Boxes chosen by left-handed users (A) and right-handed users (B) to retrieve the screws.

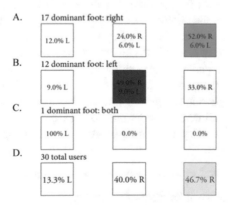

Fig. 9. Pedals chosen by the users divided by foot dominance.

Users with left footedness preferred to use the central pedal with their non-dominant foot (Fig. 9B). The user with no footedness pushed the left stopping pedal with the left foot (Fig. 9C). Finally, the aggregated results show that the use frequencies of central and right pedals are similar.

5 Discussion

The experimental study in this paper allowed the definition of four principles for the design of assembly workstations for both left-handed and right-handed users. This section describes such principles and the results that led to their definition. The comparison between the choice of the box for the components retrieval and the user laterality suggests the position of the most frequently used components. Figures 6A and 6B show the boxes chosen by left-handed users (A) and right-handed users (B), and the preferred hand. Left-handed user prefer to use their dominant hand to retrieve the left boxes, and vice-versa for right-handed users. Furthermore, Figs. 4A and 4B show different results depending on the initial position of the carter, at the beginning of the assembly operation. Such results suggest that the design of the direction of the assembly flow may affect the user performance, as it is related to his handedness. Similarly, Fig. 5B shows that the use frequency of the two central boxes is similar. However, the sample size is limited and this assumption may be verified with an extended test that includes a bigger sample of users. The similar use frequencies of the central boxes in Fig. 7 suggest that the components may be in central symmetrical position, i.e. the front position is preferred than the lateral position. The choice of the hand adopted for screwing provides confirms the previous assumptions about the components positioning in the assembly workstation. Left-handed users preferred the dominant hand for screwing the left screws and vice-versa. Right-handed users generally preferred to use the right hand. The left hand was preferred to screw the further screw from their dominant hand. Consequently, the symmetrical design benefits both left-handed users, who showed more fluid laterality, and right-handed users, who showed a strong preference for the components at their right.

Finally, the results of the choice of the stopping pedal show some discrepancy with the user footedness. The users with left footedness preferred to use the central pedal with the non-dominant foot (Fig. 9). Right-footed users preferred the right lateral pedal. The total number of the pedal choices show that central pedal and lateral pedal preferences are similar. Consequently, the central position may facilitate all the users.

Left-handed and right-handed people interact in different ways with the objects in the workplace. The experimental testing confirmed the principles of the ambidextrous design, e.g. central symmetry and front positioning, allowing the definition of the following four fundamental.

1. *Frequently used objects must be positioned according to workers laterality*

The preference of left-handed users for the object on their left and, vice-versa, the preference of right-handed users for the object on their right validated this principle. Objects and parts of objects with a central symmetry axis may be used either by the left hand and by the right hand, resulting ambidextrous. Furthermore, the mirroring of the functional parts of an object make it symmetrical and, consequently, ambidextrous.

Consequently, central symmetry and functions mirroring are fundamental principles for the design of ambidextrous workstations.

2. *The workflow direction must be in accordance with the worker laterality*

The direction of the workflow affects the worker performance, i.e. such parameter is critical for the design of an assembly workstation. When the workflow direction does not support the worker laterality, the worker is more likely to use the non-dominant arm. Such condition may produce assembly difficulties and a consequent reduction of the worker performances.

3. *The frontal position of the components is preferred than the lateral position*

The front position of the objects is preferred to the lateral position as it reduces the reaction time of the dominant hand. The users who took part to the test preferred to retrieve the components from the central boxes. The choice of the right or left central box was related to their hand laterality. Lateral boxes registered a low use frequency. Consequently, the front positioning of the components benefits both the laterlities as it reduces the effort required to the non-dominant arm for performing the assembly operations.

4. *Lateralized objects must be separated in order to allow the preferable arrangement according to the worker laterality*

In case the previous principles are not possible, the most lateralized parts of the object must be separated in order to allow the preferable arrangement. The possibility to modify the position of the components allows the workers to adapt the assembly workstation according to their laterality.

The application of such four principles allows the redesign of objects, process and workspaces, according to the worker laterality. The proper design of a workstation according to the preferences of the users is critical to prevent the negative effects of improper design, e.g. reduced productivity, poor quality, muscular fatigue, and to improve workers performances, satisfaction, health and safety.

6   Conclusions

Left-handed and right-handed people interact in different ways with the objects in the workplace. This study aimed to define a set of ergonomics principles for the design of assembly workstations for both left-handed and right-handed users. An experimental study was performed at the laboratory of the Department of Industrial Engineering of the University of Bologna involving both left-handed and right-handed users for the assembly of a centrifugal electric pump. The experimental study proved that left-handed and right-handed users differently interact with the objects in the space. The limited number of users involved in test may have affected the results of the study. However, the analysis of the users preferences and the study of their eye, foot and hand laterality allowed the definition of four fundamental ergonomics principles for the design of assembly workstations for both left-handed and right-handed users. The first principle states that frequently used objects must be positioned according to workers laterality. Furthermore, objects and parts of objects with a central symmetry may be used either by the left hand and by the right hand, resulting ambidextrous. The second principle states that the workflow direction must be in accordance with the worker laterality, i.e. when the workflow direction does not support the worker laterality, the worker is more likely to use the non-dominant arm and his performances may be reduced. The results of the test showed that both left-handed and right-handed users preferred to retrieve components from the frontal box, i.e. the third principle states that the frontal position of the components is preferred than the lateral position. Finally, lateralized objects must be separated in order to allow the preferable arrangement according to the worker laterality. The proposed principles help designers and practitioners during the design of assembly workstations, proving the importance of the ambidextrous design of workstations. Future developments of the study will include the analysis of a bigger sample of users, including both left-handed and right-handed workers with previous experience in assembly operations.

References

1. 11226, I.S.O.: Ergonomics. Evaluation of static working postures (2000)
2. Chim, J.M.Y.: Creating an ergonomic workplace by design. Japanese J. Ergon **53**, S376–S379 (2017)
3. International Ergonomics Association (IEA): Definition and Domains of Ergonomics. https://www.iea.cc/whats/
4. Walder, J., Karlin, J., Kerk, C.: Integrated lean thinking & ergonomics: utilizing material handling assist device solutions for a productive workplace (2007)
5. ISO: ISO 14738. Safety of machinery. anthropometric requirements for the design of workstations at machinery (2002)
6. Wright, L., Watt, S., Hardie, S.M.: Influences of lateral preference and personality on behaviour towards a manual sorting task. Pers. Individ. Dif. **54**, 903–907 (2013)
7. De Agostini, M., Khamis, A.H., Ahui, A.M., Dellatolas, G.: Environmental influences in hand preference: an African point of view. Brain Cogn. **35**(2), 151–167 (1997)
8. Singh, M., Manjary, M., Dellatolas, G.: Lateral preferences among Indian school children. Cortex **37**, 231–241 (2001)
9. Coren, S., Porac, C.: Fifty centuries of right-handedness: The historical record. Science **198**, 631 (1977)
10. Mandal, M.K., Dutta, T.: Left handedness: Facts and figures across cultures. Psychol. Dev. Soc. J. **13**, 173–191 (2001)
11. Lee, S.M., Oh, S., Yu, S.J., Lee, K.M., Son, S.A., Kwon, Y.H., Kim, YIl: Association between brain lateralization and mixing ability of chewing side. J. Dent. Sci. **12**(2), 133–138 (2017)
12. Jäncke, L., Peters, M., Schlaug, G., Posse, S., Steinmetz, H., Müller-Gärtner, H.W.: Differential magnetic resonance signal change in human sensorimotor cortex to finger movements of different rate of the dominant and subdominant hand. Cogn. Brain Res. **6**, 279–284 (1998)
13. Philip, B.A., Frey, S.H.: Compensatory changes accompanying chronic forced use of the nondominant hand by unilateral amputees. J. Neurosci. **34**, 3622–3631 (2014)
14. Porac, C.: Laterality: Exploring the Enigma of Left-Handedness. Nikki Levy, London (2016)
15. Oldfield, R.C.: The assessment and analysis of handedness: The Edinburgh inventory. Neuropsychologia **9**, 97–113 (1971)

16. Coren, S.: The lateral preference inventory for measurement of handedness, footedness, eyedness, and earedness: norms for young adults. Bull. Psychon. Soc. **31**(1), 1–3 (1993)
17. Engström, T., Medbo, P.: Data collection and analysis of manual work using video recording and personal computer techniques. Int. J. Ind. Ergon. **19**(4), 291–298 (1997)
18. Li, G., Buckle, P.: Current techniques for assessing physical exposure to work-related musculoskeletal risks, with emphasis on posture-based methods. Ergonomics **42**(5), 674–695 (1999)
19. David, G.C.: Ergonomic methods for assessing exposure to risk factors for work-related musculoskeletal disorders. Occupational medicine **55**(3), 190–199 (2005)
20. Annett, M.: A classification of hand preference by association analysis. Br. J. Psychol. **61**(3), 303–321 (1970)
21. Healey, J.M., Liederman, J., Geschwind, N.: Handedness is not a unidimensional trait. Cortex **22**(1), 33–53 (1986)
22. Steenhuis, R.E., Bryden, M.P.: Different dimensions of hand preference that relate to skilled and unskilled activities. Cortex **25**(2), 289–304 (1989)
23. Prieur, J., Barbu, S., Heulin, C.B.: Assessment and analysis of human laterality for manipulation and communication using the Rennes Laterality Questionnaire. R. Soc. Open Sci. **4**(8), 170035 (2017)

Impact of Industry 4.0 on Occupational Health and Safety

Aleksandra Polak-Sopinska[✉], Zbigniew Wisniewski,
Anna Walaszczyk, Anna Maczewska, and Piotr Sopinski

Faculty of Management and Production Engineering,
Lodz University of Technology, Wolczanska 215, 90-924 Lodz, Poland
{aleksandra.polak-sopinska,zbigniew.wisniewski,
anna.walaszczyk,anna.maczewska}@p.lodz.pl,
piotrsopinski@onet.eu

Abstract. Background: The objective of Industry 4.0 is to bring into existence smart, self-regulating and interconnected industrial value creation through the integration of cyber-physical systems into manufacturing. Industry 4.0 is a new paradigm of production and one that leads to a faster and more precise decision-making, entirely new approach to production, work organization manner of work task performance, which may have a significant influence on the health and safety of workers. Objectives: To provide an overview of potential effects (positive and negative) of Industry 4.0 on occupational health and safety and to list some of the recommendations regarding the integration of OHS into manufacturing in the Industry 4.0 context. Methods: A critical review of the literature currently available on this topic. Results: There are many risks as well as opportunities for occupational health and safety that derive from Industry 4.0. A considerable challenge, especially in the transitional period, is posed by insufficiency of initiatives with respect to occupational health and safety including standards and regulations, which may render them incommensurate in the face of newer and newer threats as Industry 4.0 technologies emerge. Furthermore, it may lead to forfeiting the proactive approach towards occupational health and safety that has been established in the most industrialized countries. Further research is required to enhance integration of occupational health and safety into manufacturing in the context of Industry 4.0. To achieve this, an interdisciplinary approach needs to be adopted drawing on the expertise of a team comprising engineers, IT experts, psychologists, ergonomists, social and occupational scientists, medical practitioners, and designers. The overview was carried by the research group IDEAT.

Keywords: Occupational health and safety · Occupational risk ·
Occupational hazards · Organization of work · Management ·
Industry 4.0 · Smart technologies · Smart factory

1   Introduction

The wave of digitalization that dates back to the era of the third industrial revolution has prepared the way for Industry 4.0. The term 'Industry 4.0' was first used in 2011. It is derived from an initiative launched by the German government for safeguarding the long-term competitiveness of the manufacturing industry [1]. By integrating cyber-physical systems (CPS) into manufacturing, Industry 4.0 is to bring into being smart, self-regulating, and interconnected industrial value creation. CPS includes inteeligent machines, storage systems, and production machinery which can exchange information, initiate actions and control one another. Their interconnection by way of the Internet, which has also been designated as the Industrial Internet of Things (IIoT), generates a technological leap for engineering, production, flow of materials, and sypply chain management [1].

In many regards, Industry 4.0 is much the same as the previous industrial revolution: compared to the preceding era, what drives Industry 4.0 is the transformation of the production goods and services by increasing its efficiency [2]. Nevertheless, Industry 4.0 provides a new paradigm of production that leads to a faster and more accurate decision-making, an entirely new approach to production [2], work organization and manner of job task performance, which may have a meaningful influence on the health and safety of workers. Therefore, the aim of the article is to provide an overview of potential effects (positive and negative) of Industry 4.0 and its components on occupational health and safety (OHS) and to list some of the recommendations regarding the integration of OHS into manufacturing in the Industry 4.0 context.

2   The Concept of Industry 4.0

The idea of Industry 4.0 represents the 4th industrial revolution in which cyberphysical systems play a primary role. CPS are defined as the connections between and coordination of computational and physical resources. It should be understood that computer-based algorithms control and manage many aspects of contemporary industry. CPS enable application of smart solutions in factories and warehouses [3]. Apart from cyberphysical systems, the most important components of Industry 4.0 are [3, 4]:

– *Internet of Things (IoT)* – a network of physical devices and applications connected and exchanging information in real time;
– *cloud computing* – IT technology enabling ubiquitous access to system resources and data processing via the Internet;
– *edge computing* – a method of optimizing cloud computing systems where the data is processed on the edge of the network, as close to the source as possible;
– *Big Data analytics* – processing large and varied data sets;
– *autonomous production or autonomous assembly* – self organizing computer assisted production planning;
– *the IT/OT convergence* – increase of the role of IT solutions in the operational area;
– *additive manufacturing* – 3D printing and fast prototyping;

- *advanced robots and co-robots (cobots)* – collaborative robots physically interacting with workers in production areas;
- *augmented reality* – system connecting a physical, real-world environment with computer-generated ones;
- *horizontal and vertical integration* – a kind of business expansion strategy;
- *rapid application development* – enabling support of various aspects of human activities;
- *digital twin simulation models* – a technology based on digital model of machines or production systems that enables real-time monitoring and designing new businesses.

The 4th Industrial Revolution is happening globally and concurrently. In Europe, Germany (Industrie 4.0), France (the Nouvelle France Industrielle), Sweden (Produktion 2030), Italy (Fabbrica Intelligente), Belgium/Holland (Made Different), Spain (Industria Conectada 4.0), and Austria (Produktion der Zukunft) are all actively taking an interest [5]. The EU initiated a public-private partnership under the title Factories of the Future, designed to ensure sustainable and competitive production. In the U.S., similar ideas are encouraged through the Industrial Internet Consortium with founding members such as AT&T, CISCO, GE, IBM, and INTEL. In China, the Internet Plus or Made in China 2025 initiative integrates current technological developments such as cloud computing and big data to enable state-of-the-art manufacturing. South Korea announced the Manufacturing Innovation 3.0. In all of these regions as well as in countries such as Japan and Singapore high investments are anticipated, whereas in Germany they are expected to exceed two billion Euro annually between 2018 and 2020 [2].

The rate at which technological advances are implemented is increasing yet, unfortunately, it often is rather erratic due to insufficient preparation of organizations for the implementation of new technologies [6], which fills us with misgivings as to whether we are going to forfeit a proactive character of OHS action. Whether we are going to miss the moment when preventive actions become primarily corrective again due to insufficient knowledge, skills, and competences prerequisite for predicting risks brought about by Industry 4.0 technologies. Importantly, in the recent years, the most industrialized countries have managed to develop a proactive approach to OHS. Industrial business has started to comprehend that the health and safety of workers is one of the paramount elements in the financial success of the enterprise, which is why it is so meaningful to know and understand the threats and opportunities that Industry 4.0 and its component parts bring about for OHS.

3   Research Method

In order to provide an overview of potential effects (positive and negative) of Industry 4.0 and its components on occupational health and safety and to list some recommendations regarding the integration of OHS into manufacturing in the context of Industry 4.0, a review of the literature was performed with the aid of certain keywords. First, the combination of the keywords "occupational health and safety" and "Industry 4.0" were combined (or, and). Next, "occupational health and safety" was put together with (or, and) technological categories related to Industry 4.0 such as "Internet of things", "Big data", "cobotics", "computer simulations", "artificial intelligence", "augmented reality". In the third step, the search was expanded to include other terms associated with OHS and Industry 4.0 such as "occupational health", "occupational safety", "occupational risk", "work organization", "workplace organization", "smart factory", "smart technologies", "smart production", "smart manufacturing", "smart industry", "advanced manufacturing". The literature review was conducted in the period from November 2, 2018 to January 15, 2019. The search was done in three databases: the Web of Science, Scopus and PubMed. The time range for the search was defined as 2010 through 2018. Publications of this year were not included due to continuous changes in their number. All titles and abstracts collected in the computer-based search were looked at, which allowed the authors to select those articles that appeared relevant for the purpose of the review according to the inclusion criteria. They included peer reviewed research papers, review articles, and conference papers. In the following stage of the review, documents developed by international governmental agencies which were published in English and dealt with worker health and safety issues related to Industry 4.0 technologies were included as were articles on the topic published in Polish.

4 Results

The concept of "Industry 4.0" is now one of the most discussed subjects in manufacturing technology circles, business groups, researchers and experts in several fields [7, 8]. Although there has been an enormous increase in the number of scientific publications on the subject of Industry 4.0, few of these raise OHS issues in any helpful way. The authors found 31 publications/documents that meet the inclusion criterion. All the included publications and documents were critically evaluated in order to point out the opportunities and benefits but also any concerns and issues deriving from the application (development) of Industry 4.0 for the OHS.

4.1   Industry 4.0 Opportunities and Benefits for Occupational Health and Safety

Many of the advances in technology that form the foundations of Industry 4.0 have already been used in manufacturing however, with Industry 4.0, they are expected to transform production: isolated, optimized cells will come together as a fully integrated, automated, and optimized production flow leading to greater efficiency and changing traditional production relationships among suppliers, producers, and customers—as well as between the human and the machine [4]. Some companies will be able to set up 'lights out' factories where automated robots continue production without light or heat after staff has gone home. For example, in the Netherlands, Philips produces electric razors in a 'dark factory' with 128 robots and just nine workers who provide quality and safety assurance [9]. This way the number of workers will be reduced. The remaining manufacturing jobs will contain more knowledge work as well as more short-term tasks. The workers increasingly have to monitor the automated equipment, are being integrated in decentralized decision-making, and are participating in engineering activities as part of the end-to-end engineering [10]. This may mean that workers will be involved in more creative, interesting, value-added activities and will have the opportunity to leave routine tasks and achieve a grater autonomy and self-development [11, 12].

It is worth emphasizing that the advancing computerization of industry makes it possible to create work structures that so far have only functioned in its innovative sectors. It has been pointed out that the importance of issues such as flexible working time and remote work will be growing for the manufacturing sector. To a much greater degree that before will project-based work involving the participation of individuals from and outside the organization be significant [13]. The potential consequences of the process of the blurring organizational boundaries, growing flexibility of working time, and remote work have been evaluated differently in the subject matter literature. Many researchers [1, 14] underscore the fact that increased working time flexibility should allow the worker to achieve greater work-life balance. The transformed work environment will allow workers to regulate their rhythm of work on their own.

Industry 4.0 may render labour safer and healthier owing to early and ongoing risk analysis and management based on smart safety technologies, virtual engineering, Big Data, and the Internet of Things [15]. Wireless sensor networks along with properly designed and integrated technical support may prevent accidents in autonomous and smart industrial environments [16].

Machines equipped with technical means of monitoring all parameters that have any bearing on the process will be more apt to respond appropriately the instant any dysfunction occurs [17, 18]. They will have more and more self-monitoring capability as well as the ability to monitor their surroundings and send information to diagnostic centres that will determine whether or not further intervention is necessary.

According to Gisbert et al. in [19] order to ensure the reliability of these systems, common technological platforms capable of monitoring the functioning and performance of all networks and linking sensors to remote control centres need to be implemented. These platforms will reduce occupational risks by facilitating the integration of general surveillance applications, which will be complemented with appropriate risk management. Gralewicz proposed a new concept of risk management in what is called intelligent work environment with respect to occupational health and safety using new technologies and solutions developed for the needs of safety-related areas of work. The approach to intelligent workplace management he advances relies on new technologies and solutions for which he sets the following tasks related to occupational safety: monitoring work environment worker health factors, monitoring machinery and technologies, monitoring personal protective equipment, warning workers and facilitating their timely information, facilitating decision-making, virtual 3D simulations. The performance of these tasks should proceed according to the hierarchy of risk management. This approach to organizational risk management should enable real-time responses to changes in work environment factors and personalization of risk assessment for individual worker profiles to include their psychological state, specific work environment factors at play, and location with regard to machines (including robots and cobots) [20].

There is a large range of personal protective devices that use smart technologies [17], which can help employees stay safe in dangerous workplace environments where they may be exposed to extreme noise, heat, toxic gases, chemicals, harmful elements. Similarly, technologies monitoring worker well-being (e.g. pulse, emotions, activity, temperature, etc.) can provide real-time alerts that indicate the need to adopt preventive measures designed to stop hazardous behaviors, restore safety procedures, avoid injuries, and enable an injured worker to reach for help [18, 21, 22].

Furthermore, self-aware, self-learning, self-healing, self-configurating, self-protecting machines capable of advanced analysis, fit with advanced programming, sensors, cameras, will be able to predict potential workplace hazards and manage unexpected conditions, which will facilitate prevention of worker accidents and injuries [23].

Digital factories use more and more industrial robots that replace the human in the performance of various job tasks, especially dangerous [24], monotypical, overstraining the musculoskeletal system, and excessively physically demanding for ones [23]. One type of robots that is increasingly popular is the collaborative robot (cobot). Thanks to special sensors and control methods related to the theory of cognitive machines, these robots autonomously and actively take cognizance of their surroundings (share the workspace) and analyze their activity to eliminate non-typical situations – which is indispensable if their interaction with the human is to be safe. This way productivity, and product quality can be enhanced and at the same time, occupational health complaints and diseases caused by the manner of work task performance, injuries and accidents can be prevented [25].

According to the German Federal Institute for Occupational Safety and Health, musculoskeletal injuries are the reason for 23% of sick leave days in Germany and cause production loss worth an estimated 17 billion euro annually. The situation is much the same for other European countries. These disorders are mainly attributable to physical loads during job tasks requiring lifting and carrying, which leads to muscle, ligament, bone, and cartilage damage. In some cases, common static aids such as forklifts and hoisting devices may not be used or are not flexible enough. This is when exoskeletons may prove useful designed to reduce stress/compression force on the lower back, shoulders, elbows, and wrists and therefore, protect the user against injuries to this parts of the musculoskeletal system. They provide support for the body during the positioning or using tools, handling objects, etc. [26]. The design of exoskeletons includes micromechanical elements and an ultra-light, ergonomic bearing system. In the future, the retrofitting of the structure with a sensory data transmission system is expected, which will enable the introduction of machine learning and artificial intelligence to the exoskeleton controller [27]. Therefore, these devices can provide safer and more ergonomic work conditions for the workforce which is increasingly diverse in terms of age, sex, cultural background, and the level of fitness. Furthermore, they can improve the quality of life of the disabled and the elderly [28].

In the age of European society aging, support for workers with disabilities will also be provided by humanoid robots which will be able to accurately interpret human emotions [29]. They will be able to act as personal assistants or job coaches especially in the initial stages of employment, which will reduce job related stress levels, shorten the time required for learning a new task, decrease the number of mistakes made by the worker, etc.

All of the above Industry 4.0 innovations will make the workplace a safer, more ergonomic, and comfortable environment, owing to which people will work more productively.

4.2   Concerns and Issues Deriving from the Application of Industry 4.0 Relevant for the Occupational Health and Safety

Generally, organizations are not quite ready for the implementation of Industry 4.0. Only 20% of the respondents assess themselves as being ready for the application of new supply models. The level of preparation for blurred lines between industries (17%) is even lower as is the level of preparation for the implementation of smart and autonomous technologies (15%). The latter is also connected with poor readiness for reaping the benefits generated by these technologies. Only 22% of manufacturers have a good grasp of the way new technologies change their workforce and organizational structure. Similar percentage share of respondents are aware of the influence new technologies have on the way goods and services are delivered by them. Only 16% of manufacturers know how to integrate their own solutions with external infrastructure, whereas a meager 8% have solid business foundations for the implementation of novel technologies [6]. Unsatisfactory preparation and readiness to gain from these technologies may, in the transitional period, lead to a deterioration in the quality of work, increase in the number of injuries, accidents, and other human errors.

Industry 4.0 production systems steadily grow in complexity, which is particularly evident in the context of the interplay between job content (variety, complexity, skills, uncertainty, exposure, etc.), organization (team planning, overtime, rush orders, etc.), management (duties, communication, roles, relationships, problem solving, etc.), and other organizational factors (promotions, pay increases, occupational safety, social value of work, etc.). These interactions give rise to a number of types of workplace hazards, in particular in the psychosocial category. Engineers and designers of advanced production systems often fail to notice this type of risk although it may be of utmost significance for management. Also worth pointing out is the fact that psychosocial hazards have already been recognized as considerably challenging with respect to OSH legislation and management systems BHP [12].

Workers whose job tasks include monitoring smart machines and robots or partaking in decentralized decision-making and complex engineering projects will need to act more autonomously, have excellent communication and digital skills, and an ability to organize their own work and bear greater responsibility. Unfortunately, many research findings show [6, 30] that there is a shortage of qualified staff and low level of digital culture especially among the aging workforce and people with disabilities. For this reason, workers will need to be more motivated and open to change. They will have to exhibit greater flexibility in order to cooperate more effectively and will have to embrace lifelong learning [31]. The importance of lifelong learning is growing in particular in Western Europe where there is a negative birth rate and where a majority of the economically active population is growing old. That is why workforce shortage is increasingly conspicuous for any job, which may lead to excessive fatigue, sickness absences, increased number of accidents.

Information and communications technologies (ICT) whose importance is ever increasing in Industry 4.0 widen the competency gap and extend the distance between young workers who have just recently obtained their qualifications and mature workers who exited the system of education a significant time ago [32]. Due to a decreasing share of physical labor in manufacturing and an increasing share of intellectual work, low-wage workers who are not provided with additional training will be at risk of losing their job [33], which appears to be unacceptable from a socially-inclusive occupational perspective [34].

Worker well-being monitoring technologies raise many concerns and can be perceived as an invasion of privacy which is usually experienced as a stressor. Since their use is reasonably justified, the implementation of worker well-being monitoring technologies should provide the worker with full control of her/his data. The worker should be able to choose which information is to be monitored, which could reduce the sense of having his/her privacy violated. Nevertheless, the disturbing fact is that many people may not be fully aware of how their data may be used, what it reveals about them, and whether it can be further circulated. The employer may try to persuade employees to disclose more information. However, information needs to be placed in a context (e.g., the worker's personal circumstances) before it can be properly analyzed. Most likely, this will require at least some human intervention as context is very hard to grasp and interpret. In any case, the employer will need to have been trained because the employer will bear the responsibility for actions taken based on this information [22].

The most severe constraint on machine learning is that artificial intelligence is devoid of contextual self-awareness (therefore, it is useless to expect it to understand the meaning of reality) – it merely recognizes frequent scripts but only a human/an operator can know their meaning. For non-standard events (unique) that do not match any previous scripts, it will not be able to appraise the situation effectively, in which case its behavior would be difficult to predict [27]. Innovative technologies can generate new mechanical, electrical, thermal, chemical, and other hazards. Furthermore, they can lead to new types of accidents due to incommensurate guidelines/regulations/ standards concerning their proper design as well as application. However, technological solutions quickly become obsolete, which does not conduce to the development of standards. Once again, regulatory framework and standards will not be ready on time to safeguard all workers against the consequences of the implementation of new manufacturing systems based on autonomous, smart, interconnected machines [35].

Vigorous increase in the number of devices with Internet connectivity as well as widespread exchange and processing of data over the Internet network entails a growing threat of a potential cyber attack which could pose a risk to the health and safety of workers.

5 Recommendations Regarding the Integration of OHS into Manufacturing in the Industry 4.0 Context

The review of the literature has shown that one of the potential effects (positive and negative) of Industry 4.0 and its components on occupational health and safety could be, especially in the transitional period, insufficiency of OHS initiatives including standards and statutory regulations, which could render them incommensurate in the face of newer and newer threats as Industry 4.0 technologies emerge. Furthermore, it may lead to forfeiting the proactive approach to OHS that has been established in the most industrialized countries. To maintain or improve the level of OHS in manufacturing in the context of Industry 4.0, the following recommendations have been made:

- Further research is required to improve the integration of human work and smart solutions. Design and configuration of intelligent machines still need to concentrate on physical, social, mental, and cognitive capabilities of the human being. An interdisciplinary approach to these issues should be adopted drawing on the expertise of teams comprising engineers, IT experts, psychologists, ergonomists, social and occupational scientists, medical practitioners, and designers [1].
- Further studies on psychosocial risks brought about by the new model of work organization are required.
- There is a need to continue research on collaborative robots in order to ensure a higher level of safety and accommodate physical and cognitive ability of the worker.
- New international standards should be developed or existing ones should be revised to protect workers against any and all potential physical and psychosocial risks arising from novel technologies.
- It is imperative that appropriate strategies of organizational management be established and implemented that take into consideration the protection of the worker [36].
- Cooperation with trade unions is recommended as is collective bargaining [2] with respect to the replacement of humans with robots, artificial intelligence, implementation of technologies for ongoing monitoring of worker well-being and performance.
- It is recommended that researchers, experts on advanced technologies of Industry 4.0, OSH specialists, HR specialists, and industrialists collaborate on the implementation of solution based on an all-encompassing vision of change management to ensure smooth and safe transition to the new paradigm [12].
- Adoption of socio-technical approach to the implementation of Industry 4.0 solutions is recommended so that technical innovation, models of work organization and professional development could be closely coordinated with economic and social circumstances.
- Proactive approach towards risk assessment already at the stage of design or in the early stages of the implementation of Industry 4.0 innovation be adopted, which should be feasible if simulation tools based on workplace virtualization are utilized in the design phase, whereas augmented reality techniques in the stage of prototype verification. Risk assessment that focuses on the identification of threats in the

initial stages of the manufacturing process should consider, among others, data management process, maintenance process, manufacturing technologies in use, machines, tools and materials, human error, physical and psychological load on the worker [37].

- Further research on the infallibility of personal protective devices that use smart technologies and devices for ongoing monitoring of the worker well-being is requisite.
- OHS specialists should be provided with opportunities for continual professional development, e.g. advanced training on novel technologies and effects of Industry 4.0 on workplace safety.
- Lifelong learning and continual professional development should be promoted.
- Virtual reality and augmented reality tools should be utilized during worker occupational health and safety training.
- Good practice platform should be provided showcasing examples of integrating OHS into manufacturing in the Industry 4.0 context. One example of such platform is " Plattform Industrie 4.0" created by the German Federal Ministry of Economic Affairs and Energy and the Federal Ministry of Education and Research [38].
- Measures safeguarding against unauthorized access to enterprise data and information used over the Internet. Protection against cyber threats is provided primarily through security systems (data encryption) and enterprise security architecture.

6 Summary

The review of the literature has revealed many opportunities for as well as many threats to occupational safety and health deriving from the application of Industry 4.0. A major threat, in particular during the transitional period, is insufficiency of initiatives related to occupational health and safety, including standards and statutory regulations, which may render them incommensurate in the face of emerging Industry 4.0 technologies. Furthermore, it may lead to forfeiting the proactive approach to occupational health and safety that the most industrialized countries have managed to establish. To prevent this from happening, further research is required to strengthen integration of occupational health and safety into manufacturing in the context of Industry 4.0. An interdisciplinary approach should be adopted to this end drawing on the expertise of teams made up of engineers, IT experts, psychologists, ergonomists, social and occupational scientists, medical practitioners, and designers. Apart from that, a proactive approach to risk assessment already at the stage of design or in the early stages of the implementation of innovation provided by Industry 4.0 and promotion of lifelong learning and continual professional development is also requisite.

References

1. Kagermann, H., Wahlster, W., Helbig, J.: Recommendations for implementing the strategic initiative INDUSTRIE 4.0 final report of the Industrie 4.0 Working Group. Acatech, Frankfurt am Main, Germany (2013)
2. Müller, J.M., Buliga, O., Voigt, K.I.: Fortune favors the prepared: how SMEs approach business model innovations in Industry 4.0. Technol. Forecast. Soc. Change **132**, 2–17 (2018)
3. Wrobel-Lachowska, M., Polak-Sopinska, A., Wisniewski, Z.: Challenges for logistics education in Industry 4.0. In: Nazir, S., Teperi, A.M., Polak-Sopińska, A. (eds.) Advances in Human Factors in Training, Education, and Learning Sciences. AHFE 2018. Advances in Intelligent Systems and Computing, vol. 785, pp. 329–336. Springer, Cham (2019)
4. Rüßmann, M., Lorenz, M., Gerbert, P., Waldner, M., Justus, J., Engel, P., Harnisch, M.: Industry 4.0: The Future of Productivity and Growth in Manufacturing Industries. Boston Consulting Group (2015)
5. Slusarczyk, B.: Industry 4.0: are we ready? Pol. J. Manage. Stud. **17**, 232–248 (2018)
6. Deloitte: The Fourth Industrial Revolution is here – are you ready? (2018) [online]. https://www2.deloitte.com/content/dam/Deloitte/tr/Documents/manufacturing/Industry4-0_Are-you-ready_Report.pdf. Accessed 8 Dec 2018
7. Rojko, A.: Industry 4.0 concept: background and overview. Int. J. Interact. Mob. Technol. **11** (5), 77–90 (2017)
8. Badri, A., Boudreau-Trudel, B., Souissi, A.S.: Occupational health and safety in the industry 4.0 era: a cause for major concern? Saf. Sci. **109**(11), 403–411 (2018)
9. ERPS- European Parliamentary Research Service. Industry 4.0 - Digitalisation for productivity and growth (2015) [online]. http://www.europarl.europa.eu/RegData/etudes/BRIE/2015/568337/EPRS_BRI(2015)568337_EN.pdf. Accessed 4 Dec 2018
10. Stock, T., Seliger, G.: Opportunities of sustainable manufacturing in Industry 4.0 – 13th global conference on sustainable manufacturing - decoupling growth from resource use. Procedia CIRP **40**, 536–541 (2016)
11. ILO, International Labour Office: The Futur of Work We Want: A Global Dialogue (2017) [online].   http://www.ilo.org/wcmsp5/groups/public/—dgreports/—cabinet/documents/publication/wcms_570282.pdf. Accessed 4 Dec 2018
12. Leso, V., Fontana, L., Iavicoli, I.: The occupational health and safety dimension of Industry 4.0. Med. Lav. **110**(5), 327–338 (2018)
13. Spath, D., Ganschar, O., Gerlach, S., Hämmerle, M., Krause, T., Schlund, S. (Hg.): Produktionsarbeit der Zukunft – Industrie 4.0. Frauenhofer Institut für Arbeitswirtschaft und Organisation, Stuttgart (2013)
14. Mas-Machuca, M., Jasmina Berbegal-Mirabent, J., Ines Alegre, I.: Work-life balance and its relationship with organizational pride and job satisfaction. J. Manag. Psycho. **31**(2), 586–602 (2016)
15. ABB Group: Connecting the world – Industry 4.0. (2014) [online]. http://new.abb.com/docs/librariesprovider20/Contact-magazine/contact_middle-east-industry-4-0-dec2014.pdf. Accessed 4 Dec 2018
16. Palazon, J.A., Gozalvez, J., Maestre, J.L., Gisbert, J.R.: Wireless solutions for improving health and safety working conditions in industrial environments. In: IEEE 15th International Conference on eHealth Networking, Applications and Services, Healthcom, pp. 544–548 (2013)

17. Podgórski, D., Majchrzycka, K., Dąbrowska, A., Gralewicz, G., Okrasa, M.: Towards a conceptual framework of OSH risk management in smart working environments based on smart PPE, ambient intelligence and the Internet of Things technologies. Int. J. Occup. Safe. Ergon. **23**(1), 1–20 (2017)
18. Mattsson, S., Partini, J., Fast-Berglund, A.: Evaluating four devices that present operator emotions in real-time. Procedia CIRP **50**, 524–528 (2016)
19. Gisbert, J.R., Palau, C., Uriarte, M., Prieto, G., Palazón, J.A., Esteve, M., López, O., Correas, J., Lucas Estañ, M.C., Giménez, P., Moyano, A., Collantes, L., Gozálvez, J., Molina, B., Lázaro, O., González, A.: Integrated system for control and monitoring industrial wireless networks for labor risk prevention. J. Netw. Comput. Applicat. **39**(1), 233–252 (2014)
20. Gralewicz, G.: Zarządzanie bezpieczeństwem w inteligentnym środowisku pracy (2). Bezpieczeństwo pracy **8**, 18–20 (2015)
21. Orji, R., Moffatt, K.: Persuasive technology for health and wellness: State-of-the-art and emerging trends. Health Informatics J. **24**, 66–91 (2018)
22. EU-OSHA- European Agency for Safety and Health at Work. Monitoring technology: the 21st century's pursuit of well-being? (2017) [online]. https://osha.europa.eu/en/tools-and-publications/publications/monitoring-technology-workplace/view. Accessed 4 Dec 2018
23. Beetz, M., Bartels, G., AlbuSchaffer, A., BalintBenczedi, F., Belder, R., Bebler, D., Haddadin, S., Maldonado, A., Mansfeld, N., Wiedemeyer, T., Weitschat, R., Worch, J. H.: Robotic agents capable of natural and safe physical interaction with human co-workers. In: IEEE International Conference on Intelligent Robots and Systems, pp. 6528–6535 (2015)
24. National Institute for Occupational Safety and Health (NIOSH): NIOSH Update: NIOSH seeks proposals on robotics technologies for assisting in underground mining rescue efforts. U.S. National Institute for Occupational Safety and Health (2014) [online]. http://www.cdc.gov/niosh/updates/upd-04-28-14.html. Accessed 4 Dec 2018
25. Chiabert, P., D'Antonio, G., Maida, L.: Industry 4.0: technologies and OS&H implications. Geoingegneria Ambientale e Mineraria **154**(2), 21–26 (2018)
26. Bogue, R.: Robotic exoskeletons: a review of recent progress. Ind Robot; Int. J. **42**, 5–10 (2015)
27. Szulewski, P.: Integracja informatyczna kluczowym aspektem środowiska wytwórczego w Przemyśle 4.0 (IT integration is a spirit of the Industry 4.0 manufacturing environment). Mechanik **91**(8–9), 630–636 (2018)
28. Reinert, D.: The future of OSH: a wealth of chances and risks. Ind. Health **54**, 387–388 (2016)
29. Ejdys J., Halicka, K.: Sustainable adaptation of new technology—the case of humanoids used for the care of older adults. Sustainability, MDPI **10**(10), 3770 (2018)
30. Lorenz, M., Rüßmann, M., Strack, R., Lasse Lueth, K., Bolle, M.: Man and Machine in Industry 4.0: How Will Technology Transform the Industrial Workforce Through 2025? The Boston Consulting Group (2015)
31. Moniri, M.M., Valcarcel, F.A.E., Merkel, D., Sonntag, D.: Human gaze and focus-of-attention in dual reality human-robot collaboration. In: 12th International Conference on Intelligent Environments, IE 2016, pp. 238–241 (2016)
32. Wrobel-Lachowska, M., Wisniewski, Z., Polak-Sopinska, A.: The role of the lifelong learning in logistics 4.0. In: Andre, T. (eds.) Advances in Human Factors in Training, Education, and Learning Sciences. AHFE 2017. Advances in Intelligent Systems and Computing, vol. 596, pp. 402–409. Springer, Cham (2018)

33. Wrobel-Lachowska, M., Wisniewski, Z., Polak-Sopinska, A., Lachowski, R.: ICT in logistics as a challenge for mature workers. knowledge management role in information society. In: Goossens, R. (eds.) Advances in Social & Occupational Ergonomics. AHFE 2017. Advances in Intelligent Systems and Computing, vol. 605. Springer, Cham (2018)

34. EFFRA: Factories of the future – Multi-annual roadmap for the contractual PPP under Horizon 2020 (2013) [online]. http://www.effra.eu/attachments/article/129/Factories%20of%20the%20Future%202020%20Roadmap.pdf. Accessed 8 Dec 2018

35. Jones, D., With the IEC/ISO 17305 Safety Standard Delay, What's Next? Rockwell Automation (2017)

36. Schulte, P.A., Salamanca-Buentello, F.: Ethical and scientific issues of nanotechnology in the workplace. Env. Health Persp. **115**, 5–12 (2007)

37. Tupa, J., Simota, J., Steiner, F.: Aspects of risk management implementation for Industry 4.0. In: 27th International Conference on Flexible Automation and Intelligent Manufacturing, FAIM 2017, 27–30 June 2017, Modena, Italy. Procedia Manufacturing, 11 (2017)

38. Plattform Industrie 4.0 – Anwendungsbeispiele [online]. https://www.plattform-i40.de/I40/Navigation/EN/Home/home.html. Accessed 8 Dec 2018

Lean Production Management Model for SME Waste Reduction in the Processed Food Sector in Peru

José Chávez[1(✉)], Fernando Osorio[1(✉)], Ernesto Altamirano[1(✉)], Carlos Raymundo[2(✉)], and Francisco Dominguez[3(✉)]

[1] Escuela de Ingeniería Industrial,
Universidad Peruana de Ciencias Aplicadas (UPC), Lima, Peru
{u201610172, u201320868, pcinealt}@upc.edu.pe
[2] Dirección de Investigación
Universidad Peruana de Ciencias Aplicadas (UPC), Lima, Peru
carlos.raymundo@upc.edu.pe
[3] Escuela Superior de Ingeniería Informática,
Universidad Rey Juan Carlos, Mostoles, Madrid, Spain
francisco.dominguez@urjc.es

Abstract. The reduction of waste is a constant concern for companies that form part of a supply chain. In industrial processors, these are related to logistics solutions, because the production process of the different products is highly automated. In the case of the Peruvian potato, this model is not applicable due to its irregular characteristics. In this context, this paper proposes an improvement in the process of elaboration of processed potatoes in order to reduce or eliminate waste in food sector companies. Identification tools are used for activities that do not generate value, such as the VSM, and other continuous improvement tools such as Kaizen and 5S, as well as a simulation model. In the validation, an 89% increase in the product yield, as well as a 72% efficiency increase, is obtained.

Keywords: Time reduction · Waste reduction · Production management · Lean manufacturing

1 Introduction

Currently, processes related to manufactured products are highly automated. Market competitiveness requires companies to be more profitable and efficient, either by reducing costs or saving time when carrying out their activities. This is much more evident in the processed food sector, where high levels of waste and production losses bring about higher costs, and activities with extended periods are detrimental because in-process and finished products are highly perishable. In addition, these companies work as part of a food supply chain [1]. Each of these companies faces specific problems, such as yield per hectare, optimal management and transportation conditions by wholesalers, and a demanding greater input and product yields and cleanliness from industrial processors. However, one common problem is that food is perishable.

W. Karwowski et al. (Eds.): AHFE 2019, AISC 971, pp. 53–62, 2020.
https://doi.org/10.1007/978-3-030-20494-5_5

Because the Peruvian potato has irregular shapes, automated production models cannot be applied because they would generate great losses of raw material. In addition, SMEs do not see a short-term benefit in investing in these systems. In this context, this paper proposes a production management model for waste reduction of SME potato processors that will allow for process optimization. The study applies the concepts defined by lean methodology. The proposed process reduces and eliminates waste and time-related changes. To identify these production wastes, the current VSM is drawn up. By identifying the types and amounts of waste, its causes are analyzed. After they are identified, the improvements in each process activity are considered and visualized in the future VSM. Through Kaizen and 5S tools, continuous improvement will be established to reduce waste to a greater extent. It is also expected that this study can be used by other companies in the same sector that face similar problems.

2 State of the Art

2.1 Lean Manufacturing

Lean manufacturing (LM) principles are designed to eliminate waste [2]. By waste, we refer to long cycle times or unnecessary waiting times between value-added work activities. It can also include rework or scrap. Likewise, lean applications are used along with industrial engineering tools as simulation models, since they are usually very good predictors of the performance of a new or redesigned system. In a lean production model, the following four dimensions are emphasized: internal aspects, chain value, work organization, and the impact of the geographical context [3].

However, lean implementation is not an easy task for SMEs since there are no lean experts who have years of experience in project implementation in this area. In other aspects, the framework becomes more complicated and implementation costs increase. This complicates the design of a general lean framework due to the lack of flexibility and the variety of lean tools as part of the methodology [4]. Likewise, there is no detailed description of challenges to these models, since companies want to protect and not disclose the failures of their investments. The main causes for failures include the lack of resources such as labor, capital, and communication, as well as implementation barriers. They are classified into seven categories: management, resources, knowledge, conflicts, employees, finances, and past experiences. Resistance to change by employees is a common practice. This resistance can be rooted in fear of the unknown and of failure, as well as complacency [5].

2.2 Quality Management Models

Quality management refers to the creation of an organization system that fosters cooperation and learning, and focuses on process management practices, resulting in the continuous improvement of processes, products, and services. Note that the ideas that provide quality management should not only be customized for small and medium enterprises but also for specific situations [6]. Total quality management (TQM) is one of the best-known quality management systems because there is an agreement that by implementing a TQM system, the performance and overall effectiveness of the organization will improve. TQM components include organizational management, human resource management, technological applications, and customer relationships [7, 8].

In the food processing industry, quality management models have also been developed for the organizational performance of SMEs. The components of these models focus on product design and manufacturing. Seven components are clearly identified: leadership, planning, human resource development, customer focus, supplier focus, information management, production management, and quality assurance.

Among the reasons why companies implement a TQM model, a study defines and groups the following three: (I) external market reasons, which refer to improving the image of the company and its competitive position; (II) external reasons from requirements; and (III) internal reasons, with elements such as the decision-making process at the corporate level, and productivity and quality improvement. These are the reasons why models of excellence, such as EFQM, are being increasingly applied by companies globally. This model was inspired by the TQM philosophy, although it later evolved through different revisions and incorporated aspects such as social responsibility that were not originally specified in such philosophy. Therefore, some studies consider that EFQM has been regarded as a pseudo-official TQM model. A specific case that is pointed out is the Green Lean TQM Islamic Process Management, which combines structures from LM, TQM, and environmental management system. It explained that these systems are not limited exclusively to the automotive industry but can be applied to any sector.

Many of the previous models are not designed for SMEs. In cases where there is an adapted model, such models assume that the raw material (as well as the product in process) is easily manageable and has no disadvantages. The reason why this research is conducted is to be able to solve the general problem of food processing companies that cannot add additional value to their finished products or to their customers, due to the lack of correct techniques for foods with irregular shapes. This complicates automated work with this raw material and does not prevent considerable raw material from being lost or reduced. That is why this research focuses on waste reduction that may occur in all activities of the production process of an SME.

3 Contribution

3.1 Proposed Model

This model is modified to improve the parts where companies are not currently adding value to their finished products, and therefore, to their customers. All dimensions start from the customer perspective (Fig. 1).

3.2 Model Components

The first component that this model modifies is the process values, which aims to ensure that, during the entire production process of a company, activities that do not add value to the finished products, as well as to the customer, are reduced to the

Fig. 1. Proposed model.

minimum possible. It also seeks to ensure that other types of wastes are completely eliminated from the process to guarantee the correct flow of tasks.

The second component that this model would like to change is work organization, which aims at making the flow of activities equitable, and finding and constantly eliminating the bottleneck of the process to reduce its production times.

Other components of the model that have not been modified are, with regards to planning, the technological perspective, and with regards to the process, the internal perspective. In terms of support, this model includes the components of human resource management and the financial perspective.

The food industry is related to waste. One of the most common ways to evaluate raw material and product process performance is using a block diagram. However, one of the limitations of this tool is the inability to determine the type of waste and the economic impact of its losses. For this reason, the use of a form to evaluate these aspects is proposed.

The performance evaluation form first includes all product details to be evaluated, the process, the quantity acquired, the purchase and the weight, the price per unit, and the total price. Afterward, the number of operations that make up the process is recorded. By calculating the difference between the gross and the net weight, the amount and type of waste detected are specified, whether due to deterioration, shrinkage, or waste. Finally, a section is used to describe the observations found (Fig. 2).

On the other hand, regarding the Red Card used for the application of lean tools, one of the first steps of 5S is the organization (Seiri) of the work areas. In this stage, color cards are used. This card is not only used for raw material but also for other components (Fig. 3).

EVALUACIÓN DE RENDIMIENTO DE INSUMO								Versión 1 de 1 Página 1 de 1	
Insumo: **Proceso:** **Cantidad comprada:**				Peso por unidad: Precio por unidad:					
OPERACIÓN	Tiempo Estandar (min)	Peso bruto (Kg)	Peso Neto (Kg)	TIPO					COSTO
				Deterioro (Kg)	Merma (Kg)	Desperdicio (Kg)	% Desecho		
Selección Lavado Pelado Remate Corte Zarandeado Enjuagado Oreado Empaquetado									
TOTALES									
OBSERVACIÓN									

Fig. 2. Performance evaluation form.

TARJETA ROJA	
Nombre del artículo:	
Categoría:	
1. Maquinaria	6. Inventario en Procesos
2. Accesorios y herramie	7. Producto Terminado
3. Instrumental de medic	8. Equipo de oficina
4. Materia Prima	9. Papelería
5. Refacción	10. Limpieza o pesticidas
Fecha	**Localización**
Cantidad:	.
Razón:	
1. No se necesita	4. Material de desperdicio
2. Defectuoso	5. Contaminante
3. No se necesita pronto	6.Otro
Consideraciones:	**Forma de desecho:**
1. Fragil	1. Tirar
2. Toxico	2. Vender
3.Ambiente a ____ C	3 Mover
Fecha de desecho	4. Reparar
	5. Regresar a proveedor
	6. Otros

Fig. 3. Red card.

3.3 Model Process

In the implementation process, the model begins by defining the steps to be improved. After the process to be improved has been identified, the diagnosis of the situation is made and variables are measured using the previously proposed forms. After validating that there is a possibility of improvement, the proposal is implemented based on the activities described. Finally, the proposal is validated, both technically and economically, ending with an analysis of the results obtained (Fig. 4).

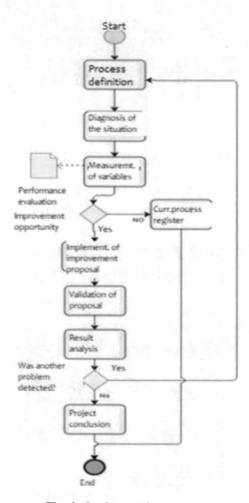

Fig. 4. Implementation process.

4   Verification

4.1   Description of the Case Study

Food Supply Peru has a single plant, located in Lima, Peru, which aims to reduce the waste generated in the process of producing potato bags.

The company production process begins with the reception and selection, where the workers separate and place the tubers in 20-kg buckets. Then, washing takes place, where dirt is removed from the product. After this, potatoes are peeled and placed in the peeling machine, and by abrasive peeling, the surface and skin are removed by friction. In the final activity, workers are in charge of removing the potato "eyes" and inspecting the peeling results. Then, during the cutting process, an operator manipulates a machine and places the potatoes in a vertical position for cutting. Later, when potatoes are shaken, potato sticks that do not meet the required dimensions are removed. After this, potatoes are rinsed and a preservative is added to guarantee the preservation of the product, and after they are vented, excess moisture is eliminated. Finally, to maintain the organoleptic characteristics of the potato, the operator packages the tuber with a shovel in polypropylene bags. For this process, an industrial scale is used for the packaging activity.

4.2 Validation of the Proposal

To validate the proposal, a mixed method will be applied, where both field and simulation data will be collected. For the simulation part, the Arena program will be used, with the variables presented above (Fig. 5).

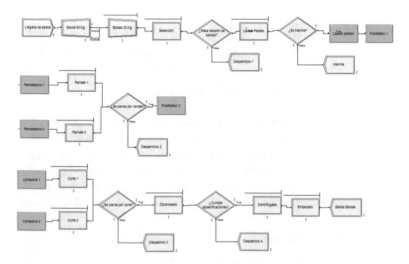

Fig. 5. Simulations of both the current and proposed situations performed to compare them under the same conditions.

4.3 Result Analysis

According to the technical and economic data, the viability of the proposal is positive. It is important to consider that there is a 6% difference in the technical validation that can affect the final results, although not excessively.

The results obtained through the case study and the simulation include a reduction of 70% in the maximum pressing time, from 5 to 2 min. Process efficiency increased from 59% to 74%. On the other hand, raw material performance of the production process increased from 79% to 89%.

5 Analysis

5.1 Description of the Scenarios

For the scenarios, two groups of initial values were established. The first group is focused on the distribution of innovative operations included in the new process (washing-peeling, centrifugation, and packaging), while the second group is the percentage of raw material and/or product that is lost as waste or decreased products in operations with higher losses (Chart 1).

Aspecto	Operación	Escenario 1	Escenario 2	Escenario 3
	Lavado-pelado	TRIA(1,2,2.5)	TRIA(2,3,3.5)	TRIA(3,4,5.5)
Distribución	Centrifugado	TRIA(0.5,1.5,2.2)	TRIA(1.5,2.5,3.5)	TRIA(2.5,3.5,4.5)
	Empaquetado	TRIA(0.5,1.,2)	TRIA(1.5,2.,3)	TRIA(2.5,3.,4)
% de merma/desperdicio	Lavado-pelado	4	8	12
	Remate	0.5	1.5	2
	Corte	1	3	4
	Zarandeado	0.5	1.5	2

Chart 1. Variables of the scenarios

The first scenario is the most favorable one, where variables with better values than those of the case study are presented. The second scenario is a slightly more realistic one, where distributions as well as losses of raw material and of in-process products are the closest to the results obtained in the case study. Finally, in the third scenario, the most pessimistic case is presented, where the waste level percentage does not vary in operations. Production time distributions have higher averages in this scenario.

5.2 Analysis of the Scenarios

To perform an analysis, a chart is drawn (Chart 2):

	Cantidad de unidades producidas (diarias)	Tiempo total del proceso (min)	Eficiencia del proceso	Rendimiento de materia prima
Resultado	200	10.33	74%	89%
Escenario 1	224	10.33	74%	94%
Escenario 2	194	12.50	59%	86%
Escenario 3	188	14.40	68%	80%
Promedio	202	11.89	69%	87%
% variacion max y min	16%	28%	20%	15%

Chart 2. Comparison of net flows in scenarios

On average, the number of units produced is 202, with a total processing time of 11.89 per unit. The average process efficiency is estimated at 69% with an expected 87% yield. The largest difference between these indicators is the total time of process, since any delay has a great impact on their calculations.

In addition, an economic evaluation of the scenarios is performed (Chart 3):

Scenarios	Current	Scenario 1	Scenario 2	Scenario 3
Savings	S/ 154,764.02	S/ 100,596.61	S/ 162,502.22	S/ 201,193.23
Expenses	S/ 23,163.90	S/ 15,056.54	S/ 24,322.10	S/ 30,113.07
Net flow	S/ 131,000.12	S/ 85,540.08	S/ 138,180.13	S/ 171,080.16
Variation		65%	105%	130%

Chart 3. Comparison of net flows in scenarios

According to data from the previous chart, the variation with the pessimistic scenario is 65% with regards to the current situation or scenario. In the normal scenario, the expected net flow variation does not change significantly. This is because the conditions remain the same. As for the optimistic scenario, there is a positive improvement of 30% in the net flow. This is mainly because the savings can increase if the results are better than expected, owing to the positive acceptance by workers, the implementation speed, and an increased fulfilled demand.

5.3 Future Processes

For future processes, other important issues to consider to generally improve the proposed model, which have not been discussed by the present study, include, but are not limited to,

- Joining of manual activities
- Inclusion of process validation forms
- Process automation systems

We add the Kanban technique to the proposed LM list.

6 Conclusion

According to the analysis that was performed, the company, on average, can increase the number of units produced to 202 units per day, with a production time of 11.89 min. Also, on average, process efficiency is expected to increase up to 69% with a performance of 87%.

One of the main problems faced by potato processors in Peru is losses due to the irregular shape of the native potato. In the case study of one of the leading SMEs in the sector, this problem is reflected in the raw material yield of 79% and a process efficiency of 59%. After having analyzed the operations of the process, a new production procedure is proposed taking into account these indicators. The implementation of the improvement proposal shows that it is possible to reduce the amount of raw material wasted and time spent by food processing SMEs, without having to automate processes. Through a performance evaluation form, we can quantify the losses of raw material and in process products, as well as determine bottlenecks. Consequently, the yield of the material is increased to 89% and the process efficiency to 74%. To achieve this result, a color card format is applied to determine what to do with the inputs. Finally, the proposal is profitable for SMEs since the investment is recoverable in the first year, with a resulting VAN of S/6,531.27 and a TIR of 135%.

References

1. Montgomery, D.C., Borror, C.M.: Systems for modern quality and business improvement. Qual. Technol. Quant. Manag. **14**(4), 343–352 (2017)
2. Moyano-Fuentes, J., Sacristán-Díaz, M.: Learning on lean: a review of thinking and research. Int. J. Oper. Prod. Manag. **32**(5), 551–582 (2012)
3. Rishi, J.P., Srinivas, T.R., Ramachandra, C.G., Ashok, B.C.: Implementing the lean framework in a small & medium & enterprise (SME) – a case study in printing press. IOP conference series. Mater. Sci. Eng. **376**(1) (2018)
4. Almanei, M., Salonitis, K., Xu, Y.: Lean implementation frameworks: the challenges for SMEs. Procedia CIRP **63**, 750–755 (2017)
5. Kumar, R.: Barriers in implementation of lean manufacturing system in Indian industry: a survey. Int. J. Latest Trends Eng. Technol. **4**(2), 243–251 (2014)
6. Das, A., Paul, H., Swierczek, F.W.: Developing and validating total quality management (TQM) constructs in the context of Thailand's manufacturing industry. Benchmarking: Int. J. **15**(1), 52–72 (2008)
7. Talib, H.A., Mohd Ali, K.A., Idris, F.: A review of relationship between Quality Management and organizational performance: a study of food processing SMEs. In: Proceedings - 2009 2nd IEEE International Conference on Computer Science and Information Technology, ICCSIT 2009, pp. 302–306 (2009)
8. Talib, H.H.A., Ali, K.A.M., Idris, F.: Quality management framework for the SME's food processing industry in Malaysia. Int. Food Res. J. **20**(1), 147–164 (2013)

Qualitative Features of Human Capital in the Formation of Enterprise Agility. Research Results in Polish Enterprises

Hanna Wlodarkiewicz-Klimek[✉]

Faculty of Engineering Management, Poznan University of Technology,
Ul. Strzelecka 11, 60-965 Poznan, Poland
hanna.wlodarkiewicz-klimek@put.poznan.pl

Abstract. Unpredictability is a crucial feature of the modern environment. It is a characteristic that gives the organization both opportunities and threats. The enterprises' success is determined by their ability to adjust efficiently their activity and potential to changes and challenges that appear in the environment. Companies that aim to stay on the market and act efficiently must become agile organizations. It means that they must be able to identify opportunities fast and use them for their own development. The estimation whether the appearing situation is recognized and assessed as favorable by the organization is determined by the potential and assets that it possesses or can potentially possess. Human capital is the superior factor that generates the potential of the organization and that integrates all other assets. Each action initiated by the organization is connected with involving human capital both in a causative and executive approach. The more the organization provides a good configuration of human capital features for the maximization of its quality, the better the results of its actions will be.

The aim of this paper is to present the results of research on a group of Polish enterprises that examined the qualitative features of human capital that affect the formation (stimulation) of enterprise agility. It has been chosen from a group of factors describing qualitative features of the human capital to point at characteristics that significantly contribute into the upgrading the level of agility of enterprises by identifying and reaching for opportunities. Qualitative features have been grouped into four groups: competences, interpersonal relations and contacts, culture and organizational climate. The results obtained from the research initiated a selection of a set of qualitative features of human capital that represent crucial importance for enterprise agility.

Keywords: Turbulent environment · Human capital · Agile enterprise

1 Introduction

The new reality in which agile organizations are developed is shaped by the primacy of knowledge. Knowledge in an organization must be taken heed of alongside human capital because people are not only carriers and conveyors of knowledge, but also its creators [1]. Human capital has the resources of information and knowledge as well as

© Springer Nature Switzerland AG 2020
W. Karwowski et al. (Eds.): AHFE 2019, AISC 971, pp. 63–72, 2020.
https://doi.org/10.1007/978-3-030-20494-5_6

wisdom. It also has predispositions, capabilities and skills to acquire, disseminate and use new knowledge in business processes. It is a stimulating, motivating and determining factor for undertaking any changes in an organization. The influence of human capital on shaping the behaviour of an organization should be considered in a broad context, taking into account a social dimension [2, 3], an organizational dimension [4–6] and an individual dimension [7–9]. Therefore, it can be assumed that an organization's ability to survive and develop in a dynamically changing business environment is possible primarily due to the potential of human capital. Properly shaped human capital significantly affects the efficiency of perceiving and using opportunities, which essentially contributes to the improvement of an enterprise's agility.

2   Factors Determining the Level of Enterprise Agility

Assuming that an agile enterprise is a knowledge-based organization capable of quickly identifying opportunities and using them for their own development, any relations that appear at the interface between an enterprise and the environment should be treated as a potential source of opportunities. The structure and dynamics of the environment in which an agile enterprise functions is of key importance. A variable and turbulent environment in its essence is a natural source of opportunities in both dimensions of stimulated and unexpected surprising events. Whether situations appearing in the environment will be recognized and assessed by an organization as conducive and beneficial for functioning and development depends on the potential that these entities have or can have at their disposal. The potential of enterprises is made up of all tangible and intangible resources necessary to achieve intended effects, which are in the possession of an enterprise or are possible to be obtained from the environment. From the point of view of an agile enterprise, the key potential for identifying, capturing and implementing opportunities is knowledge.

In the agile enterprise model, knowledge is considered to be inseparable from human capital. Man is treated as a carrier, creator and conveyor of organizational knowledge and is a key factor in an organization's success. From this point of view, it is important to recognize the qualitative characteristics of human capital expressed in competences, interpersonal relations, culture and organizational climate.

The context of the described conditions resulting from the adopted assumptions of shaping an agile enterprise is the basis for shaping and stimulating the development of dimensions (features) of agility, which are: brightness, flexibility, intelligence and shrewdness. Raising the level of agility in individual dimensions leads to an increase in the overall agility level of an enterprise.

In the description of individual dimensions of agility, particular attention was paid to their essence, as well as to actions that characterize them in the context of an organization's response to opportunities. The characteristics of individual dimensions of agility together with the features allowing to identify the general level of agility of an enterprise are as follows [10, 11]:

– **brightness**, the ability to quickly perceive market opportunities and threats in the environment, is associated with an enterprise's ability to:

- penetrate diverse segments of the environment (thanks to internal diversity and specificity of internal units),
- configure events and phenomena occurring in the environment into opportunities,
- comprehensively assess opportunities in terms of their value and risk.

An enterprise's key activities enabling it to achieve a high level of brightness are:

- constant monitoring of changes in macroenvironment and competitive environment,
- conducting a detailed assessment of potential benefits and risks for all new ventures undertaken by an organization;

- **flexibility**, understood as a feature of available resources consisting in extending the scope of their use and thus undertaking various tasks, can be determined by:
  - resource flexibility obtained as a result of their redundancy, diversification and universality,
  - flexibility of organizational structures, including network and organic structures. The most important actions taken by organizations to achieve a higher level of flexibility are:
  - purposeful maintenance of reserve production/technological/service capacities in the area of production resource management, which may be implemented at the appearance of a new venture,
  - employing staff (white-collar employees) who, thanks to their competences, are able to flexibly change and carry out various tasks,
  - making the organizational structure of management more flexible, which can each time adapt to emerging changes;

- **intelligence**, being an enterprise's ability to learn and understand the situation and to respond to it purposefully, is expressed through its ability to:
  - shape the strategic, tactical and operational management system,
  - develop the intellectual potential of human capital,
  - saturate the management system with creative staff,
  - develop the space of intelligent enterprise behaviours (methods and tools supporting particular business line innovations),
  - recognize the relationship between events in the environment,
  - obtain a fast flow of information in management systems,
  - concurrently implement processes,
  - standardize and automate information-decision processes,
  - maintain the efficiency of the information system,
  - increase the cost-effectiveness of the management system.

The increase in the level of intelligence by enterprises is mainly achieved by undertaking the following activities:

- stimulating and supporting the employee continuous development at all organizational levels,
- defining a binding (formally described) organizational strategy,
- fast and effective creation of the business relations network adequate to the needs reported by clients;

- **shrewdness**, which is characterized by the ability to use knowledge in new situations in practice, can be defined as an organization's ability to:
  - adapt in the environment understood as reactive adaptation to changes taking place in it,
  - make an impact on the environment in order to bring about beneficial changes in it,
  - change the environment understood as a transition to another environment.
    An enterprise's key activities enabling it to achieve a high level of shrewdness are:
  - organizational knowledge management,
  - ability to trigger off the activity of other entities, which results in the appearance of new opportunities,
  - skilful and fast adaptation of an enterprise's operations to the needs determined by the environment (in relation to the ability to configure and use both one's own and others' resources).

3 The Concept and Structure of Research on the Impact of Qualitative Features of Human Capital on Shaping the Agility of Enterprises

Due to the need for transparency and logic in the research procedure, the following assumptions were made [12]:

- agility is the ability of an enterprise to perceive and use opportunities;
- an enterprise's agility is described by: brightness, flexibility, intelligence and shrewdness;
- there are qualitative features of human capital, whose shaping leads to improving the agility of an enterprise. These include: competences, culture and organizational climate features as well as interpersonal communication features;

The main objective specified in the research process in the context of the assumptions of research has been operationalized to the following specific issues:

1. What behaviours regarding agility characterize modern enterprises?
2. What are the competences of human capital conducive to enterprise agility?
3. What level of the competence of human capital is important for achieving a high level of agility in the dimensions of: brightness, flexibility, intelligence and shrewdness?
4. What are the features of culture and organizational climate conducive to enterprise agility?
5. What level of culture and organizational climate is important for achieving a high level of agility in the dimensions of: brightness, flexibility, intelligence and shrewdness?
6. What are the features related to establishing and maintaining interpersonal contacts that support the agility of an enterprise?

7. What level of the competence of human capital is important for achieving a high level of agility in the dimensions of: brightness, flexibility, intelligence and shrewdness?

Due to the fact that the research problem was not consistently described and structured, neither theoretically nor empirically, quantitative research was adopted as the basic research method. The research procedure was based on conducting surveys aimed at diagnosing the existing state in terms of defining the level of agility of enterprises and assessing the implementation of qualitative features describing human capital (i.e.: competences, organizational culture, organizational climate), as well as features characterizing interpersonal communication.

The substantive basis for formulating the questionnaire was a detailed theoretical analysis of the problems of enterprise agility and human capital culminating in both areas with own original models based on knowledge and shaping human capital in an enterprise.

With regard to the area of diagnosis of the agility level of enterprises, the following structure of the survey was adopted: (1) The starting point for formulating questions were activities aimed at increasing agility in particular dimensions of enterprise agility (i.e. brightness, flexibility, intelligence and shrewdness) described in the model of an agile knowledge-based enterprise; (2) In the model, three activities favourable for the improvement of enterprise agility were assigned to each of the dimensions; (3) Further, each of the activities was described by three types of situations: (a) a situation corresponding to the behaviour of an enterprise with a high level of agility; (b) a situation characterizing enterprises with a medium level of agility and (c) a situation that identifies enterprises with a low level of agility.

When creating the structure of a research questionnaire related to the area of the qualitative features of human capital, based on literature studies, key elements were distinguished that characterize: competences, interpersonal contacts, organizational culture and organizational climate. The separated elements constituted the basic part of individual survey questions. The questionnaire was constructed in such a way as to obtain two basic categories of data. The first one referred to general information on the subjects studied and intended to create a research sample. The second part of the survey was the substantive part of the research, in which respondents rated the following six areas: (I) behaviours defining an enterprise's agility, (II) competences, (III) organizational culture and organizational climate, (IV) interpersonal communication.

The analytical part, apart from traditional induction-deduction methods, also made use of statistical methods implemented in the Statistica software package. To describe the actual state of the occurring phenomena, descriptive statistics tools were used, giving the possibility of a clear order and presentation of the data set. To determine the relationships occurring in the studied phenomena between variables characterizing the analyzed subjects, a chi-squared χ test between the analyzed pairs of variables was used. For determining the dependence, the level of significance was $\alpha = 0.05$. If there was a statistically significant relationship between the pairs of studied variables, an analysis of correspondence was carried out, the purpose of which was to indicate mutual interrelationships between them contained in the initial contingency table in the common space.

Aiming to determine the level of features favourable for the agility of enterprises, the analysis of average values depending on the dimension of agility was used.

The survey was conducted in 2017 on a sample of 109 enterprises located throughout Poland. The study covered mainly enterprises involved in industrial processing, a large group were also enterprises involved in transport and warehouse management. The proper part of the research was an online survey, which was preceded by contact and consultation by phone. The telephone conversation was conducted with a representative of the board or another authorized person, who then was asked to complete a questionnaire. The survey was conducted using the CAWI (Computer-Assisted Web Interview) method.

4   Conclusions from Empirical Research

The empirical research which aimed at getting to know and determining the impact of human capital features on the agility of an enterprise allowed to formulate conclusions. The following conclusions referred to the grouped categories indicated in the detailed research problems describing the main objective:

- **Behaviours characterizing agile enterprises**

Few (in the study only 11%) enterprises are characterized by a fully high level of agility, understood as achieving the highest degree of agility in the following dimensions: brightness, flexibility, intelligence and shrewdness. It looks different in the analysis of individual dimensions. The research shows that the largest group of enterprises is characterized by a high level of brightness, a relatively high percentage of enterprises also declare a high of shrewdness. However, in the case of flexibility and intelligence, the largest group of enterprises declares reaching the medium level. The observations made lead to the reflection that the surveyed enterprises have well-developed predispositions to monitor and assess the environment, as well as to use knowledge in practice in order to adapt to new situations taking place in the environment. However, the potential of efficient multidimensional management of internal resources and continuous learning of an organization is not yet fully exploited.

- **Qualitative features of the human capital**

Competences of human capital in achieving the agility of an enterprise
In terms of the impact of competences on enterprise agility based on statistical dependencies, it was identified that competences have the greatest impact on shaping flexibility and shrewdness, and to a lesser extent on intelligence. In addition, the study showed that brightness is the dimension of agility, the achievement of which does not depend on competence. A detailed set of competences, together with an indication of their impact on particular dimensions of agility, is presented in Table 1.

It was also observed that enterprises with a high level of agility achieve the highest average competence ratings. Therefore, it should be assumed that increasing the agility level of an enterprise will involve ensuring the competence of human capital in particular dimensions at an equal or higher level than theirs.

Table 1. The impact of competences on shaping agility in the dimensions of: brightness, flexibility, intelligence and shrewdness

Competences in the enterprise	Impact on shaping agility			
	Brightness	Flexibility	Intelligence	Shrewdness
Substantive / specialist knowledge relevant to the specific character of the enterprise	No	No	Yes	No
General knowledge	No	Yes	No	Yes
Education necessary for the proper performance of work	No	No	yes	Yes
Confirmed professional qualifications	No	Yes	Yes	Yes
Professional experience determining success in the industry	No	No	Yes	No
Creativity and innovativeness	No	Yes	No	Yes
Independence	No	Yes	No	Yes
Responsibility	No	Yes	No	Yes
Self-motivation	No	Yes	No	Yes
Entrepreneurship	No	Yes	No	Yes
Ability to establish business relationships	No	Yes	No	No
Professionalism	No	No	No	No
Decision-making	No	No	No	Yes
Self-discipline	No	Yes	No	Yes
Effectiveness and efficiency	No	Yes	No	Yes
Communication and interpersonal skills	No	No	No	Yes
Ability to cooperate in a team	No	Yes	No	Yes
Ethical conduct	No	Yes	No	No
Civility	No	No	No	No
Work culture	No	Yes	No	Yes

Culture and organizational climate in achieving enterprise agility

The results of the research procedure in the context of defining the features of culture and organizational climate affecting the attainment of the dimensions of brightness, flexibility, intelligence and shrewdness allow to formulate the conclusion on statistically significant connections between selected features and flexibility, intelligence and shrewdness. Brightness is the dimension the achievement of which is not dependent on culture and organizational climate, while the dimension with the greatest number of connections turned out to be flexibility. The analysis of the impact of individual features of culture and organizational climate on achieving agility in particular dimensions is presented in Table 2.

A dependency was also observed in relation to flexibility and intelligence indicating that enterprises with a high level of agility obtained the highest scores of statistically dependent features. This rule is not confirmed at the level of shrewdness.

Table 2. The influence of culture and organizational climate on shaping agility in the dimensions of: brightness, flexibility, intelligence and shrewdness

Organizational culture and climate	Impact on shaping agility			
	Brightness	Flexibility	Intelligence	Shrewdness
Conscious increase in employees' independence and responsibility	No	Yes	No	No
Employees' broad access to information and knowledge that can be important for work	No	Yes	Yes	No
Convincing employees of their special importance for the company's success	No	Yes	No	No
Developing teamwork	No	No	No	No
Striving for continuous development and improvement of various areas and activities of the organization	No	No	Yes	Yes
Perception of employees' competences as one of the most important resources of the enterprise	No	Yes	Yes	No
Joint resolution of conflicts and disputes	No	Yes	No	No
Coherent perspective (management's and employees') of looking at the organization and understanding of its activities	No	Yes	No	No
Openness and reactivity to the changing environment	No	Yes	No	No
Understanding the importance of the client in achieving success	No	Yes	No	No
Well-defined and implemented organizational strategy	No	Yes	No	No
Employees' identification with the goals of the organization	No	Yes	No	No
Shared vision of the organization's future	No	Yes	No	No
Employees' independence in decision-making processes	No	Yes	No	No
Willingness to share risk	No	Yes	No	No
Trust in honest treatment in the organization	No	Yes	No	No
Determining operating standards (timeliness, efficiency)	No	Yes	No	No
Forbearance and tolerance of mistakes	No	Yes	No	No
Justice and equal organizational policy (decisions taken independently of emotions and prejudices)	No	Yes	No	No
Stimulation for innovativeness and development	No	Yes	No	No

Interpersonal communication in achieving enterprise agility

The features of interpersonal communication show the lowest degree of connections with the agility dimensions of an enterprise. Individual features affect brightness and shrewdness, three out of nine affect flexibility and two shrewdness. A detailed analysis of the relationships is presented in Table 3.

In the case of flexibility, intelligence and shrewdness, the regularities observed in the previous areas of analysis are confirmed and refer to the highest level of statistically dependent features achieved by enterprises with the highest level of agility. This rule is not confirmed with respect to brightness.

Table 3. The impact of interpersonal communication features on shaping agility in the dimensions of: brightness, flexibility, intelligence and shrewdness

Features of interpersonal communication	Impact on shaping agility			
	Brightness	Flexibility	Intelligence	Shrewdness
Having clearly defined and commonly known rules and ways of communicating	No	No	No	No
Ensuring constant access to current knowledge about the current state and condition of the enterprise	No	No	No	Yes
Undertaking activities aimed at increasing the effectiveness of communication (meetings, discussions, forums, presentations)	Yes	No	No	No
Ensuring daily communication of the superior with the subordinate	No	No	No	No
Awareness of barriers and limitations in communication	No	No	No	No
Use of electronic systems supporting interpersonal communication	No	Yes	Yes	Yes
Sense of security and trust (employees are not afraid to share key knowledge)	No	No	No	No
Building positive interpersonal relations	No	Yes	No	No
Supporting and developing teamwork	No	Yes	No	No

5 Summary

Increasing interest in the concept of an agile enterprise from both the world of science and business has contributed to the reflection and research into qualitative factors shaping agility and affecting its improvement.

The research results presented in the article clearly indicate the important role of qualitative features of human capital in shaping enterprise agility. A special interest should be given to the dimension of flexibility, which shows the highest degree of sensitivity to the impact of qualitative features of human capital. Flexibility, as the ability of resources to expand the possibilities of their use, increases considerably with the development of personal competences and as a result of proper building of the organizational climate and culture. Also interesting are the results regarding the impact of interpersonal communication, which, contrary to expectations, has a rather insignificant influence on shaping agility.

To summarize, it is essential for enterprises wanting to consciously shape and raise their level of agility to focus, in a particular way, on creating conditions for developing and achieving those individual competences and employees' behaviour as well as intra-organizational culture and climate that indicate a significant relationship with agility.

References

1. Stańczyk-Hugiet, E.: Strategiczny kontekst zarządzania wiedzą. Wydawnictwo Akademii Ekonomicznej we Wrocławiu, Wrocław (2007)
2. Domański, S.R.: Kapitał ludzki i wzrost gospodarczy. PWN, Warszawa (1993)

3. Kałkowska, J., Pawłowski, E., Włodarkiewicz-Klimek, H.: Zarządzanie organizacjami w gospodarce opartej na wiedzy, Monografia. Wydawnictwo Politechniki Poznańskiej, Poznań (2013)
4. Wlodarkiewicz-Klimek, H.: The analysis and assessment of the degree adaptation of human capital in polish enterprises' to the knowledge-based economy requirements. In: IFKAD 2015—10th International Forum on Knowledge Asset Dynamics, Culture, Innovation end Entrepreneurship: Connecting the Knowledge Dots, Proceedings, pp. 143–154 (2015)
5. Baron, A., Armstrong, M.: Human Capital Management: Achieving Added Value Through People. Kogan Page, London and Philadelphia (2008)
6. Wlodarkiewicz-Klimek, H.: The changes of human capital structure in condition of adaptation the enterprises' management systems to the knowledge-based economy requirements. In: Advances in the Ergonomics in Manufacturing: Managing the Enterprise of the Future: Proceedings of the 5th International Conference on Applied Human Factors and Ergonomics, 19–23 July 2014, AHFE 2014, pp. 70–81 (2014)
7. OECD: The Knowledge-Based Economy. OECD, Paris (1996)
8. Fitz-enz, J.: The ROI of Human Capital: Measuring the Economic Value of Employee Performance. AMACOM American Management Association (2009)
9. Włodarkiewicz-Klimek, H.: Human capital in the development of mechanisms improving the agility of organizations. In: Trzcieliński, S. (ed.) Advances in Ergonomics of Manufacturing: Managing the Enterprise of the Future. Proceedings of the AHFE 2017 International Conference on Human Aspects of Advanced Manufacturing, July 17–21, 2017, The Westin Bonaventure Hotel, Los Angeles, California, USA, pp. 88–105 (2017)
10. Trzcieliński, S.: Zwinne przedsiębiorstwo. Wydawnictwo Politechniki Poznańskiej, Poznań (2011)
11. Włodarkiewicz-Klimek, H.: Agility of knowledge-based organizations. In: Schlick, C., Trzcieliński, S. (eds.) Advances in Ergonomics of Manufacturing: Managing the Enterprise of the Future. Proceedings of the AHFE 2016 International Conference on Human Aspects of Advanced Manufacturing, July 27–31, 2016, Walt Disney World®, Florida, Springer International Publishing, pp. 375–384 (2016)
12. Włodarkiewicz-Klimek, H.: Kapitał ludzki w kształtowaniu zwinności przedsiębiorstw opartych na wiedzy. Wydawnictwo Politechniki Poznańskiej, Poznań (2018)

Agile Manufacturing

Reflections on Production Working Environments in Smart Factories

Sebastian Pimminger[1(✉)], Werner Kurschl[2], Mirjam Augstein[2],
Thomas Neumayr[1], Christine Ebner[3], Josef Altmann[2],
and Johann Heinzelreiter[2]

[1] Research and Development, University of Applied Sciences Upper Austria,
Hagenberg, Austria
{sebastian.pimminger, thomas.neumayr}@fh-hagenberg.at
[2] School of Informatics, Communications and Media, University of Applied
Sciences Upper Austria, Hagenberg, Austria
{werner.kurschl, mirjam.augstein, josef.altmann,
johann.heinzelreiter}@fh-hagenberg.at
[3] School of Management, University of Applied Sciences Upper Austria,
Steyr, Austria
christine.ebner@fh-steyr.at

Abstract. The fourth industrial revolution has brought new challenges to the production workers in so-called smart factories. The flexible production process and many different product variants call for well-trained human workers. This paper contributes towards a better understanding of manual assembly workers in their everyday environment, since most advances in the production environment are technologically driven. Findings based on the Contextual Design methodology will be discussed and further analyzed in the crosscutting aspects of user, work, social, and environmental context.

Keywords: Human factors · Human-Systems integration ·
Systems engineering

1 Introduction

Modern production working environments allow fully customized products in response to customer needs that are manufactured on demand with lot sizes down to one. However, this trend, often referred to as the fourth industrial revolution, is not enabled by fully automated systems and robots, but by highly skilled, well-trained and well-practiced human workers [14].

This paper contributes towards a better understanding of manual assembly workers in their everyday environment. We want to gain a better understanding of their real needs and identify contextual influences from a holistic user-centered perspective. For our reflections on production working environments, we follow the Contextual Design methodology described by [10]. We draw on data gathered in four Contextual Inquiries and four additional interviews conducted at two companies. One company is a manufacturer of automation technology, the other is a manufacturer of welding machines.

© Springer Nature Switzerland AG 2020
W. Karwowski et al. (Eds.): AHFE 2019, AISC 971, pp. 75–85, 2020.
https://doi.org/10.1007/978-3-030-20494-5_7

We focused on humans carrying out manual assembly tasks and stakeholders (e.g. production line scheduler and shift supervisor) in close contact with them.

The following sections cover our Contextual Design process with Contextual Inquiries and the Contextual Analysis, which resulted in an affinity diagram. The contextual data was further analyzed by categorizing the data into user, work, social, and environmental context, which gives a new perspective on the data. We also show how these more than 600 affinity notes are distributed among the four context types.

2 Background

The Contextual Design (CD) methodology describes a framework for planning and implementing a user-centered design process throughout all phases of a project. CD was first invented in 1988 and published as an "emergent method for building effective systems" by Wixon et al. in [17]. In 1993, Holtzblatt and Jones published a more detailed description of the Contextual Inquiry (CI) method, which is the recommended method for early project phases [9]. The CI is usually structured as an interview in which the researcher observes the user's day-to-day activities and discusses those activities with the user. The first book with an in-depth description and an extensive guide to prepare and implement a CD process was provided in [3]. This work explains all CD phases and provides practical examples. A more dense guide for practitioners that focuses on core techniques was introduced as Rapid Contextual Design in [10]. In the latest iteration, CD was overhauled in order to account for changes in technology and its influence on humans' live (e.g. "always-on, always-connected and always carried devices") [11].

Although the application of CD involves a wide range of advantages (e.g. uncover hidden information and identify actual user needs), it is often not used, especially in time-critical projects. Nevertheless, some examples for applications of CD in research projects are as follows: The General Motors UX design team describe the first of several CD projects in [6]. They focused on gaining deeper understanding of how drivers interact with today's entertainment, communication, navigation and information systems in their vehicles. In [4], Coble et al. describe a CI process conducted to gather physicians' requirements for a clinical information system. Another example for a user-centered clinical IT system design is described in [16]. Fouskas et al. applied CI for gathering users' requirements in the context of mobile exhibition services [5].

The fourth industrial evolution is changing the industrial production process again. We move from mass production of few items to an ever greater number of variants of products, taking into account even more specific consumer needs [2, 12]. This change has an impact on the production process and the involved production workers, because assembly tasks vary in a smart production depending on the lot size and the specific product. This requires flexible and highly skilled human production workers. This kind of flexibility adds additional complexity to the production process, which can lead to cognitive load and the risk of more errors, which must be addressed by assisting humans with appropriate tools and systems [7, 13]. Since full automation is currently not feasible, because production facilities cannot be fully automatized due to the rising number of product variants and short lifecycles, it is therefore very important to understand the needs of the human production workers in a smart factory to be able to address them accordingly with appropriate measures and assisting tools.

3 Method

For our reflections on production working environments, we draw on data gathered in four Contextual Inquiries and four additional interviews gathered at two companies. One company is a manufacturer of automation technology; the other is a manufacturer of welding machines. We focused on humans carrying out manual assembly tasks and stakeholders in close contact with them. The CI participants were production employees, the four standalone interview partners were a production line scheduler, a shift supervisor, a team leader and a manager. The data gathered in the interviews and observations was consolidated and used to create an affinity diagram and personas. A detailed description of our contextual design process and experiences can be found in [1].

3.1 Contextual Inquiries

The CI sessions and interviews were conducted directly at the production facilities of the respective company. We focused on the daily work of the human assembly workers, aiming to explore how they experience and execute their assembly tasks. The CI sessions comprised of (1) an interview with open questions (60 min), (2) an observation at the participants' workplace carrying out their daily tasks (about 45 min). This involved the assembly of a complete product and was done in a production line with several workstations (by a single employee). One observation took place in the context of a training where the participant had not manufactured the product before and got step-by-step instructions from an experienced colleague. The final part of the CI session was (3) a wrap-up to clarify open questions and make sure there were no misinterpretations (15 min).

The full CI sessions were done with four assembly workers and interview-only with five so-called indirect users (a production line scheduler, a shift supervisor, two team leads, and a manager). Participants' mean age was 32 years (SD = 12.00), ranging from 21 to 59 years (three participants without exact indication of age and therefore not taken into account), 7 participants were male and 2 female. Their average working experience in their current position was 10.67 years (SD = 12.56) ranging from 2 to 44 years.

All interviews were recorded with audio and the observations were recorded with video. Additionally, important situations and artefacts were photographed. In total, we gathered about 13 h of audio, 10 h of video recordings, about 150 photographs and 35 pages of written notes.

3.2 Contextual Analysis

During the analysis phase, the recorded material was reviewed and transformed into affinity notes. Affinities notes define the basic building blocks of the CD methodology and describe a meaningful event, problem or issue in the context of our users. The material included audio and video recordings and the notes taken during the interviews and observations.

In the next step we transformed the affinity notes into a so called affinity diagram. The affinity diagram tries to identify different issues within the collected data by organizing the notes in groups under common labels (blue labels). These labels are further grouped together (pink labels) and summarized into whole areas of concerns (green labels). This creates a hierarchical structure of four levels.

The affinity building was done in several team workshops with in-depth discussion of the observed material resulting in 682 organized notes and labels (see Fig. 1). Overall six main areas of concern (green labels) were identified. Based on these areas and the underlying data originating directly from the actual work context of our users, the findings were interpreted in relation to our research and design goals.

Fig. 1. Final affinity diagram.

3.3   Context Type Classification

The affinity diagram organizes a large number of ideas. However, with the increasing size of affinity diagrams, it becomes more difficult to understand and comprehend them in their entirety. Yet, this is necessary in order to deduce requirements, visions and other design-informing models. Although the Contextual Design process provides methods such as the Wall Walk to handle this complexity, it may still lead to difficulties for non-experienced team members to work with this amount of information. Another difficulty we encountered is that the affinity diagram mixes different context types all over the diagram. If you are working with data in question, you always have to keep the entire information of the diagram in mind.

Therefore, we pick up the idea of [18] and assign different context types to each blue label. This will generate a new view on the data of the affinity diagram. For example, if you want to focus on the design of workplaces, you may simply access the environmental context and get all relevant data across the different areas of the affinity diagram. In our approach, we pick the first level groupings (blue labels) for the classification with context types. As highlighted in [11], the blue labels summarize the key points of the underlying data without abstracting away too much information (like the overlying level of pink labels).

For the classification, we use the following four context types based on the work of Wurhofer et al. [18]. Although they are well known from literature there is no common definition and understanding. Therefore, we use the context types as follows:

User context. Characteristics of the user like skills, values, or knowledge.
Work context. Characteristics of work like work organization, tasks, and staff.
Social context. Characteristics of the social environment like role, hierarchy, or relationship to other people.
Environmental context. Characteristics of the physical environment like noise, light conditions, or temperature.

Multiple experts should do the classification individually. Possible different ratings can be resolved in a group discussion afterwards.

4 Findings

In the following, we describe our findings on how manual assembly employees experience their day-to-day work. The first part describes the findings grouped by the main categories of the affinity diagram. The second part creates a new view on these data by categorizing the blue labels of the affinity diagram in four context types. All quotations and technical terms have been translated from German to English.

4.1 Main Areas of the Affinity

The affinity diagram consists of 682 affinity notes, which describe events, problems and issues in the context of manual assembly workers' daily work routines. These affinity notes are grouped into 154 blue labels. Further, the blue labels are aggregated to 45 pink labels and six green labels which identify the top areas of the diagram.

Assembly Worker. Assembly workers do not need a special training for their job. Often career changers are hired. A minimum qualification for new employees is to speak German. Since many people from different cultural backgrounds work together in this area, this is important for personal interaction and for reading and understanding assembly instructions. Currently, these are written in German exclusively. Hiring technically experienced employees often leads to shorter and more efficient training. These workers are also more likely to be faster in the assembly and training process. However, finding well-qualified employees is difficult.

The average assembly worker is between 20 and 50 years old and female. There is a high turnover rate and often employees are leased laborers or part-time employees. Assembly work is carried out alone or in pairs manufacturing various products (with multiple configurations) on various assembly stations depending on the order situation.

Training. Experienced colleagues usually train new employees. However, this means their manpower is missing in the day-to-day routine. Preferred learning methods include observations, collaborative work and working alone in a controlled environment. Learning only by instructions, paper or digital, is usually not enough. Furthermore, experienced colleagues pass on tips and tricks, which are often not documented.

The training usually happens directly at the workplace. A dedicated training workplace for practicing difficult work steps would be desirable, but is often not available. The supervisor or team lead decides when an employee is ready to work independently. The key factor is that a certain level of production and quality is reached at the end of the training.

Workplace. Assembly workers prefer a well-organized and clean workplace. All tools are at hand and are taken from and returned to the designated place. All materials and components are arranged sequentially, resulting in an efficient production flow. Ergonomic aspects, such as adjustable table heights are also important.

At the workplace, assembling happens while standing. Sitting would restrict the freedom of movement too much. Individual tasks also require repeated stooping and kneeling. The tasks may also be physically exhausting or require lots of strength. To protect components and assembly parts, the electrostatic discharge (ESD) concept must be strictly adhered to. Therefore, employees need to wear special clothing such as work coats, shoes and gloves. Most tasks require some sort of hand or power tools (e.g. pliers or tweezers). One of the most important tools are electronic tightening systems. Some work steps require bolts to be tightened to their proper torque and angle specification. These are regulated in the guideline VDI/VDE 2862 [15]. Other assembly aids (e.g. assembly jigs, mounting rails, clamps) serve as a third hand. Some assistance systems are necessary (e.g. display for work instructions); other assistance systems facilitate work. For example, Pick-by-Light is a valuable support, but may not meet expectations (complex and expensive to operate). Assembly workers are critical of new technologies and feel insecure in dealing with them.

Assembly Instructions. The initial assembly instructions are prepared at the production facilities. They are based on documentation (e.g. technical manual) by the engineering department and improved with additional and more detailed information like screwing order torques and illustrations. Assembly workers use these instructions for training and guidance during their actual work. The instruction manuals are provided either analog or digital. In general, assembly instructions should be short, clear and understandable. Since digital instructions provide interactive capabilities such as zooming, videos and automatically displaying the current instruction, they can often be better understood. Workers have problems with instructions if they are not up-to-date, inaccurate, incomplete, or contain too much information. In this case, it is often easier and more efficient to ask a colleague for help. Another challenge is the manufacturing of many product variants. This requires special attention as the instructions may vary depending on the selected product variant and may have to be created on the fly. In addition, multi-variant instructions tend to be more complex and, therefore, may cause more assembly errors and problems in the manufacturing process. The modification of instructions is subject to a defined process. Workers file change requests or give feedback to their supervisors and team leads. An interactive editing feature on the display would be desirable.

Assembly Tasks. The production line consists of several workstations that are connected one way or another. Test equipment and packaging can be located in the line, but also spatially separated. The production line goes through a specified lifecycle and is constantly being optimized and changed. Multiple products can be produced in one production line, each with several variants. The assembly process starts with a placed production order. All data is tracked and documented in the manufacturing execution system (MES). Typically, orders with different quantities are processed, ranging from quantities of one to 70, with an average of three pieces per order.

In the first assembly step, each product receives a unique identification (e.g. label with bar code). At each workstation, the product (and possibly also parts and sub-assemblies) must be scanned for documentation. In addition, the assembly process is started or stopped via a terminal (with touch display). Each assembly step must be confirmed (via touch display or foot switch). During the assembly process, the workers take the materials and parts from Kanban bins or a Kitting Rack [8]. All materials are prepared beforehand and are available in the line (Kanban system). The worker usually carefully lays all the necessary parts on the assembly area. Often these must be cleaned with special cleaning supplies. However, this step is rarely found in the instructions. For a single screwing task, sometimes several steps are needed. There are different types of screw heads (e.g. Torx and Hex key) and they have to be changed multiple times. The torque may also be adjusted according to instructions. Some screwing tasks require by law special electronic tightening systems that are suitable for safety-critical tightening jobs in accordance with the guideline VDI/VDE 2862 [15]. This system may also specify a particular screwing sequence. These systems require the work piece to be always in the same position in order to work correctly. Craftsmanship and manual skills (e.g. fixing the piece of work with arms and screwing at the same time) and certain tricks (e.g. to avoid bending of cables while assembling) are also necessary for assembling products. At the end of the assembly step, the workers should check their activities against the instruction and with their tactile and auditory sense. Then, the product is taken from one workstation to another or placed in a buffer between them.

If workers have any questions or need help, they should contact their team lead. However, they often hesitate and do not dare to ask. The workers receive feedback on key production figures and their performance as team. They can also submit their own suggestions for improvement via the continuous improvement process (CIP).

Assembly Error. In general, assembly errors happen rarely. The errors occur due to incorrect positioning of parts, incorrect mounting and wrong tool usage. Reasons for this can be for example, fatigue, lack of concentration, ambiguous instructions and complex assembly tasks.

Errors happen regardless of the experience and qualifications of the workers. However, for freshmen and inexperienced employees, errors occur more frequently. They are especially prone to error in the manufacturing process (e.g. wrong login to the station terminals or wrong testing equipment and procedures). However, faulty materials like upstream processes (pre-assembly of modules or design errors) may also cause errors. Workers often cannot recognize these errors because they are hidden to some extent (e.g. a supplier delivers wrong parts).

Workers discover their faults through independent testing during the assembly process or through automated test systems (with specified test procedures). The personal competence is an important factor when conducting independent tests. There is often special training, which helps the employees finding and assessing assembly errors. Unfortunately, not all errors can be detected during the assembly process either by humans or by test systems. Some errors only occur at the final product test or even, in the worst case, at the customer. Since errors mean negative consequences (e.g. additional costs for repairing, reputation with customers), employees must pay special attention to error prevention.

Depending on type and classification, assembly workers may repair errors on-site in the production line or the product is transferred to dedicated repair stations. Whenever errors happen, the workers document them and report to their supervisor. Depending on the nature and type of the error, there are different responsibilities and ways in which these are processed (e.g., construction errors are reported back to the engineering department). For assembly workers, feedback on errors is important. It will help them in the future to avoid them. Often, this feedback comes directly from colleagues, their supervisor or as general information posted on bulletin boards.

4.2   Crosscutting Context Types

Our analysis is based on the 154 blue labels of the affinity diagram. Each label has been categorized in one of the four context types *user context* (UC), *work context* (WC), *social context* (SC) and *environmental context* (EC) by three experts individually. Out of 154 blue labels, there were 89 classifications with exact matches, 57 classification with two matches out of three, and 8 classifications with no matches at all. In a group discussion, all classifications were merged into a final result.

As shown in Table 1, most labels are assigned to WC (63%), followed by UC (22.7%), SC (8.4%), and EC (5.8%). This helps us to identify crosscutting topics that are otherwise scattered throughout the affinity and may easily be overlooked.

In the following, we present our findings with these four context types. We do not repeat the actual contents of the affinity (they can be found in Sect. 4.1), but describe the respective context type and which questions it can answer.

Table 1. Classification of the blue labels of the affinity diagram to the context types *user context* (UC), *work context* (WC), *social context* (SC) and *environmental context* (EC). The number depicts the quantity of labels in the particular context types and the distribution of blue labels in the affinity diagram and context type classification in absolute numbers and percentages.

Area/Context type	UC	WC	SC	EC	$\sum$	%
Assembly worker	11	4	5		20	13.0
Training	6	4	2		12	7.8
Workplace	3	13		9	25	16.2
Assembly instructions	3	15	1		19	12.3
Assembly tasks	5	36	2		43	27.9
Assembly errors	7	25	3		35	22.7
$\sum$	35	97	13	9	154	
%		22.7	63.0	8.4	5.8	

User Context. The UC represents the characteristics of assembly workers. It collects requirements and minimum qualifications for this type of jobs as well as skills and knowledge that is necessary to carry out daily assembly tasks. This includes the training of assembly tasks, basic tool usage and also tips and tricks from colleagues that are not included in official instructions. The UC also gives answers to the questions of what causes workers feel overloaded and what expectations do they have of their jobs.

As our findings in the affinity diagram suggests, the finding and fixing of assembly errors are linked with the skills of the workers and thus with the UC. Furthermore, the UC indicates which assembly errors are caused by the workers themselves. This is also supported by the data in Table 1, as assembly errors contribute to the UC as the second largest issue of the affinity.

Work Context. WC is by quantity the most significant context type and is represented in all areas of the affinity (see Table 1). The WC is centered on all aspects related to assembly tasks. This includes how tasks are carried out, what tools and equipment are used and how parts and components are handled. For assembly workers it is also important to know how to deal with assembly errors as they happen. The necessary steps include the detection, repair and documentation. Furthermore, WC covers the whole work organization like shift work and training that is relevant for workers to know about. This is especially true for the training, as the WC describes who will train freshman workers and how and where this training will happen. Generally, WC seems to be closely related to SC. The execution of work tasks are closely linked with knowledge and skills of the workers.

Social Context. The SC describes the social setting in which the assembly work takes place. As Table 1 highlights, all areas (except workplace) in the affinity diagram describe aspects of social relationships and influences on the assembly workers. With help of the SC, we can identify all social interfaces that interact with the workers and answer questions like who are my colleagues and how do I interact with them. It also illustrates how collaboration in the production line takes place. The SC may also identify problems while interacting with others. Another important aspect of the SC is feedback and communication between workers and their supervisors. For example, the team receives feedback and general information (e.g. common assembly errors and key production figures) from their supervisor. On the other hand, workers provide feedback on encountered problems and issues to their supervisor.

Environmental Context. In our analysis, the EC is built exclusively from data of the area of the workplace. The EC characterizes the physical structure and environmental conditions of the workplace. It provides an overview of objects the worker interacts with (besides the actual work piece). For assembly workers, ergonomic aspects are important since they make their daily work easier and less physically demanding. The EC also gives indications on safety regulations that have to be respected by the assembly workers.

5   Discussion

In general, we can encourage other researchers in using CD. With help of the CD/CI methodology, we could uncover numerous requirements, constraint and influences that would otherwise most probably have stayed hidden. For example, assembly tips and tricks play an important role in effective manufacturing, but are nowhere documented and only passed on from colleagues to colleagues. We believe, that the additional effort pays off in the long run, as we hope to get better results than with other methodologies and techniques (e.g. traditional interviews and focus groups). For example, the onboarding of new project employees seems more effective with tools like the affinity diagram.

The analysis of context types helps us to extract specific contextual data with relative low effort. In our perspective, the newly generated views help us to better summarize and integrate the data. For example, filtering the contextual data by UC, reveals characteristics of specific work roles that can be easily turned into personas. Another example would be SC, which identifies human interfaces that interact and collaborate with users in question. Additionally, this approach allows some quantitative analytics, for instance, how many affinity notes contribute to a specific context. It also shows the distribution of areas like worker, training, workplace, work instructions, assembly tasks, and assembly errors among the different contexts. For example, 16% of the affinity notes in our data are linked to the workplace area, while only 8% are linked to the training area.

6 Conclusion

We started with an in-depth contextual inquiry in a smart factory, which lead to 682 affinity notes. Although the Contextual Design process provides methods such as the Wall Walk to handle this vast number of affinity notes, it can still lead to difficulties for non-experienced team members to work with this amount of information. Therefore, we picked up an idea loosely based on [18] and assign different context types (e.g. WC, EC) to labels of the affinity diagram. This generates a new view on the data for specific contexts that helps us to better understand and analyze the data. This new view with separated different contexts allows us to focus on a specific context, which is by default intermingled in the affinity diagram. This separation can for example be useful, if someone wants to design a new assisting tool for a smart factory worker. In this case, the work context may lead directly to the relevant affinity notes.

Acknowledgments. Our work has been conducted within the scope of the project Human-Centered Workplace 4 Industry (HCW4i), funded through the COIN program and managed by the Austrian Research Promotion Agency (FFG).

References

1. Augstein, M., Neumayr, T., Pimminger, S., Ebner, C., Altmann, J., Kurschl, W.: Contextual design in industrial settings: experiences and recommendations. In: Proceedings of the 20th International Conference on Enterprise Information Systems—Volume 2: ICEIS. INSTICC, SciTePress, Portugal, pp. 429–440 (2018)
2. Bauernhansl, T., Ten Hompel, M., Vogel-Heuser, B.: Industrie 4.0 in Produktion, Automatisierung und Logistik: Anwendung, Technologien und Migration. Springer Vieweg, Wiesbaden (2014)
3. Beyer, H., Holtzblatt, K.: Contextual Design: Defining Customer-Centered Systems, 1st edn. Morgan Kaufmann, San Francisco (1997)

4. Coble, J., Maffitt, J.S., Orland, M.J., Kahn, M.G.: Contxtual inquiry: discovering physicians' true needs. In: Proceedings of the Annual Symposium on Computer Application in Medical Care (1995)
5. Fouskas, K.G., Pateli, A.G., Spinellis, D.D., Virloa, H.: Applying contextual inquiry for capturing end-users behaviour requirements for mobile exhibition services. In: Proceedings of the 1st International Conference on Mobile Business, pp. 8–9 (2002)
6. Gellatly, A., Hansen, C., Highstrom, M., Weiss, J.P.: Journey: general motors' move to incorporate contextual design into its next generation of automotive HMI designs. In: Proceedings of the 2nd International Conference on Automotive User Interfaces and Interactive Vehicular Applications, pp. 156–161. ACM, New York, USA (2010)
7. Gorecky, D., Schmitt, M., Loskyll, M.: Mensch-Maschine Interaktion im Industrie 4.0-Zeitalter. In: Industrie 4.0 in Produktion, Automatisierung und Logistik. Springer Fachmedien Wiesbaden, Wiesbaden, pp. 525–542 (2014)
8. Hanson, R., Medbo, L.: Kitting and time efficiency in manual assembly. Int. J. Prod. Res. **50** (4), 1115–1125 (2012)
9. Holtzblatt, K., Jones, S.: Contextual inquiry: a participatory technique for system design. In: Schuler, D., Namioka, A. (eds.) Participatory Design: Principles and Practices. Chapter 9. Lawrence Erlbaum Associates (1993)
10. Holtzblatt, K., Wendell, J., Wood, S.: Rapid Contextual Design. Elsevier/Morgan Kaufmann, San Francisco (2005)
11. Holtzblatt, K., Beyer H.: Contextual Design: Design for Life. Morgan Kaufmann, Cambridge (2017)
12. Koren, Y.: The Global Manufacturing Revolution: Product-Process-Business Integration and Reconfigurable Systems. Wiley, New York (2010)
13. Peissner, M., Hipp, C.: Potenziale der Mensch-Technik Interaktion für die effiziente und vernetzte Produktion von morgen. Fraunhofer Verlag, Stuttgart (2013)
14. Pfeiffer, S.: Robots, industry 4.0 and humans, or why assembly work is more than routine work. Societies **6**, 16 (2016)
15. VDI-Richtlinie: VDI/VDE 2862 Blatt 2 Minimum requirements for application of fastening systems and tools - Applications in plant construction, mechanical engineering, equipment manufacturing and for flange connections in components under pressure boundary (2015)
16. Viitanen, J.: Contextual inquiry method for usercentered clinical it system design. Stud. Health Technol. Inform. **169**, 965–969 (2011)
17. Wixon, D., Holtzblatt, K., Knox, S.: Contextual design: an emergent view of system design. In: Proceedings of the SIGCHI Conference on Human Factors in Computing Systems, pp. 329–336. ACM (1990)
18. Wurhofer, D., Buchner, R., Tscheligi, M.: Research in the semiconductor factory: insights into experiences and contextual influences. In: 7th International Conference on Human System Interactions (HSI), pp. 129–134. IEEE (2014)

The Relation of Proexploitation Attributes with Selected Criterion of Agility of Public Transport Vehicles Manufacturing

Joanna Kalkowska[(✉)]

Faculty of Engineering Management, Poznan University of Technology,
Strzelecka 11 Str., Poznan, Poland
joanna.kalkowska@put.poznan.pl

Abstract. The paper presents the relation of proexploitation attributes with selected criterion of agility of public transport vehicles manufacturing. Moreover, it describes the expert method of inferring about future states of public transport vehicles (exploitation and service phases as well as elimination and utilization phases) based on earlier process of product development and manufacturing. A numerical formalization of the expert intuition of twenty specialists is a starting point of the method. The examined dependency regards the relation between pro-exploitation attributes of vehicle with the selected criterion of agility associated with product development process of vehicles. Hypothesis about the essential relation of selected criterion of agility with proexploitation attributes of vehicle was verified with the method of progressive inference.

Keywords: Agility · Exploitation · Lean manufacturing

1 Introduction

The goal of this paper is to present some research results concerning the relation of proexploitation attributes with selected criterion of agility of public transport vehicles manufacturing process. Public transport vehicles are usually ranked among the group of transport vehicles within the notional category machines. The research analysis involves three large enterprises producing buses, trams and local metropolitan coaches. The problem-related presented in the paper has the character of a relation linking attributes of the proexploitation of a given group of public transport vehicles selected criterion of agility. Agility is a concept which is relatively widely presented and discussed in literature. Early definitions of agility were studied by Kidd, Goldman, Nagel and Preiss, Yusuf and many others [1–3]. Agility is understand as a capability to use market opportunities according to the brightness, flexibility, intelligence and shrewdness [4]. Agility is a concept which is dedicated to enterprises, however, agile enterprise needs a changeable environment as only such environment generates opportunities. Moreover, according to the one of the newest approach to the agility represented by Trzcieliński [5], the existence of an agile enterprise is justified by the presence of opportunities [6]. However, in agile production management, first and foremost, the key is to strive for excellence, which defines production standards at the

© Springer Nature Switzerland AG 2020
W. Karwowski et al. (Eds.): AHFE 2019, AISC 971, pp. 86–91, 2020.
https://doi.org/10.1007/978-3-030-20494-5_8

level called the world class (World Class Manufacturing) [7]. Concepts and models of agility were considered in the monograph as a continuation of previous concepts and models of leanness. In addition, they were treated as the culmination of approaches known as World Class Manufacturing. Agility has become (rooted in models of leanness and in world class standards) a response to the situation in which the paradigm of liquidation of waste, associated with the concept of Lean proved to be the insufficient defence of an enterprise producing transport vehicles against further accelerations of business environment volatility.

2 Exploitation and Proexploitation Attributes

Exploitation definition involves the stage of using and operating the vehicle, in addition in the system of his use an use is a basic process [8]. However, functional features are providing about directing at improving manufacturing processes, according to the current state of technology. Generally, the contemporary problems of the exploitation concern five thematic areas [8, 9]:

- the description, the modelling and improving objects and systems of the exploitation,
- strategy of the reconstruction and reconstructions of the exploitation systems,
- predicting the permanence and reliabilities,
- providing the efficient productive maintenance,
- economic of the exploitation and services.

Based on the exploitation meaning, next, the definition of the proexploitation approach to improving the processes of public transport vehicles creation was specified as: a set of purposeful construction, organizational, technological and economic activities undertaken in relation to the production process of microbuses, buses, trams and passenger rolling stock of regional railways (including its conception, maturing and stabilization) in order to make the most efficient and appropriate (from the point of view of the customer) use of manufactured vehicles at the stages of their use, servicing, liquidation and disposal after the end of operation [10].

On the basis of thirteen so-called good exploitation practices [11, 12], the following seven attributes of exploitation of minibuses, buses, trams and passenger rolling stock of regional railways were established:

- E1-effectiveness and functional efficiency (including the ease of starting exploitation) and economical use and maintenance (exploitation proper).
- E2-reliability and durability, resistance to atmospheric conditions and damage.
- E3-exploitation safety (including elimination of accidental start-ups, unauthorized use, maintenance and possible intentional damage).
- E4-ergonomics (including general aesthetics of a vehicle, configuration and ease of operation, resistance to operating errors and instantaneous overloading).
- E5-ecological operation.
- E6-diagnostic and service susceptibility; repairability and audit control susceptibility.
- E7-recycling and disposal susceptibility [10].

3 Criterion of Agility and Proexploitation Attributes – Issue and Relation

In order to check the titled relational hypothesis based on the Delphi method, an expert method for predicting the dependence of specific attributes of the proexploitation of public transport vehicles on selected agility criterion was developed.

The inference about later exploitation states of public transport vehicles (use and maintenance phase and liquidation and disposal phase) was carried out using the author's method of numerical formalization of expert intuition of specialists. For the needs of the conducted research, the following eleven criteria of lean production and organizational agility of an enterprise producing public transport vehicles were defined (with the unchanged number of proexploitation attributes) [13]:

- A1-Rapid growth of production capacity and productivity due to the flexibility and easy reconfiguration of production systems and thanks to its own assembly supported by external production processes and outsourcing.
- A2-Maximizing and improving the effectiveness of machines and technical devices (Total Productive Maintenance, Overall Equipment Effectiveness) and systematic and continuous improvement of technical preparation and implementation of production, carried out in a short time following the concept of: Kaizen, Just in Time, 5S, SMED, FMEA, Ishikawa Diagram, zero stocks, low capital involvement and conducting simulation experiments.
- A3-Automation of proper production processes of public transport vehicles as well as their integration and autonomy (Jidoka) using information and communication technologies after previous re-engineering of these processes (Business Process Reengineering, Kaikaku).
- A4-Effective task and activity time management, continuous improvement of processes (PDCA), using Outsourcing and the rules of supply chain management.
- A5-Production of public transport vehicles with a short life cycle but reliable in operation (in case of product failure immediate repair or replacement with a new one).
- A6-Customer satisfaction and loyalty as the company's goal (Mass Customization manufacturing systems integrated with quality - Total Quality Management - and quick response to customer needs).
- A7-Employees' identification with the interest of the company (Identity) and showing willingness to raise the level of knowledge and improve own competences (Open Book Management).
- A8-Flat organizational structure of an enterprise (empowerment and relative autonomy of the operational level - Employee Empowerment); management through team-based activities and leadership (Leadership) based on participation and susceptibility to change and refinement.
- A9-Targeting enterprise cost management towards productivity and quality.
- A10-Flexibility of the scope of changes in business processes, enabling the introduction of these changes in a quick and economical way.
- A11-Fulfilling by an enterprise the social requirement of environmentally friendly company management - providing employees with personal and social satisfaction.

In the research carried out for the purposes of this paper, the dichotomous relationship (existence of dependence or lack thereof) linking the area of seven exploitation attributes with the individual eleven criteria of lean production and organizational agility of public transport vehicle creation processes was adopted. It was considered that the existence of a dependence relationship can be mentioned when the numerical value of the median along with the lower limit of the interquartile range exceeds the lower level of significant dependence [10]. In a matrix form, the final results of the study of the dependence of seven attributes of the pro-operation of public transport vehicles on the eleven criteria of lean production and organizational agility of the processes of creating public transport vehicles are presented in the form of a graphic model (Fig. 1).

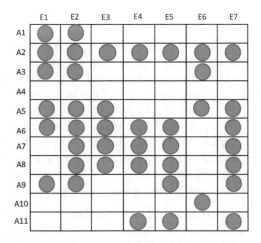

Fig. 1. The relation of proexploitation seven attributes of public transport vehicles on the eleven criteria of organizational agility of the processes of creating public transport vehicles

4 Closing Remarks

The most important conclusions (of the nature of cognitive and utilitarian achievements) from the conducted research were formulated as follows:

- The median value along with the lower limit of the interquartile range in as many as forty one cases, the study covering seventy seven potential relationships, exceeded the adopted level of significant dependence.
- The following two attributes of proexploitation regarding reliability and usability and service durability, and the resistance of a public transport vehicle to atmospheric conditions and damage as well as to recycling and usability susceptibility proved to be the most dependent on the possible implementation of the organizational criteria of the agility of the production processes of public transport vehicles.
- Possible implementation of organizational criteria for the agility of production processes of public transport vehicles is also very beneficial from the point of view

of improving their efficiency and functional effectiveness (including ease of commencement of operation) and the economics of use and operation (proper operation). A similar high level of expert assessments was obtained by positive effects of organizational agility on the eco-friendliness of use and servicing of the vehicles in question.

– The remaining three attributes of proexploitation public transport vehicles regarding the safety and ergonomics of their use and maintenance as well as diagnostic and repair-related susceptibility of these vehicles also remain in a significant (but slightly weaker than the above) relation with eleven criteria of organizational agility of their production processes.

– If, on the basis of the obtained results of expert opinion research, one should try to create the hierarchy, from the point of view of proexploitation, of the most recommended undertakings of organizational agility of production processes of public transport vehicles, two priorities should undoubtedly be considered as priority:

(a) maximizing and improving the effectiveness of machinery and technical equipment (Total Productive Maintenance, Overall Equipment Effectiveness) and systematic and frequent improvement (Continuous Improvement) of technical preparation and implementation of production processes for the creation of public transport vehicles, carried out in a short time following the concept of: Kaizen, Just in Time, 5S, SMED, FMEA, Ishikawa Diagram, zero inventory, low capital involvement and conducting simulation experiments,

(b) orientation towards customer satisfaction and loyalty as the goal of an enterprise (Mass Customization manufacturing systems integrated with quality - Total Quality Management - and quick response to customer needs).

– At the next level of the hierarchy of recommendations dictated by the requirements of production proexploitation, four organizational agility measures of the production processes of public transport vehicles should be located:

(a) actions to identify employees with the interest of the company (Identity) and showing willingness to raise the level of knowledge and improve own competences (Open Book Management),

(b) striving to flatten the organizational structure of an enterprise (empowerment and relative autonomy of the operational level - Employee Empowerment) and management through team-based activities and leadership (Leadership) based on participation and susceptibility to change and improvement,

(c) production of public transport vehicles with a short life cycle but reliable in operation (in case of product failure, immediate repair or replacement with a new one),

(d) targeting business cost management towards productivity and quality.

An important confirmation of the professional competence of the surveyed experts and the tool justification for the correctness of the research method adopted for the paper was a relatively low assessment of the dependencies of the operational value of public transport vehicles on the other five potential organizational undertakings of agile technical improvement and preparation and implementation of the proper production processes. These undertakings have only productive orientation. Their relationship with proexploitation is therefore indirect. The obtained research results can be found out as a significant results (especially for practice).

The customer's expectations (using, servicing, liquidating and disposing of a vehicle) concerning almost all seven attributes of the exploitation of minibuses, buses, trams and passenger rolling stock of regional railway require the innovative improvement of construction, technology and organization as part of the technical preparation and implementation of the production of these vehicles. The model presented at Fig. 1 shows that the practice of the production phase of creating processual engineering products requires their gradual testing which "confronts" design with realities. Testing variants of solutions of this type is possible, among others, by means of a computer simulation [14]. Such solutions will be the subject of further research in the future.

References

1. Kidd, P.T.: Agile Manufacturing: Forging New Frontiers. Addison-Wesley, Reading (1994)
2. Goldman, S., Nagel, R., Preiss, K.: Agile Competitiors and Virtual Organizations. Van Nostrand Reinhold, New York (1994)
3. Yusuf, Y., Sarhadi, M., Gunasekaran, A.: Agile manufacturing: the drivers, concepts and attributes. Int. J. Prod. Econ. **62**, 33–43 (1999)
4. Trzcieliński, S.: Zwinne przedsiębiorstwo (Agile Enterprise). Wydawnictwo Politechniki Poznańskiej, Poznań (2011)
5. Trzcieliński, S.: Changeability of environment – enemy or ally? In: Management Science in Transition Period in South Africa and Poland. Cracow University of Economy, Cracow-Stellenbosch (2013)
6. Wlodarkiewicz-Klimek, H.: Agility of knowledge-based organizations. In: Proceedings of the AHFE 2016 International Conference on the Human Aspects of Advanced Manufacturing (2016)
7. Hayes, R.H., Wheelwright, S.C.: Restoring Our Competitive Edge: Competing Through Manufacturing. Wiley, New York (1984)
8. Żółtowski, B.: Doskonalenie systemów eksploatacji maszyn (Improvement of operational systems of machines). Problemy Eksploatacji **2**, 7–20 (2012)
9. Legutko, S.: Eksploatacja maszyn (Machines Operation). Wydawnictwo Politechniki Poznańskiej, Poznań, p. 376
10. Kałkowska, J.: The Pro-operational Approach in the Creation of Public Transport Vehicles, p. 158. Poznan University of Technology, Poznan (2018)
11. Olearczuk, E.: Warunki konieczne Dobrej Praktyki Eksploatacyjnej (DPE) obiektów technicznych Polskie Naukowo-Techniczne Towarzystwo Eksploatacyjne Standard eksploatacyjny (2000)
12. Kasprzycki, A., Sochacki, W.: Wybrane zagadnienia projektowania i eksploatacji maszyn i urządzeń Wydawnictwo Politechniki Częstochowskiej Częstochowa, p. 260 (2009)
13. Sindhwani, R., Malhortra, V.: Twenty criteria agile manufacturing model. Int. J. Emerg. Technol. Adv. Eng. **5**, 182–185 (2015)
14. Pacholski L., Pawlewski P.: The usage of simulation technology for macroergonomic industrial system improvement. In: Goossens, R.H.M. (ed.) Advances in Social & Occupational Ergonomics, Advances in Intelligent Systems and Computing, 487, pp. 3–14. Springer, Cham (2017)

Management System of Intelligent, Autonomous Environment (IAEMS). The Methodological Approach to Designing and Developing the Organizational Structure of IAEMS

Edmund Pawlowski[✉] and Krystian Pawlowski

Faculty of Management Engineering, Poznan University of Technology,
Strzelecka 11, 60-965 Poznan, Poland
{edmund.pawlowski,krystian.pawlowski}@put.poznan.pl

Abstract. This paper concerned with designing, building, and developing of Intelligent Autonomous Environment (IAE). IAE is a system of integrated industrial buildings, or a residential district, or a shopping district, etc., defined as an intelligent environment that has all systems of self-steering and adaptation. This paper focuses on IAEMS's organizational structure designing and developing. The main problems of designing are complexity and changeability of IAS. The response to such complexity and changeability context of IAE is the presented methodological conception of organizational structure designing, based on agile approach assumes: 1. Continuously scanning the environment and adopting the organizational strategy to the requirements of external and internal stakeholders. 2. Changing the list of business processes of IAE (function tree of IAE) in accordance with the stage of IAE cycle. 3. Flexibility of organizational structure forms in different stages of IAE life cycle. 4. Flexibility transformation organizational structures from one stage of the life cycle to the next one

Keywords: Intelligent autonomous system ·
Organizational structure, designing · Management system

1 Introduction

This paper concerned with designing, building, and developing of Intelligent Autonomous Environment (IAE). IAE is a system of integrated industrial buildings, or a residential district, or a shopping district, etc., defined as an intelligent environment that has all systems of self-steering and adaptation. Adaptation concerns reaction on changes both in external and internal environment, as well as ability to keep up with the changing technology and social progress. IAE combines Smart Economy, Smart People, Smart Governance, Smart Mobility, Smart Environment, and Smart Living. Intelligent Autonomous Environment is a systemically understand environment for people life, both private and social, and business life. This paper is a part of a larger research project called "Innovative Management System of Designing, Building,

© Springer Nature Switzerland AG 2020
W. Karwowski et al. (Eds.): AHFE 2019, AISC 971, pp. 92–99, 2020.
https://doi.org/10.1007/978-3-030-20494-5_9

Maintenance and Development of IAE". The project, undertaken at the construction enterprise WPIP Ltd. (Wielkopolskie Przedsiebiorstwo Inzynierii Przemyslowej SP. z o.o.) and Faculty of Engineering Management of Poznan University of Technology, just started in 2018. The research framework of the project has been presented at the ISHE 2018 Conference [1, 2]. Model of IAE Management System (IAEMS) (Fig. 1) includes [1]:

1/ Reference Functional Model of IAE (Function Tree), contains a list of business processes unfolded in form of a function tree diagram,

2/ Reference Model of Organizational Structure of IAE MS, contains organizational charts, tables of functions deployment to organizational units and managerial positions, job description cards as well as processes' owners and leaders,

3/ Reference Model of Processes and Standards of IAE MS, includes the classifiers of processes developed based on the IAE MS function tree, maps of processes, and standards of processes' realization included in the process description cards and manuals,

4/ Reference Methods and Technics of Management supporting designers and managers, Each method and technique should be adjusted to particular needs of management in each phase of IAE lify cycle. Preliminary research already implicates the usefulness of the following methods: Quality Function Deployment (QFD), Concurrent Engineering (CE), Failure Modes and Effects Analysis (FMEA), Methods of Strategic Analysis.

5/ Computer Systems Support for designers and managers. Designed IT systems are divided into two groups:

- Technical support systems (Meta-BMS and other SMART Building management systems)
- Support systems for IAE MS in the phase of building design, exploitation and development referring to particular elements of the management system (unfolding of a function tree system, designing and development of the organizational structure, and chosen methods and techniques of management)

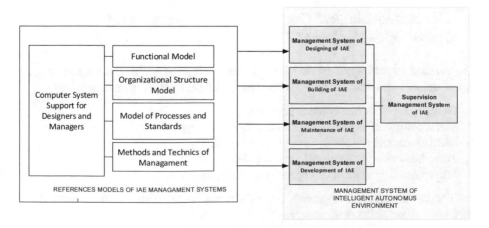

Fig. 1. The structure of IAE Management System

This paper focuses only on one of IAE Management System, it focuses on organizational structure designing and developing the structure from one of IAE phase to another one.

2 Intelligent Autonomous Environment as a Subject to Organizational Structure Designing

The main problems of designing the Management System of IAS are complexity and changeability of IAS:

1. IAE can be simply a few buildings, or a city district.
2. The management system of IAE covers the whole cycle of IAE life (designing, building, maintenance and development).
3. The duration time of the whole cycle of IAE can be a few or several dozen years.
4. The environment of IAE is important from economical, legal, sociological and technological points of view.
5. The each stage of the cycle engage different groups of internal and external stakeholders.

The response to such complexity and changeability context of IAE is the presented methodological conception of organizational structure designing. The conception methodology, based on agile approach assumes:

1. Continuously scanning the environment and adopting the organizational strategy to the requirements of external and internal stakeholders.
2. Changing the list of business processes of IAE (function tree of IAE) in accordance with the stage of IAE cycle.
3. Flexibility of organizational structure forms in different stages of IAE life cycle.
4. Flexibility transformation organizational structures from one stage of the life cycle to the next one.

3 The Methodological Conception of Organizational Structure Designing of IAE

The methodological approach is based on the concept of multi-dimensional space of organizational structure design [3]. The space is described by five dimensions:

1. Dimensions of organizational structure,
2. Methodology of organizational structure modelling,
3. Methodological approach to organizational design,
4. Procedures of organizational structure design,
5. Strategy and principles of organizational structure design.

The methodological approach description we begin with the fifth dimension – principles of organizational design.

3.1 Agile Effect as the Main Strategy of Designing the Organizational Structure of IAE

Based on S. Trzcielinski concept we have distinguished four dimensions of agility [2, 4]:

1. Brightness - the ability to perceive the market opportunities,
2. Flexibility - the feature of resources that enables to expand the repertoire of undertaken opportunities,
3. Intelligence - the ability to learn and adapt to anticipated changes,
4. Shrewdness - the ability to run operational activities in practical way and cope with difficulties to utilize the opportunities.

Depending of the contract for IAE project (construction size, financial budget, duration time) the model of agility can be used on four levels of agility [2]:

(1) The reactive model consists in identifying and using opportunities, which are the needs reported by existing and potential customers.
(2) The pro-active model consists in creating opportunities by offering previously unknown functionalities of products and services.
(3) Network model - in which the company focuses on key processes and products from the point of view of generating value for the client, while the remaining ones are carried out by subcontractors treated as virtual organizational units of the enterprise.
(4) The virtual enterprise model, in which the key resource is knowledge about the demand and supply markets, and the enterprise implements a business model based on its role as a market broker.

3.2 Dimensions of Organizational Structure of IAE

There are two main approaches of organizational structure interpretation: classic (traditional) and contemporary. Classic interpretation reduces the issue of shaping the structure to organizational chart and the scope of job duties and responsibilities. Contemporary interpretation claims that organizational structure encompasses everything which limits the behavior latitude of elements in organization. Manipulating the latitude of elements can be done in many different ways, and each of them is treated as a new dimension of organizational structure. Multidimensional concept of organizational structure is commonly accepted in literature although a number of suggested dimensions differs [5–9]. The broadest interpretation was suggested by "Aston school" [5]. Their concept includes five dimensions: configuration, specialization, centralization, standardization, formalization. Classis organizational structure theory focuses on the dimensions of configuration and specialization, treating all other as an obvious results of those two. In contrast, the research proved low coherency between all five dimensions [5, 6]. Therefore it is possible to manipulate them relatively separately and thus create a wider area of organizational solutions exceeding the stiff classic structures.

In accepted Agile model of IAE, dimensions of configuration and specialization define the characteristic of resource and organizational processes flexibility. It leads to favoring flat structures, based on horizontal forms of coordination and process based management. By introducing the notions of shrewdness, business intelligence and swiftness we appeal to beyond-standard business behavior (standardization dimension) and usage of workers intelligence by flexible allocation of decision-making authorities, according to the postulate of hybrid centralization and decentralization [10]. A model of setting variables (structure dimensions) favoring the agile characteristics of IAE is presented on Fig. 2. To conclude, we notice a positive relation between deliberate usage (by designers and managements) of contemporary concept of multidimensional interpretation of organizational structure and implementation of Agile concept.

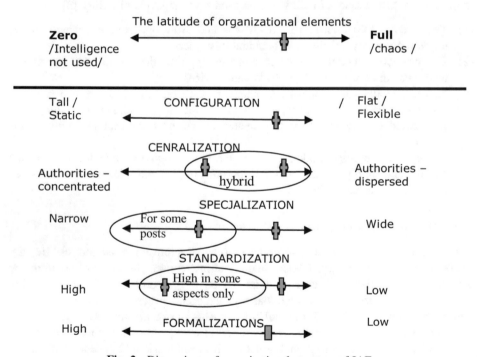

Fig. 2. Dimensions of organizational structure of IAE

Kast and Rosenzweig [11] divided all types of organizational structures into two classes: static-mechanistic structures and adaptation-organic structures. The class of adaptation-organic structure is characterized by: openness to environment and ability to adapt in conditions of uncertainty, low level of standardization and formalization of structure, low level of differentiation and specialization of activities – sometimes even overlapping activities, loosely defined functions and tasks, multiple means of coordination, dispersed structure of authority based on knowledge and not on position in the structure, decentralization, responsibility shared by many participants. These characteristics are still valid and correspond with the dimensions of organizational structure set the most far right (in the direction of maximum latitude of elements). At the end of 70s it meant the support to matrix structures and set the direction for network structures. From the point of view of contemporary taxonomy of organizational structures, the above characteristics correspond with network and virtual structures. It is worth mentioning though, that the essence of transformation from classic linear structure or line and staff structure into network and virtual structure is not only the change of organizational chart (configuration dimension), but the activation of all dimensions in order to "blend into the environment" and virtual usage of environment for own business purposes. It leads to blurring the borders between the company and environment. The "architecture of organization" appears, which includes internal structure and structures existing outside but used for internal objectives of organization [12, 13].

3.3 Methodological Assumption to Modelling of IAE's Organizational Structure

Our design and consulting experiences indicate the following characteristics of a model favoring Agile effects:

- keeping the division and allocation of functions at the level of teams – organizational units (not workplaces)
- connecting the organizational structure with strategy in one organizational model (ex. coherent development of process based structure and parameters of Balanced Scorecard)
- focusing on coordination through flexible but precise model of communication defining: process, problem, decision making process members, scope of decision making latitude, form of document, frequency and form of contact
- electronic version of documentation placed in intranet, enabling direct access to all structural data (not only one's own but of all organization members)
- Introduction of structural rules into the logic of functioning computer systems of Work Flow, DDM, QMS.

3.4 Methodological Approach to Organizational Design of IAE

Methodological approaches are diversified based on accepted paradigms of organizational designing: starting point to designing paradigm, subject of designing paradigm, designing latitude paradigm [14]. A starting point to designing may be an existing state corresponding with diagnostic method or postulated state corresponding with prognostic methods. Because of the level of novelty of Agile concept, it seems more efficient to use a prognostic approach suggested by G. Nadler within the framework of the IDEALS concept. Considering the subject of design we separate two further approaches: traditional – functional, focusing on an organizational improvement of functional organizational units, and process approach encompassing the design of entire business process. Process approach, which is a cornerstone of reengineering [10], is environment and customer oriented, and thus directly corresponds with Agile assumptions.

3.5   Procedures of Organizational Structure Design of IAE

At the level of detailed procedures and rules of organizational structure designing a number of methods was elaborated. A majority of them is an output of consulting companies and therefore they remain confidential. The part that has been published includes several dozen of methodologies. The most commonly known are Shlezinger, Mintzberg, Butler and Hall methods. Below we present a proposal of a framework methodology of a design of an organizational structure of IAE. Procedure consist of eight phases.

1. Phase I – Recognition of the company's strategy and defining the IAE' projects portfolio (number of actual and future projects, market segmentation, size of the projects and duration time)

2. Phase II – Analysis of internal and external conditions for the company and particular IAE projects:

a/ Assessment of surroundings' turbulence (on sales market, delivery, competition behavior and the level and changeability of technology).

b/ Assessment of necessary and achievable flexibility of resources (technology and equipment, financial resources, human resources, information resources).

3. Phase III – Elaboration of the company's organizational strategy (goals for IAE projects and level of agility: reactive, proactive, network, virtual enterprise)

4. Phase IV – Designing of a structure of business processes (function tree) for each IAE project and supervisory system.

5. Phase V – Designing of subject structures for particular processes (organizational charts)

6. Phase VI – Integration of process and subject structures of a company as a whole. Designing of communication and coordination mechanisms. Structure integration with Balanced Scorecard.

7. Phase VII – Designing and implementation of structural rules in information systems Work flow, DDM, QMS.

8. Phase VIII – Implementation and control of strategic goals achievement and Agile effects.

4   Conclusions

Organizational structure designing is a complex and multiaspect process. Ultimate sequence of designing actions and decisions is a resultant of former decisions related to methodology selection. Structures are not the objective as such, but they are a tool which facilitates management Changes in structure should be subordinated to defined business objectives included in company's strategy. This methodology is based on the literature review and our hitherto designing and consulting experience, however it is still a theoretical model as it has not yet been verified in this sequence. Presented concept of methodology requires further operationalization. Practical verification of the methodology is planned for the next year.

References

1. Pawlowski, E., Pawlowski, K., Trzcielinski, S., Trzcielinska, J.: Designing and management of intelligent, autonomous environment (IAE): The research framework. In: Ahram, T., et al. (eds.) IHSED 2018, AISC 876, pp. 381–386. Springer Nature, Switzerland AG (2019)
2. Trzcielinski, S., Trzcielinska, J., Pawlowski, E., Pawlowski, K.: Methodology of shaping the agility of the intelligent autonomous environment management system. In: Ahram, T., et al. (eds.) IHSED 2018, AISC 876, pp. 652–658. Springer Nature, Switzerland AG (2019)
3. Pawlowski, E.: Designing the organizational structure of a company. A concept of multidimensional design space. In: Csatth, M., Trzcielinski, S., Management Systems. Methods and Structures. Monograph, pp. 107–122. Publishing House of Poznan University of Technology (2009)
4. Trzcielinski, S., Trzcielinska, J.: Some elements of theory of opportunities. Hum. Factors Ergon. Manuf. Serv. Ind. 21, 124–131 (2011)
5. Pugh, D.S., Hinings, C.R.: Organizational Structure. Extensions and Replications, pp. 3–11. The Aston Programme II, Saxon House (1976)
6. Mrela, K., Pankow, M.: Organizational Structure's Effectiveness in Their Context, PAN Warszawa, pp. 41–52, 53–81 (1980)
7. Mintzberg, H.: Structure in Fives: Designing Effective Organizations, pp. 1–24. Prentice-Hall International Editions, Englewood Cliffs, NJ (1983)
8. Hall, R.H.: Organizations: Structures, Processes and Outcomes, 4th ed., pp. 101–115. Prentice-Hall, Englewood Cliffs (1987)
9. Strategor, Zarządzanie firmą, PWE Warszawa, pp. 281–287 (1996)
10. Hammer, M., Champy, J.: Reenginnering the Corporation: A Manifesto for Business Revolution, pp. 33–36, 93. Nicholas Brealey Publishing Ltd. (1995)
11. Kast, F.E., Rosenzweig, J.E.: Organizations and Management: A System and Contingency Approach, pp. 234–242. Mc Graw Hill Inc., New York (1979)
12. Kay, J.: Podstawy sukcesu firmy, PWE W-wa, pp. 99–115 (1996)
13. Kozuch, B.: Nauka o organizacji, Cedewu.pl, pp. 259–263 (2007)
14. Pawlowski, E.: O wspolczesnych paradygmatach projektowania organizatorskiego. In: Materiały V Międzynarodowej Konferencji Naukowej nt. „Zarzadzanie Organizacjami Gospodarczymi", Politechnika Lodzka, Lodz, pp. 421–427 (1998)

Lean Manufacturing Model in a Make to Order Environment in the Printing Sector in Peru

Adriana Becerra[1]([⊠]), Alessandro Villanueva[1], Víctor Núñez[1],
Carlos Raymundo[2], and Francisco Dominguez[3]

[1] Escuela de Ingeniería Industrial,
Universidad Peruana de Ciencias Aplicadas (UPC), Lima, Peru
{u201316223,u201412080,Victor.nunez}@upc.edu.pe
[2] Dirección de Investigaciones,
Universidad Peruana de Ciencias Aplicadas (UPC), Lima, Peru
Carlos.raymundo@upc.edu.pe
[3] Escuela Superior de Ingeniería Informática,
Universidad Rey Juan Carlos, Mostoles, Madrid, Spain
francisco.dominguez@urjc.es

Abstract. The printing sector in Peru constantly faces the need to reduce production time because late deliveries to the customer owing to high manufacturing time are critical problems. Previous studies have proposed using the lean manufacturing philosophy to reduce idle times by improving the flow of information within the production processes in large manufacturing companies but not printing SMEs using these tools in a make-to-order environment. This document focuses on implementing an affordable lean manufacturing model in an SME company with a make-to-order environment to reduce its manufacturing times. Therefore, the Kanban, Single Minute Exchange of Dies (SMED), and value stream mapping (VSM) processes will be adapted for implementation in these companies. Then, the model is validated through its application in a case study; through process simulation, production times were reduced by 24% for an SME, delivering all orders on time, and eliminating 100% penalty costs for late orders.

Keywords: Make to order · Simulation · Value stream mapping · Kanban · SMED

1 Introduction

In the last decade, the need for shorter production times has increased in all types of industries. Additionally, the ever-changing market needs create demands for higher flexibility and cost reduction against market globalization. Further, markets demand more product variety in terms of personalization so that companies may remain competitive within the sector. One sectors thus affected is the printing sector; printing is part of the non-primary manufacturing activities. In Peru, this industry has reported a decrease in their GDP growth rate in recent years, reaching −2.7% in 2015.

© Springer Nature Switzerland AG 2020
W. Karwowski et al. (Eds.): AHFE 2019, AISC 971, pp. 100–110, 2020.
https://doi.org/10.1007/978-3-030-20494-5_10

The printing sector includes activities such as printing books, newspapers, magazines, advertising collaterals, labels, and any other type of printed products. The Ministry of Production [1] indicates that at 40%, the printing sector is one of the sectors with the lowest economic efficiency. Late deliveries owing to high production times have been identified as one of the main factors contributing to this situation.

Currently, effective production without delays in delivery is critical for companies that wish to remain active in the market, and this requires quick responses and strict compliance with the quality, quantity, and agreed delivery terms. Therefore, the implementation of a more efficient production management model has become a crucial factor for manufacturing companies that seek efficiency and efficacy in their processes. In addition, while there are several methodologies aimed at improving production processes, the Japanese lean manufacturing techniques have become an important reference because of the results that can be achieved through their application. Previous studies have revealed that by using tools such as SMED, 5S, and TPM under this technique, production-time-related issues were reduced by more than 95% [2]. Further, the use of these tools may improve efficiency through value-added processes, such as VSM analysis and the Kanban tool [3]. These tools improve the flow of information and materials by reducing delivery times, thus increasing business profitability [4].

This paper is structured as follows: first, a literature review is presented, where previous studies are examined through states of art. Second, the contribution adapted and justified in a case study. Third, the validation of the proposed model is presented. Fourth, the discussion of the results obtained and recommendations for future research studies. Finally, the conclusion of the project based on the main objectives and accounting for the contributions made and the model validation is presented.

2 State of the Art

2.1 Lean Manufacturing Philosophy

In the following paragraph, we analyze the studies performed through the implementation of the lean manufacturing philosophy in case studies. In the current scenario, the printing industry is considered to be a part of the worldwide manufacturing sector. This sector plays a vital role in terms of changing basic manufacturing ideas and thoughts [5]. For example, many manufacturing companies have already changed their productive processes through different lean manufacturing methodologies. These methodologies were utilized to reduce the production times for the fulfillment of customer orders. Author [2] achieves a reduction of 7.6% of the total time, solving 95% of production problems thanks to tools such as SMED, 5S, and VSM in a confectionery company. In this particular case, the Kaizen philosophy was additionally used to secure the continuous improvement of the contribution. Similarly, authors [6] and [7] reported reductions in production times through SMED, VSM, 5S and line of balance. Their main results are reflected in the reduction from 145 to 54s in preparation time at a label production company [6] and a 40% reduction in production times coupled with an increase in cycle efficiency from 71.24% to 81.18% in an automotive company [7]. Finally, author [8] used different lean tools to reduce cycle times from 8.5 to 6 days in a paint manufacturing company. From these references, it can be concluded that the different lean manufacturing tools have been useful for the different authors because they have helped solve the many time-related problems identified in their production processes.

2.2 SMED

Over the years, different authors began implementing lean manufacturing tools individually to solve specific problems within their production processes. For this reason, this section assesses the studies reported by the authors who used the Single Minute Exchange of Dies (SMED) tool to solve their problems. This section presents the case of author [9] who implemented SMED to reduce the setup times for machinery-based processes at a toll manufacturing company. Further, authors [10] and [11] implemented this tool within their respective companies, resulting in a 37% reduction in preparation times for blow molding machines at a brewing company [10] and a 2.06 h reduction in preparation times at a shaft production company [11]. Finally, author [12] reduced machining setup times for valve production, achieving a total delivery time of 3.6 d for their products. Therefore, SMED has been considered one of the most effective tools within the lean philosophy for reducing process setup times.

2.3 Kanban

Another tool used by various authors for the solution of specific problems has been Kanban. Therefore, this section reviews papers in which problems were resolved through Kanban. Author [13] managed to solve order fulfillment delays and reduce customer dissatisfaction due to incomplete orders and excess inventory through Kanban. Regarding excess inventory, a 56.8% reduction of stored products was achieved. In addition, the adoption of the Kanban system provided strategic benefits and improved service quality. Further, author [3] was able to reduce high levels of inventory caused by the inefficient flow of materials and lack of process specifications in the company through a Kanban-based implementation proposal for each product line. However, author [4], after implementing Kanban at a bearings manufacturer, significantly reduced delivery times from 26 to 21 d besides improving the flow of information and materials and inventory times. In addition, prompt product deliveries increased from 76 to 99%. For these reasons, the Kanban tool has been considered one of the most effective within the lean philosophy for improving material and information flows and reducing production times.

3 Collaborations

3.1 Proposed Model

The model presented is an adaptation on the traditional lean manufacturing model adapted to a make-to-order (MTO) environment and company size according to the application market. This is reflected in Fig. 1.

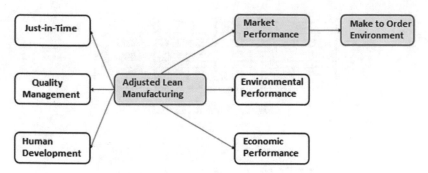

Fig. 1. Production management lean manufacturing model.

3.2 Proposal Development

The proposal includes implementing changes to lean manufacturing tools to adapt them to specific markets, such as the printing industry, with adjustments for an MTO production environment.

The proposed model applies changes to lean manufacturing tools aimed at reducing production times. Within the established procedures, the changes implemented are the use of diagnostic tools, value stream mapping (VSM), operational tools, SMED, Kanban, and follow up tools with KPIs. The following paragraphs explain the changes introduced by each lean manufacturing tool, adjusted to the production management model proposed.

The VSM diagnostic tool is used as a visual tool to understand the current working conditions through a current VSM, where activities that generate and do not generate value to the process are determined in order to identify opportunities for improvement. Here production processes as well as the times that add and do not add value to the process, are clearly exposed. Based on this, a future VSM in which where the implemented improvements are reflected may be developed. For these purposes, the general, process, flow of information, and material set of symbols are taken into account.

Once the current VSM has been developed, the "mudas" or wastes from times that do not add value to the process are determined. The proposed model is applicable if the wastes identified are related to high waiting times, transportation, and machinery setup.

To resolve high waiting times and transportation problems, the Kanban tool is used. The change implemented via this tool is the application of two different types of Kanban: one for production (waiting times) and the other for supermarket restocking (transportation time).

As shown in Fig. 2, the original Kanban matrix has been modified to adapt it to an MTO environment. In the proposed model, Kanban cards are used for processes (production), adding a Kanban column for supermarkets (restocking).

For Kanban production cards, as shown in Fig. 3, changes are made in their design. In these cards, the date, product, customer, quantity of products, process owner, and specifications are recorded.

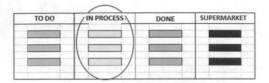

Fig. 2. Kanban Matrix.

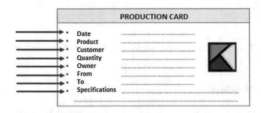

Fig. 3. Kanban production card.

The Kanban cards will be used for all production process activities and will be provided to all area operators. This will help to reduce waiting times owing to a lack of product knowledge as it optimizes the flow of information throughout the production process. In addition, this will help organize the production area since each operator will know what the previous and next processes of each product are, as shown in Fig. 4.

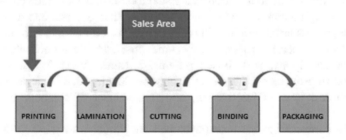

Fig. 4. Use of the production Kanban card.

However, the Kanban card used for the supermarket has undergone modifications in its design and information. This card records the date, material, quantity, area, and owner, as shown in Fig. 5. In addition, a shelf will be used to store materials within the production area. In other words, this shelf will be a supermarket within the production area for restocking materials. To manufacture these shelves, the size of the production area and the ergonomics of the operators must be considered.

This card will be placed on each shelf in order to determine moment when the material must be restocked. This method decreases the travel time for area operators since they will not have to travel to the warehouse to look for the necessary materials to conduct their work anymore, as shown in Fig. 6.

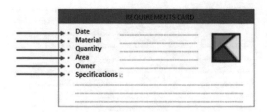

Fig. 5. Kanban requirements card.

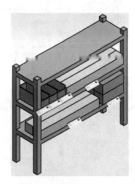

Fig. 6. Supermarket shelf.

For problems related to high setup times, SMED is used. This tool categorizes all the internal and external configuration activities, as shown in Fig. 7. Internal activities are those that can only be performed when the machine is off, whereas external activities are those that can be performed during normal machine operation, i.e., while it is on. The application of the SMED methodology comprises four stages. The preliminary stage is where the internal and external configuration conditions are identified. In the next stage, more configurations are separated. Then, internal activities are converted to external activities. Finally, the change is applied, and all the aspects of the configuration operation are rationalized.

Finally, to implement the proposed new tools and ensure their proper use, pre-implementation and standardization trainings, which are performed by subject matter experts, are required. Training schedules and implementation costs depend on the company in which the proposed model will be applied. The defined process shown in Fig. 8 will be followed.

Setup Operations	Internal	External
Remove Roll from machine	x	
Place Roll on the floor		x
Internal Cleaning	x	
External Cleaning		x
Go to the warehouse		x
Look for materials		x
Transport materials		x
Take banner roll from box		x
Place new roll	x	
Adjust roll	x	
Verify Ink Levels		x
Go to the warehouse		x
Look for inks		x
Transport inks		x
Place inks	x	
Control color quality		x
Calibrate machine	x	

Fig. 7. Printing process SMED.

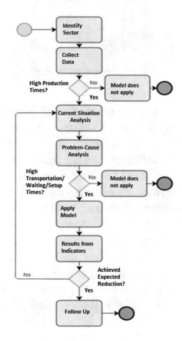

Fig. 8. Process workflow.

4 Validation

To validate the proposed production management model, a simulation will be used. For these purposes, times will need to be recorded at workstations.

The company under study is Krea Visual S.A.C., an SME in the printing sector located in Lima. This company prints advertising products (banners, vinyls, banners, posters, backlite and rollscreen) of various sizes, based on the requirements and specifications of their customers. In recent years, this company has faced problems with the timely delivery of their products owing to the high transportation, product waiting, and machine setup times. Therefore, it was chosen to validate the proposed model.

In order to understand the process that is to be simulated, an order process flow was created once the improvements had been implemented. The process begins once the production order is placed within the company. As per the current process, after being placed, the order is transferred to the design area. After the design work is finished, the process follows to the printing area. This area now performs their production activities using the production Kanban. To do this work, the materials required, banners and inks, need to be checked out from the supermarket, and this is when the product Kanban is activated. Both banners and inks are purchased from the supermarket; if there are not enough supplies, they must be restocked as required. The subsequent lamination process follows the same steps. The lamination sheets, blades for cutting, supplies for bookbinding, and fill paper for packaging must be purchased from the supermarket (Fig. 9).

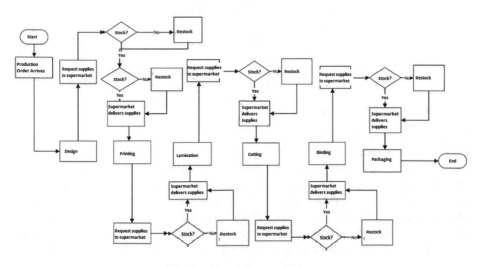

Fig. 9. Simulation workflow.

4.1   Variables

For the validation of the proposal, a simulation is performed using the ARENA Simulator program, for which the input and output variables must be determined, as shown in Fig. 10.

To perform the simulation, the time data collected in the production area of the company were considered. For this simulation, the following variables are assessed:

- Time between orders: There is a database where the times between production orders are recorded.
- Service time (Process n): There is a database in which the service time is recorded for each workstation (design, printing, lamination, cutting, binding, and packaging).

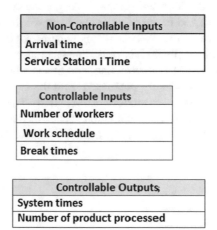

Non-Controllable Inputs
Arrival time
Service Station i Time

Controllable Inputs
Number of workers
Work schedule
Break times

Controllable Outputs
System times
Number of product processed

Fig. 10.   Controllable and non-controllable variables.

4.2   Analysis of Indicators

Through the ARENA software simulation, it has been confirmed that the proposed model does not present errors. Therefore, when executed, it yields a report with the corresponding data and required statistical results. There is an average time of 30 runs per system of 103 min. This is the average production time for a banner unit. Therefore, for the production of 9.2 banners, a production order will take a total of 946.3 min for completion. This result evidences a 300.3 min or 24% initial time reduction, which makes the project operationally viable since this reduction eliminates delivery delays by adhering to the production times agreed with the customer, as shown in Fig. 11.

	CURRENT	RESULTS	REDUCTION
Production Time	1246.6 min	946.3 min	24%
Transportation Time	179.4 min	124.1 min	30.80%
Waiting Time	312.8 min	207.2 min	33.70%
Setup Time	349.6 min	210.2 min	39.80%

Fig. 11.   Comparison table.

In addition to the reduction of the total production time, the main causes of the initial problem reported were reduced. According to the results of the simulated system, the total transportation time for the production of 9.2 banner units is 124.1 min, which represents a 30.8% reduction in current transportation times. Furthermore, the total waiting time of 207.2 min, which represents a reduction of 33.7% from the current waiting times. Furthermore, the total setup time is 210.3 min, which represents a reduction of 39.8% from current setup times.

5   Conclusions

Using SMED as the only time reduction tool provides a reduction of 11.1%. If the Kanban tool is used, a 12.9% reduction is reported. However, if both tools are used together, the company is able to reduce production times by 24%. This completely eliminates late delivery penalties for the company.

With the help of the case study, the functionality of the model was successfully demonstrated. The use VSM as a diagnostic tool identified the times that do not add value, where the setup, transportation, and waiting times stood out. Based on this, an improvement point was presented, which was adjusted for operator restrictions owing to their nature.

With the proposed model, the total transportation time for the production of 9.2 units was reduced by 30.8% of the current times. In addition, a 33.7% reduction was reported for total waiting times. Finally, the setup times were reduced by 39.8% of the current times.

References

1. PRODUCE Estudio de la Situación Actual de las empresas peruanas [En línea] Ministerio de la producción: Perú. Recuperado de (2015). http://demi.produce.g.,ob.pe/images/publicaciones/publi81171136fe74561a7_79.pdf
2. Djekic, I., Zivan ovic, D., Dragojlovic, S., Dragovic, R.: Lean manufacturing effects in a serbian confectionery company - case study. Organizacija 47(3), 143–152 (2014). http://dx.doi.org/10.2478/orga-2014-0013
3. Tomas Rohac, M.J.: Value stream mapping demostration on real case study. Procedia Eng. 100(2015), 520–529 (2015)
4. Ranjan, R.U., Mahesh, B.P., Sandesh, S.: On-time delivery improvement using lean concepts - a case study of Norglide bearings. Int. J. Innov. Res. Sci., Eng. Technol. 3(6), 13349–13354 (2014)
5. Womack, J.P., Jones, D.T.: Lean Thinking, Banish Waste and Create Wealth in Your Corporation, Simon & Schuster. Top of Form, New York, NY (1996)
6. Azizi, A.: Thulasi a/p Manoharan: designing a future value stream mapping to reduce lead time using SMED-a case study. Procedia Manuf. 2, 153–158 (2015)
7. Nallusamy, S.K., Ahamed, A.: Implementation of lean tools in an automotive industry for productivity enhancement - a case study. Int. J. Eng. Res. Afr. 29, 175–185 (2017)
8. Jafri, M., Seyed, M.: Production line analysis via value stream mapping: a lean manufacturing process of color industry. Procedia Manuf. 2, 6–10 (2015)

9. Díaz-Reza, J.R., García-Alcaraz, J.L., Martínez-Loya, V., Blanco-Fernandez, J., Jimenez-Macías, E., Avelar-Sosa, L.: The effect of SMED on benefits gained in maquiladora industry. Sustainability **8**(12), 1237 (2016). http://dx.doi.org/10.3390/su8121237
10. Lopes, R.B., Freitas, F., Sousa, I.: Application of lean manufacturing tools in the food and beverage industries. J. Technol. Manag. & Innov. **10**(3), 120–130 (2015)
11. Sabadka, D., Molnar, V., Fedorko, G.: The use of lean manufacturing techniques - smed analysis to optimization of the production process. Adv. Sci. & Technol.-Res. J., 187–195 (2017)
12. Ashif, M., Goyal, S., Shastri, A.: Implementation of lean tools-value stream mapping & SMED for lead time reduction in Industrial Valve Manufacturing Company. Applied Mechanics & Materials, 813/8141170-1175 (2015). 10.4028/www.scientific.net/AMM.813-814.1170
13. Papalexi, M., Bamford, D., Dehe, B.: A case study of Kanban implementation within the pharmaceutical supply chain. Int. J. Logist.: Res. & Appl. **19**(4), 239–255 (2016). https://doi.org/10.1080/13675567.2015.1075478

The Influence of Macroenvironment Changes on Agility of Enterprise

Stefan Trzcielinski[✉]

Faculty of Engineering Management, Poznan University of Technology,
Strzelecka Str. 11, 60-965 Poznan, Poland
stefan.trzcielinski@put.poznan.pl

Abstract. In spite of different understanding of agility the dominating view-point is that agile enterprise is focused on using opportunities. Opportunity is a situation existing in the enterprise's environment that favours the enterprise achieving its goals with use of disposable resources. On the base of open systems theory commonly accepted in strategic management the business environmental changes can be recognized by the enterprise either as opportunities or threats. In this paper the focus is on influence of macroenvironment on opportunities. The undertaken problem is if turbulences in macroenvironment result in enterprise shift to agility. As a measure of this shift, some of the company's activities were taken that are symptoms of its agility. To get an answer for the research problem the macroenvironment have been segmented according to PEST analysis and for each segment factors potentially influencing the enterprise's opportunity has been defined. Statistical data about the value of these factors and the symptoms of opportunities were used, their reliability was checked as a measure of changes in the environment and occurrence of opportunities and their correlation was analyzed. The study has been run internationally however in this paper only data from two countries are presented. The results indicate the crucial macroenvironmental factors affecting the shift to agility. Such results can be useful for managers who are interested to exploit market opportunities by shaping their businesses agile.

Keywords: Agility · Agile enterprise · Opportunity ·
Symptoms of opportunity · Business environment

1 Introduction

The concept of an agile enterprise was developed as an alternative and competitive in relation to the concept of a lean company developed by Toyota. While the lean company defends itself against the changeability of the environment, eliminating or reducing the waste, the agile company needs this changeability, because only the changing environment generates opportunities [1]. As Goldman et al. writes "Agile companies aggressively embrace change. For agile competitors – people as well as companies – change under uncertainty are self-renewing sources of opportunities out of which to fashion continuing success" [2]. Being agile means being a master of change, and allows one to seize opportunity [3]. Agile organization is opportunity oriented [4].

© Springer Nature Switzerland AG 2020
W. Karwowski et al. (Eds.): AHFE 2019, AISC 971, pp. 111–121, 2020.
https://doi.org/10.1007/978-3-030-20494-5_11

The concept of agility is used in variety of contexts: production, manufacturing [5], enterprise, supply chain [6], supply network [7], software developing, projects, etc. [8].

Usually, agility is associated with the ability to quickly and effectively respond to changes in the business environment [9], adaptability to the changes, ability to innovate, ability to thrive in a changing and unpredictable business environment, to be capable of operating profitably in a competitive environment of unpredictably, changing customer opportunities [3], ability to speed respond to new market opportunities [5].

In this article, in the central point of agility, we put the ability to seize opportunities that appear just because the business environment is changing. In this context, the question arises about the attributes or dimensions of agility thanks to which the organization and, in particular, the company is agile. The literature on the subject does not provide an answer to this question. The authors point to the effects of the organization's agility, for example, the ability to respond quickly, innovate, adapt, etc., or the management methods that an agile company should use. Therefore before moving to the core topic of this paper we present here our view point on opportunities and attributes or dimensions of agility.

2   The Agility

2.1   The Business Opportunity

Authors writing about agility rarely define the phenomena of opportunity, although the essence of an agile enterprise is the ability to use them. More often the term of opportunity is defined on the basis of entrepreneurship. The authors usually associate opportunity with events taking place in the business environment. For example Casson appoints that opportunities are such situation in which new products, materials and organizational solutions can be introduced and sold in beneficial mode [10]. According to Baron opportunities emerge from a complex pattern of changing conditions: technological, economic, political, social and demographical [11]. Kaish and Gild see opportunities as market gaps caused by market disequilibrium [12] and Choi and Shepherd suggest that opportunities exist when there is customer demand for a new product [13].

In this article, the opportunity is understood as a situation favorable to the enterprise in achieving its objectives with use of accessible resources, that exists in the enterprise's environment or is a postulated state of features of the environment (Fig. 1). The situation is created by events that occur in this environment. It can be multi-events or single event situation [14]. It is worth noticing that whether or not the situation is favorable depends on the objectives and resources that the company can dispose of. These are both own resources and network resources. The opportunity is therefore the relationship between the goals, resources and the situation in the environment, such that it favors achieving the goals with the resources available.

There is a discussion in the literature whether or not opportunities are created or recognized. Schumpeter argued that opportunities are created as a result of innovative activities that cause market disequilibrium [15]. Kerzner, on the other hand, represented the view that opportunities exist and must be recognized. We believe that there is no contradiction between these views. Both views refer to two levels of agility of the enterprise. The first - the lower level, popularized in the literature mainly as quick response, is related to the ability of the company to recognize the opportunities [17]. The process of recognizing begins with marketing research, as a result of which customer needs are identified, run through engineering departments, where the product is modified to meet those needs and return to the marketing and sales department, which are responsible for its promotion and delivery to the customer.

The second one – higher level, consists in creating opportunities as a result of the company's creative activities. In this case, the process of creating opportunities begins with the engineering departments that develop an innovative the product, and goes through the marketing department that promotes this product. The model shown in Fig. 1 corresponds to both creation and the recognition of opportunities.

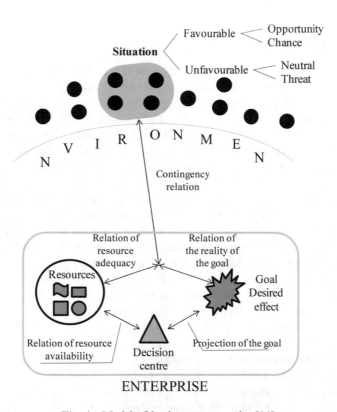

Fig. 1. Model of business opportunity [14]

Whether the given situation in the environment was an opportunity is verified by the results achieved by the enterprise. Such situations are usually complex multi-events systems. When conducting investigation on opportunities, instead of using a complicated description of the situation, you can use the symptoms of the opportunity. These symptoms are actions that the company undertakes in connection with an occurrence of the opportunity. The following symptoms of the opportunity were taken into consideration in this work.

O1. New registrations of businesses. If there is an increase of new registered businesses then we can supposed that the entrepreneurs starts the business as he see the market opportunities.

O2. Degree of customer orientation. Customers' needs and expectation create opportunities. The more the enterprise is customer-oriented, the better it satisfies his needs and thus uses opportunities.

O3. Total export growth rate. Increase of export may suggest existence of unmet demand for products or services at foreign markets. Existence of the demand means existence of opportunity. Agile enterprise gives response for a such opportunity.

O4. Foreign direct investment outward. Before the decision about foreign direct investment a detail analysis is made about its profitability. If there is an increase of FDI then we can presume that the investment project is rational as it meets some opportunities.

2.2   Dimensions of Agility

To shape agility we need to know what is its essence. This issue has not been well systematized in the literature. As a starting point, we assume that the essence of an agile organization is its ability to seize opportunities. Opportunities appear in a changing and turbulent environment. The ability to seize opportunities in such an environment is the source of a competitive advantage. By adopting this point of view, the concept of agility can be operationalized by defining its characteristics necessary to identify opportunities. Our research has led to the following characteristics and dimensions of agility: brightness [18], flexibility [19, 20], intelligence [21], and shrewdness [19]. These dimensions concern the agility of the employee, the team and the entire organization. In this study, we refer them only to agility of the entire enterprise.

Brightness is the enterprise's ability to identify business opportunities as well as threats. Shaping brightness requires systematic and methodical observation of the environment. Flexibility is a feature of the enterprise's resources thanks to which it is possible to expand the range of manufactured products and services provided, and thus to use various opportunities. Intelligence is the ability of an enterprise to learn, to predict future states of the environment, which can turn into opportunities, and to undertake targeted adaptation activities. Shrewdness is the ability to use the opportunity in a practical way. It seems that such a set of features and at the same time the dimensions of the opportunity comprehensively covers the actions by which the company is oriented on taking advantage of market opportunities.

The agility of the enterprise is not a specific state of its dimensions: brightness, flexibility, intelligence and shrewdness. An enterprise can be agile to varying degrees and in different ways. Four levels of agility can be distinguished. The lowest level is represented by the company that gives quick respond to opportunities arising on the market. It represents a reactive model of agility - manifested in its ability to notice opportunities that, according to Kirzner, exist on the market and only need to be discovered [16]. The second level represents a company that creates opportunities through innovative activities. This is a proactive model of agility, corresponding to the meaning of opportunity by Schumpeter [15]. The third level of agility is represented by a dynamic networking enterprise that acquires necessary competences and resources from partners selected for utilization of single opportunity (project) [22]. The highest level of agility is represented by a virtual enterprise, which in its advanced form turns into the market broker [23, 24].

2.3   Domain of Opportunities

Opportunities are situations that favor the company in achieving its goals with use of disposable resources. These situations arise in the environment. The theory of strategic management divides the business environment into macroenvironment and the industrial (competitive) environment. The industry environment is segmented into: new entrants, suppliers, customers, substitute products and competitors. In this work we deal with the impact of macro-environment on the agility of the enterprise. The macro-environment is divided into segments according to the PEST analysis [25]. For individual segments, the following measures of variability of changes taking place in them were adopted.

Political and Legal Segment

P1. Favoritism in decisions of government officials. To what extent do government officials show favoritism to well-connected firms and individuals when deciding upon policies and contracts?

P2. Burden of government regulation. How burdensome is it for companies to comply with public administration's requirements (e.g., permits, regulations, reporting)?

P3. Efficiency of legal framework in challenging regulations. How easy is it for private businesses to challenge government actions and/or regulations through the legal system?

P4. Transparency of government policymaking. How easy is it for companies to obtain information about changes in government policies and regulations affecting their activities?

P5. Number of procedures required to start a business.

Economic Segment

E1. Inflation. Annual percent change in consumer price index.

E2. Effect of taxation on incentives to invest. To what extent do taxes reduce the incentive to invest?

E3. Total tax rate. This variable is a combination of profit tax (% of profits), labor tax and contribution (% of profits), and other taxes (% of profits).

E4. Financial services meeting business needs. To what extent does the financial sector provide the products and services that meet the needs of businesses?

E5. Ease of access to loans. How easy is it for businesses to obtain a bank loan?

Socio-demographical Segment

S1. Tertiary education enrollment rate. (ISCED levels 5 and 6).

S2. Quality of the education system. How well does the education system meet the needs of a competitive economy?

S3. Quality of management schools

S4. Extent of staff training. To what extent do companies invest in training and employee development?

S5. Availability of scientists and engineers. To what extent are scientists and engineers available?

Technological Segment

T1. Mobile-cellular telephone subscriptions. Number of mobile-cellular telephone subscriptions per 100 population

T2. Availability of latest technologies.

T3. FDI and technology transfer. To what extent does foreign direct investment (FDI) bring new technology into your country?

T4. University-industry collaboration in R&D. To what extent do business and universities collaborate on research and development (R&D)?

T5. Government procurement of advanced technology products. To what extent do government purchasing decisions foster innovation.

3   Methodology

The impact of macro-environment factors on the agility of enterprises was analyzed in the cross-section of two countries - Poland and Finland. It was assumed that with the intensification of the occurrence of the symptoms of opportunity, the agility of enterprises increases. Data on the symptoms of the opportunity and on the factors characterizing individual segments of macro-environment were obtained from statistics published by the Global Competitive Index Report and the Global Innovativeness Index and UNCTAD Stat. Data came from the years from 2009 until 2017. Symptoms of opportunity are treated as dependent variables and factors in macroenvironment's segments - as independent variables. Their values are presented in Table 1.

In order to check the reliability of the measuring scales, the Cronbach alpha test was carried out. As the data takes values both on the ordinal and ratio scale, the standardized Cronbach alpha coefficient was used.

As a result of the analysis of reliability of measuring scales, in the case of Poland, the number of dependent variables was limited to three and independent variables to seventeen. In order to determine if there is a statistically significant relationship between macroenvironment factors and symptoms of opportunity, the Spearman's rank correlation analysis was performed. The correlation was carried out between reliable variables, and it was assumed that the opportunities caused by changes in the environment occur with a two-year delay in relation to these changes. This assumption is justified in previous studies on the impact of the organizational climate on the inflow of foreign direct investment. The time shift between the organizational climate and investment decisions is two years.

4 Results

The independent variables describing particular segments of macroenvironment were chosen following a logical but hypothetical effect on dependent variables regarding the symptoms of the opportunities and indirectly agility. The change of cumulated values of these variables for individual segments of macro-environment are presented in Fig. 2.

Table 1. Value of independent and dependent variables and theirs reliability

No.	Environment factor	Measurement scale	09.	10.	11.	12.	13.	14.	15.	16.	17.	FI	PL
P1	Favoritism in decisions of government officials	1 = show favoritism to a great extent; 7 = do not	5,9	5,0	5,0	5,1	5,3	5,3	5,4	5,6	5,6		
			2,5	(illegible)	(illegible)	(illegible)	3,1	3,1	3,1	3,0	2,8		
P2	Burden of government regulation	1 = extremely burdensome; 7 = not	4,4	4,3	4,4	4,8	5,0	4,5	4,2	4,5	4,7		
			2,3	2,7	2,6	2,6	2,7	2,9	2,8	2,7	2,8		
P3	Efficiency of legal framework in	1 = extremely difficult; 7 = extremely easy	6,0	5,5	5,7	5,9	5,9	5,6	5,8	5,8	5,6		
			2,9	3,1	3,3	3,2	2,9	2,8	3,1	3,0	2,6		
P4	Transparency of government	1 = extremely difficult; 7 = extremely easy	5,7	5,6	5,8	6,1	6,1	5,8	5,9	6,0	5,8		
			3,0	3,7	4,0	3,8	3,6	3,6	3,6	3,6	3,3		
P5	Number of procedures required to start a	World Bank methodology	3,1	3,0	3,0	3,0	3,0	3,0	3,0	3,0	3,0		
			10,0	6,0	6,0	6,0	6,0	4,0	4,0	4,0	4,0		
E1	Inflation - Annual percent change	Consumer price index (year average)	1,6	1,6	1,7	3,3	3,2	2,2	1,2	-0,2	0,4		
			2,5	3,5	2,6	4,3	3,7	0,9	0,0	-0,9	-0,6		
E2	Effect of taxation on incentives to invest	1 = to a great extent; 7 = not at all	2,8	3,0	3,1	3,2	4,0	3,9	3,6	3,9	4,3		
			2,3	3,1	3,1	3,1	3,3	3,1	3,3	3,3	3,3		
E3	Total tax rate	profit, labor, and other taxes: (% of profits)	47,8	47,7	44,6	39,0	40,6	39,8	40,0	37,9	38,1		
			38,4	42,5	42,3	43,6	43,8	41,6	38,7	40,3	40,4		
E4	Financial services meeting business needs	1 = not at all; 7 = to a great extent	6,0	6,0	5,8	5,9	6,1	6,0	6,0	5,6	5,6		
			4,0	4,8	4,8	4,8	4,9	4,9	4,9	4,6	4,5		
E5	Ease of access to loans	1 = extremely difficult; 7 = extremely easy	4,7	4,5	4,5	4,4	4,2	4,0	4,0	5,3	4,0	0,65	0,68
			3,4	2,9	2,7	2,5	2,5	2,6	2,6	4,3	4,3		
S1	Tertiary education enrollment rate	Gross tertiary education enrollment rate	93,2	94,4	94,4	93,7	95,2	93,7	93,7	88,7	87,3		
			65,6	66,9	69,4	70,5	72,4	73,2	73,2	71,2	68,1		
S2	Quality of the education system	1 = not well at all; 7 = extremely well	6,2	5,6	5,9	5,8	5,9	5,9	5,7	5,7	5,8	X	
			3,8	3,8	3,7	3,7	3,4	3,6	3,6	3,6	3,6		
S3	Quality of management schools	1 = extremely poor; 7 = excellent	5,5	5,3	5,3	5,6	5,6	5,6	5,4	5,4	5,6		
			4,5	4,2	4,0	4,0	4,0	4,0	4,1	4,2	4,2		
S4	Extent of staff training	1 = not at all; 7 = to a great extent	5,2	5,2	5,3	5,4	5,5	5,3	5,2	5,4	5,3		
			3,6	4,2	4,1	4,0	4,0	4,0	4,0	4,0	4,0		
S5	Availability of scientists and engineers	1 = not available at all; 7 = widely available	5,9	6,0	6,0	6,2	6,3	6,2	6,1	6,1	6,0		
			4,1	4,2	4,1	4,2	4,2	4,2	4,2	4,3	4,2		
T1	Mobile-cellular telephone subscriptions	Number per 100 population		144,6	156,4	166,0	172,5	171,7	139,7	135,5	134,5		
				117,0	120,2	128,5	132,7	150,0	156,4	148,7	146,2		
T2	Availability of latest technologies	not at all; 7 = to a great ex	6,6	6,6	6,6	6,6	6,5	6,6	6,6	6,6	6,6		
			4,4	4,7	4,6	4,6	4,4	4,5	4,6	4,8	4,8		
T3	FDI and technology transfer	1 = not at all; 7 = to a great extent	4,8	4,3	4,2	4,4	4,4	4,3	4,4	4,6	4,8	X	
			4,9	5,0	5,0	4,8	4,6	4,6	4,5	4,6	4,9		
T4	University-industry collaboration in R&D	1 = do not collaborate; 7 = collaborate extens.	5,5	5,6	5,6	5,6	5,8	6,0	6,0	5,7	5,6		
			3,0	3,6	3,6	3,6	3,5	3,6	3,5	3,3	3,2		
T5	Government procurement of	1 = not at all; 7 = great extent	4,7	4,7	4,7	4,5	4,2	4,1	3,8	3,8	4,0		X
			3,7	3,7	3,3	3,2	3,1	3,2	3,1	2,9	3,1		

No.	Symptom of opportunity	Measurment scale	09.	10.	11.	12.	13.	14.	15.	16.	17.	FI	PL
O1	New business density	New registrations of businesses (per 1,000)			26,20	3,40	3,60	2,30	2,30	3,40	3,40	X	
					4,10	0,50	0,50	0,00	0,50	0,50	0,50		
O2	Degree of customer orientation	1 = poorly; 7 = extremely well	5,20	5,30	5,50	5,40	5,20	5,20	5,30	5,40	5,50	0,62	0,58
			4,80	4,80	4,80	4,90	4,90	4,90	5,00	5,10	5,10		
O3	Merchandise export	Exports growth rates, annual	-34,84	10,60	13,84	-7,66	1,87	-0,15	-19,53	-3,19	17,55	X	
			-19,92	17,01	18,14	-1,76	-1,76	7,33	-9,51	2,36	13,30		
O4	Foreign direct investment outward	FDI outward flows and stock, annual	5 681,08	10 167,18	5 011,07	7 543,42	-2 401,88	1 182,12	-16 583,88	25 621,79	1 726,82		
			1 806,36	6 147,18	1 026,44	2 900,99	-1 346,10	2 898,11	4 995,54	8 073,71	3 590,71		

The measurement scale of these variables is such a one that the higher cumulative value corresponds to a better situation in the environment, that is, a situation more supportive to running a business. The more the line rises or falls on the chart, the greater the variability of the environment. The comparison of curves on the charts

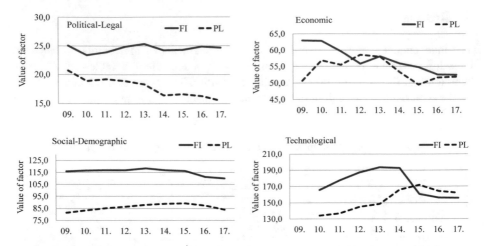

Fig. 2. The cumulative value of factors characterizing segments of macroenvironment

shows that in the analyzed period, both in Poland and Finland, the largest turbulences occurred in the economic and technological segments, with negative trends emerging in the second half of the analyzed period. In addition, in Poland, this negative trend also occurs in the political and legal segment. The most stable is the socio-demographic segment.

The reliability of these variables, in the case of Poland, even after removing three variables concerning the environment and one variable concerning the symptoms of the opportunity (Table 1), is relatively low but acceptable. The correlation of the variables referring to environment and the symptoms of occurrence of the opportunities is presented in Tables 2 and 3.

Table 2. Correlation between symptoms of opportunity (O) and factors of environment's segments (PEST) - Finland

	P1	P2	P3	P4	P5	E1	E2	E3	E4	E5	S1	S2	S3	S4	S5	T1	T2	T3	T4	T5
O1	0,14	-0,39	-0,39	-0,64	0,14	-0,56	-0,19	0,49	0,72	0,05	0,13	-0,16	-0,18	-0,49	-0,13	-0,11	-0,14	0,11	0,10	0,03
O2	0,74	-0,59	0,08	-0,44	0,52	-0,78	-0,20	0,37	0,51	-0,07	-0,54	0,02	-0,13	-0,77	-0,39	-0,44	0,21	0,48	0,13	-0,25
O3	-0,04	-0,14	0,23	-0,33	0,41	-0,16	-0,61	0,50	0,18	0,64	-0,06	-0,26	-0,21	-0,32	-0,38	-0,20	0,00	0,45	-0,67	0,56
O4	0,35	-0,74	0,23	-0,27	0,41	-0,83	-0,46	0,57	-0,04	0,31	-0,21	-0,04	-0,65	-0,67	-0,65	-0,77	0,20	0,28	-0,22	0,15

Table 2 shows that variables describing environmental segments are not statistically significantly correlated with the symptoms of an occurrence of opportunity. The exception is the variable E1 - Inflation - Annual percent change, which is correlated with O2 - Degree of customer orientation and O4 - Foreign direct investment outward. This is a negative correlation. A statistically significant relationship E1-O2 can be interpreted in such a way that if inflation increases, customer care is improved so that the drop in the value of money is compensated by a higher price for better customer service. The relationship E1-O4 can be explained by the fact that the drop in value of money increases the exchange rate of "hard" currencies, which increases the cost of foreign investment and reduces their outflow. The negative correlation S4-O2 shown in Table 2: Extent of staff training - Degree of customer orientation is difficult to rational explanation.

Table 3. Correlation between symptoms of opportunity (O) and factors of environment's segments (PEST) - Poland

	P1	P2	P3	P4	P5	E1	E2	E3	E4	E5	S1	S3	S4	S5	T1	T2	T4
O2	-0,12	0,93	-0,33	-0,12	-0,94	-0,43	0,73	0,06	0,97	-0,55	0,96	-0,44	0,06	0,66	0,93	0,00	0,13
O4	-0,73	0,22	-0,36	-0,63	-0,28	-0,93	-0,04	-0,86	0,19	0,36	0,25	0,37	-0,47	-0,32	0,71	-0,36	-0,60

As can be seen in Table 3, the symptom O2-Degree of customer orientation is correlated statistically significantly with the following environmental factors:

P2 - Burden of government regulation - this factor may induce to stabilize relations with the client through its good service in order to balance the burdensome related to government regulations;

P5 - Number of procedures required to start a business - this factor is negatively correlated and its reduction reduces of the bureaucratic burden and in this way may help devote more time to the client;

E4 - Financial services meeting business needs - improving the accessibility of financial resources allows the allocation of these funds to better meet customer needs;

S1 - Tertiary education enrollment rate - a higher percentage of university students increases the inflow of new knowledge to business and thus enables a better response to the customer's needs;

T1 - Mobile-cellular telephone subscriptions - an increase the value of this factor means enables better communication with customers and therefore potentially better service.

Symptom O4 - Foreign direct investment outward is negatively significantly correlated - with environmental factors:

E1 - Inflation - annual percent change - increase of inflation reduces the attractiveness of foreign direct investment;

E3 - Total tax rate - an increase in tax burden makes it difficult to accumulate capital needed for foreign investments.

5 Discussion and Conclusions

The Tables 2 and 3 show that only two symptoms of seizing the opportunity are statistically significantly correlated with the variables describing the company's environment. These are O2 - Degree of customer orientation and O4 - Foreign direct investment outward. Both are reliable symptoms of seizing the opportunity and as such can be a substitute for enterprise's agility. Comparing O2 and O4 curves between countries, it can be noticed that in the case of Poland, that the customer orientation as well as outflow of foreign direct investment were more systematic than in Finland (Fig. 3). This is surprising because in Finland over the period under consideration,

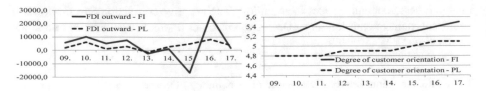

Fig. 3. The cumulative value of factors characterizing segments of macroenvironment

changes taking place in the political-legal and socio-demographic segments were less turbulent than in Poland (Fig. 2). However, as shown in Table 3, in the case of Finland, the changes taking place in these segments of the environment were not significantly correlated with the symptoms of the use of the opportunity. In fact, only one factor in the economic segment, namely E1 - inflation, is correlated statistically significantly with these symptoms. Changes in this segment were more turbulent than in the political-legal and socio-demographic segments. It is also worth noting that the weaker in Finland than in Poland influence of macroenvironment on the use of opportunities and indirectly on the agility of enterprises may be caused by the generally higher value of factors describing particular segments of macroenvironment. It can be hypothesized that a more comfortable environment is less favorable to agility than a less friendly environment.

In the case of Poland, not only the values of factors describing each segment of macroenvironment are lower, which means that it is a less friendly environment, but the dynamic of changes taking place in these segments is greater. This may be an explanation of why more factors than in Finland are statistically significantly correlated with the symptoms of the opportunity. In addition, these are factors from every segment of macroenvironment.

The research results presented in this study are preliminary. In order to be able to draw more generalized conclusions, such research should be carried out on a larger population of countries and the reliability of measuring scales of the considered variables should be improved. It seems, however, that they allow conclusions to be drawn that in a more comfortable (friendly) environment, businesses tend to use lean strategy rather than agile, and that greater turbulences of the environment leads to implementing agile strategies, because such environment generates opportunities giving the enterprise competitive advantage.

References

1. Wlodarkiewicz-Klimek, H., Kalkowska, J., Trzcielinski, S., Pawlowski, E.: External conditions of enterprise's development in a knowledge based economy. In: Marek, T., Karwowski, W., Frankowicz, M., Kantola, J., Zgaga, P. (eds.) Human Factors in a Global Society, pp. 625–633. CRC Press (2014)
2. Goldman, S., Nagel, R., Preiss, K.: Agile competitors and virtual organization: strategies for enriching the customer. Van Nostrand Reinhold, New York (1995)

3. Dove, R.K.: Design Principles for Highly Adaptable Business Systems, with Tangible Manufacturing Example, Maynard's Industrial Handbook. McGraw Hill, New York (2001)
4. Worley, C.G., Williams, T., Lawler, E.E.: The Agile Factor. Building Adaptable Organization for Superior Performance. Jossey-Bass, USA (2014)
5. Gunasekarany, A., Yusuf, Y.Y.: Agile manufacturing: a taxonomy of strategic and technological imperatives. Int. J. Prod. Res. **40**(6), 1357–1385 (2002)
6. Ambe, I.M.: Agile supply chain: strategy for competitive advantage. J. Glob. Strat. Manag. **4**(1), 5–17 (2010)
7. Purvis, L., Gosling, J., Naim, M.M.: The development of a lean, agile and lcagile supply network taxonomy based on differing types of flexibility. Int. J. Prod. Econ. **151**, 100–111 (2014)
8. Putnik, G.D., Putnik, Z.: Lean vs agile in the context of complexity management in organizations. Learn. Organ. **19**(3), 248–266 (2012)
9. Goldman, S., Praiss, K. principal investigators Nagel, R., Dove, R.: 21st Century Manufacturing Enterprise Strategy: An Industry-Led View, vol. 1. Iacocca Institute, Lehigh University (1991)
10. Casson, M.: The entrepreneur: an economic theory. Rowman & Littlefield (1982)
11. Baron, R.A.: Opportunity recognition as pattern recognition: how entrepreneurs "connect the dots" to identify new business opportunities. Acad. Manag. Perspect. **20**(1), 104–119 (2006)
12. Kaish, S., Gilad, B.: Characteristics of opportunities search of entrepreneurs versus executives: sources, interests, general alertness. J. Bus. Ventur. **6**(1), 45–61 (1991)
13. Choi, Y.R., Shepherd, D.A.: Entrepreneurs' decisions to exploit opportunities. J. Manag. **30**(3), 377–395 (2004)
14. Trzcielinski, S., Trzcielinska, J.: Some elements of theory of opportunities. Hum. Factors Ergon. Serv. Ind. **21**(2), 124–131 (2011)
15. Schumpeter, J.: Capitalism, Socialism and Democracy. Taylor & Francis e-Library, London (2003)
16. Kirzner, I.: Competition and Entrepreneurship. University of Chicago Press, London (1973)
17. Trzcieliński, S.: The influence of knowledge based economy on agility of enterprise. In: 6th International Conference on Applied Human Factors and Ergonomics, Procedia Manufacturing, pp. 6615–6623 (2015)
18. Trzcielinski, S., Trzcielinska, J.: How enterprises identify market opportunities: research results and findings. In: Trzcielinski, S. (ed.) Advances in Human Factors and Ergonomics, pp. 116–128. Springer, Cham (2017)
19. Trzcielinski, S.: Flexibility of SMEs. In: Schlick, Ch., Trzcielinski, S. (eds.) Advances in Ergonomics of Manufacturing: Managing Enterprise of the Future, Advances in Intelligent Systems and Computing, vol. 490, pp. 417–427. Springer, Switzerland (2016)
20. Trzcielinski, S.: Agile enterprise – research on flexibility. In: 10th International Workshop on Robot Motion and Control, IEEE, pp. 213–216 (2015)
21. Trzcielinski, S.: Research on intelligence of medium sized enterprises. In: PICMET 2016: Technology Management for Social Innovation, pp. 1993–2001 (2016)
22. Trzcielinski, S., Rogacki, P.: The model of virtual workshop of manufacturing company. In: Fallon, E.F., Karwowski, W. (eds.) Human and Organizational Issues in the Digital Enterprise, vol. 2, pp. 470–478. National University of Ireland Galway, Galway (2004)
23. Trzcielinski, S., Wypych-Zoltowska, M.: Towards the measure of virtual teams effectiveness. Hum. Factors Ergon. Manuf. **18**(5), 501–514 (2008)
24. Trzcielinski, S., Wojtkowski, W.: Towards the measure of organizational virtuality, vol. 17, no. 6, pp. 575–586 (2007)
25. Rothaermel, F.T.: Strategic Management. McGrawHill, USA (2019)

Agile Management Methods in an Enterprise Based on Cloud Computing

Michal Trziszka[✉]

Faculty of Engineering and Management, Poznan University of Technology,
Strzelecka 11, 60-846 Poznan, Poland
michal.trziszka@put.poznan.pl

Abstract. Globalization and development of IT systems are conducive to changes in the entrepreneurs' approach to the methods of carrying out tasks related to running a business. The growing competition and the need to quickly respond to the changing business environment mean that business owners are more willing to switch to agile business management methods. Cloud computing as a modern form of providing IT solutions can be successfully used in agile enterprises. The tools provided in the subscription model are increasingly being chosen by such enterprises. The aim of the article is to present the tools available in the cloud computing for managing a agile enterprise and to focus on security aspects of such solutions. The research conducted on several dozen companies using the software provided in the subscription model allowed to draw the conclusion that most enterprises are not aware of the threats on the Internet and can not protect themselves against them. In most cases, companies are not aware of the possibility of leakage of confidential data, business secrets and management models. Full confidence in cloud computing and data stored on the Internet makes small and medium enterprises an easy target for cyber-criminals. As a result of the conducted research, the article presents not only the most frequently chosen software for agile enterprise management, but also selected options to increase the protection of corporate data stored in the cloud.

Keywords: Systems engineering · Cloud computing management ·
Cloud computing · Cloud · Agile management · Agile in the cloud

1 Introduction

The feature of a modern company is flexibility, efficiency and speed in responding to changing environments [1]. All of these activities are aimed at increasing the company's competitiveness on the market, which is to provide it with an advantage over other entities operating in the same sector or industry. New solutions in the sphere of management and organization, tailored to the type of work performed, in addition to the benefits brought to employees and employers, enable continuing economic growth and increasing the competitiveness of enterprises [2].

Competitiveness is considered a multidimensional characteristic of the enterprise, associated with the ability to adapt to changes and transformations occurring in the environment. It is a feature that indicates the company's outstanding ability to take

© Springer Nature Switzerland AG 2020
W. Karwowski et al. (Eds.): AHFE 2019, AISC 971, pp. 122–129, 2020.
https://doi.org/10.1007/978-3-030-20494-5_12

such actions that guarantee stable and long-term development, and also contribute to the creation of high market value [3]. This approach is characterized by agile companies, distinguished by the following features: smartness, flexibility, intelligence and cunning [4]. The agility of the company is a complex condition conducive to perceiving occasions appearing in the environment and using them in accordance with the possessed resources. Due to the type of resources that dominates in shaping agility, S. Trzcielinski distinguished four types of agility: technological, financial, social and entrepreneurial. Technological agility is primarily based on the use of modern technological solutions that result in the improvement of the speed and quality of services provided. Technological changes, in turn, have a considerable impact on the optimization of costs, and hence - increased profits and better organization of the enterprise. Skilful building of competitive advantage is based on the proper and effective use of knowledge resources and intellectual capital [5]

Thanks to the Internet, more and more companies have the opportunity to use expensive and modern cloud computing technologies without investing in expensive devices, software and specialists.

2 Cloud Computing – IT Services from the Internet

Cloud computing is a term that includes the provision of IT services by an external company via the Internet. The National Institute of Standards and Technology (NIST) has developed a definition of cloud computing, according to which it is a model that enables sharing of shared configurable computing resources (servers, programs, memories, applications and services) that can be delivered quickly with minimal management effort and the interaction of the service provider [6]. Such services can not only be delivered from anywhere and any time in the world, but also from any device such as a computer, smartphone or iPad. Also, users who have access to data in the cloud using special programs can use them anywhere, anytime. The possibilities of cloud computing will make the company not limited in its operations geographically. Employees of international companies have access to ongoing projects at the same time, and small enterprises can collect and process data without having to install costly servers and software. This is confirmed by the results of the report of the Institute for Market Economics, according to which many companies appreciate the advantages of cloud computing [7]. When using it, the IT system configuration is limited to the necessary minimum, because the service provider provides space on the server and software. It also translates into better settlement, less needs in the area of office space and a reduction in the number of IT employees [8]. An important advantage of the cloud is its performance, i.e. unlimited disk space and advanced programming platforms. Gatner [9] estimates that in 2021, the cloud market will be worth $ 300 billion. Information and telecommunications capabilities as a cloud computing solution, also causes the phenomenon of virtualization, resulting in a series of profound changes in many aspects of the functioning of enterprises, not only in terms of management, but also organizational. According to T. Listwan, the concept of management processes virtualization should be understood as "the process of changing the organizational space and building relationships (including electronic ones) based on shared value systems between line managers, task force managers and managers or personal specialists, in order to provide information and carry out human resources tasks" [10]. This is a wider use of flexible forms that allow, among other things, for dislocation (e.g. teleworking), broadening the scope of duties and rights (increasing independence) and implementing new solutions in organizing teamwork processes [6].

3   Benefits and Threats Resulting from Access to Cloud Computing

The pressure associated with growing customer expectations, intensifying competition and globalization of activities in most areas of the economy, entail the need to improve the functioning of the company, in accordance with the Agile principle [11], that responding to changes is more important than acting in accordance with the plan. Changes in the needs, expectations and preferences of consumers, which are still accelerating the development of technology and technology, make companies constantly be surprised by something. This is reflected in the organization of work. Owners and managers of companies - acting as employers - often under the influence of external conditions make unexpected decisions relating to, among others, to work methods or personnel composition changes. As a result, uncertainty is now an element of the organization's life and people working in it [12].

This state affects the dynamic development of cloud services and results mainly from the benefits of cloud computing such as availability on demand, cost reduction, availability regardless of location, measurability, responsibility for infrastructure on the service provider's side. For the cloud model, it is also characteristic that users usually pay for what they use, avoiding large fixed initial costs associated with the independent configuration and operation of advanced computer hardware, which directly results from the financial agility of the company.

Three types of clouds can be identified in global use:

1. Software as a Service (SaaS) - permanent access to the software with payment only for the actual use of it: Skype, Gmail, Google Docs, ifirma.pl, nozbe.com;
2. Platform as a Sewice (PaaS) - a working environment for creating, processing, installing and running own applications: Azure from Microsoft, Elastic Compute Cloud from Amazon;
3. Infrastructure as a service (IaaS) - a service consisting in offering IT equipment via a computer network, such as a processor, operational memory or disk memory mp. virtual disk offered by dropbox.com [13].

Cloud computing guarantees security through a negligible failure of the company's management system and protects against data loss. The information collected is usually copied and stored in at least two parallel data centers. When a possible failure of one of the centers occurs, a backup from the parallel processing center is automatically enabled [14].

In a survey conducted among employees of Polish IT companies in December 2018, on a sample of 523 people, it was determined that the majority of respondents highly evaluate the implementation of cloud solutions in enterprises (82% of respondents). In the survey, the following advantages of cloud applications are listed:

- The speed of implementation;
- Low maintenance costs compared to normal licenses/equipment;
- Free technical support;
- High level of service availability;
- Frequent software updates;
- Availability of solutions via the Internet.

On the other hand, the use of cloud services is also associated with concerns about security, performance, cost increases and availability [15]. New threats or new forms of existing threats appear. Electronic communication, facilitating the flow of information, increases the risk of outflow of confidential information outside organizations and the risk of intellectual property violation. New technical solutions are also used in online crime, e.g. data theft, blocking websites [16]. It must be clearly emphasized that the cloud computing is not free from defects experienced by both employers and employees. The most important ones include:

- Work addiction - software availability even after the office;
- Security of data stored in the cloud;
- Loss of control over software/hardware;
- Difficulty in controlling the update;

Entrepreneurs are aware of the benefits of cloud computing as well as its disadvantages. Many of them (over 62%) introduce additional security methods such as:

- Security policy regarding user passwords (78%);
- Connection using VPN (42%);
- Connection with software on the Internet from the company's internal network (23%);
- Additional login authorization from a mobile device (e.g. Google Authenticator) (4%);

In the study, 92% of respondents admitted to problems with the security of their computers, the majority (64%) concerned e-mails with phishing phishing sites, 30% received a computer virus sent electronically, and 6% of respondents were attacked by ransomware.

Cloud computing, despite the multitude of protections which the suppliers of not only infrastructure equipment, but also the SaaS-based software providers themselves, can not be treated as a safe and reliable solution. However, you can not omit the key aspect that affects the security of the corporate environment - the user. Securing your computer, an appropriate password hardship policy, and encrypting your connections are essential to protect yourself against potential data leaks and attacks on the Internet.

4 Nozbe – Software Available in the Cloud for Agile Enterprises

Communication in the company and workflow is key in an agile company. Combining these two most important functions in one tool is nothing complicated and innovative. Let's analyze this type of solution on the example of the Nozbe application.

This product is easy to use and at the same time contains many advanced features. Helps teams manage time and tasks. The developers of Nozbe noticed that the e-mail is not suitable for group work. Thanks to the Nozbe application, work on projects allows the team to complete more tasks and to inform each other who and what they are currently working on. Additionally, it is possible to synchronize with other applications, i.e. Dropbox, Google Calendar. Nozbe allows you to choose one task that is the most important at the moment and mark it with a symbolic star. Other tasks are available on the calendar. Importantly, tasks marked with the current date are automatically marked with an asterisk so that the employee can remember that the deadline is approaching. Every Nozbe user can create as many different projects as he needs. He shared them with the other members of the project groups to carry out tasks that had previously been delegated to them. For simplicity, projects are labeled. In addition, contexts are assigned to them, e.g. a computer, telephone, home, office. Newly delegated tasks are instantly added to the list of immediate tasks. You can add comments to individual commands, including with guidance on the implementation of the action plan. Comments can also contain short checklists for unmarking, images, photos, drawings, documents, etc. Instead of traditional e-mail, it is better to contact by notifying about the last activities of the team. Messages are sent cyclically (every hour), but only when someone has done something in those projects that the person is co-organizing. Nozbe works on all popular browsers: Internet Explorer, Google Chrome, Mozilla Firefox, Safari, Opera. It should be emphasized that the application also works without internet. You can manage your projects without a network connection, drag and drop files to load them into Nozbe via the cloud.

The Nozbe application uses the GTD (Getting Things Done) method of David Allen. It is a way of organizing tasks based on collecting orders and managing task lists and projects. The goal of GTD is to release from the obligation to remember about all obligations and plans, while maintaining high productivity at work and in private life. According to Allen, two key elements in time management are control and perspective. Control over liabilities is ensured by the detailed GTD process, while the perspective allows the inclusion of current tasks in the context of six levels:

1. Current tasks.
2. Current projects.
3. Areas of responsibility.
4. Annual goals.
5. Vision for 5 years.
6. All my life [17].

One of the sources of the Allen method's popularity is its flexibility - GTD offers processes that can be implemented in many ways. That is why the method is often used by lifehackers who adapt existing tools to it (text files, e-mail client, wiki-based pages, time management programs) or create new ones. Some uses of GTD are based solely on paper materials [18]. There are currently many computer programs supporting GTD work. The described method fits perfectly into the agile enterprise philosophy, which requires conscientious execution of manager's orders and quick response to changes. The method allows to organize the chaos in the work plan and in the closest surroundings (on the desk, in files on the computer) [19].

5   Methods of Securing Cloud Services

Among the data security in the cloud, in terms of business management, procedures and management of access to data are applied:

- security policy - a set of procedures and rules according to which the organization manages resources,
- identity and access management of users in the cloud - Identity and Acces Management - gives the ability to monitor the actions performed by users (Role - based Access Control - defines the roles associated with the appropriate responsibilities assigned to certain functions in the organization; Discretionary Access Control - the system uses hardware and software mechanisms to identify and authorize users, Mandatory Access Control - mandatory access control is a method of access control in the system - entities are processes and users themselves, and device objects, registry keys, files and directories, and both functions have specific security attributes),
- backup copies (this is data that is a copy of the data currently in use and is used to reproduce those that have been damaged or lost [20],
- encrypted communication - encryption and decryption of messages to protect against unauthorized retrieval or reading of messages,
- Keys and certificates - bits of bits, which are input parameters for the algorithm that encrypts data. It is a form of an electronic certificate that can be used to confirm the identity of a specific person [20].

Despite the benefits and high technological level, cloud computing has been defined as the technology with which the largest business risk is associated. Responsible in companies for the area of risk management, audit, compliance with regulations and finances, pay particular attention to the possibility of data leakage and the risk of failure to meet the requirements related to GDPR (in Poland - GDP). According to Gartner's experts, unauthorized access to information contained in cloud computing will be the biggest challenge for enterprises. Already, interruptions in network access cost the world economy 2.5 billion dollars a year [21]. The weak point of the cloud is man, not technology. Gatner estimates that by 2022, at least 95 percent of security failures in the cloud will be the result of organizational and human errors.

6   Conclusion

An increasing number of enterprises are looking for savings in the area of ICT infrastructure. The increasing popularity of the method of providing IT services such as cloud computing, not only changes the existing IT resource management model, but also for the agile enterprise is a tool to modify the competitive advantage, mainly due to the reduction of operating costs and the efficiency and flexibility of operations. One of the most important issues in the area of cloud computing is the need to ensure safe data transfer and storage, ways to identify and verify users, and the legal requirements of the supplier. Despite the identified threats, the cloud computing market is growing dynamically. It is estimated that in 2019, nine out of ten companies will use solutions based in part or entirely on a computing cloud. In 2025, business will already place there 60% of all its digital information, which will force the cloud computing service providers to implement better security systems and data protection.

References

1. Trzcielinski, S.: Modele zwinności przedsiębiorstwa. In: Nowoczesne Przedsiebiorstwo, Monografia Instytutu Inzynierii Zarzadzania. Politechnka Poznanska, Poznan (2005)
2. Harasim, W.: Zarzadzanie zasobami ludzkimi i kapitalem ludzkim. In: Czlowiek i organizacja XXI wieku, Wydawnictwo Naukowe Wyzszej Szkoly Promocji. Warszawa (2013)
3. Kot-Radojewska, M.: Elastyczne formy zatrudnienia jako czynnik wzrostu konkurencyjności przedsiębiorstwa, Zeszyty Naukowe Wyższej Szkoły Humanitas. Zarządzanie (2014)
4. Trzcielinski, S.: Zwinne przedsiębiorstwo. Wydawnictwo Politechniki Poznanskiej, Poznan (2011)
5. Walczak, W.: Wiedza zrodlem budowania przewag konkurencyjnych wspolczesnego przedsiebiorstwa. In: Kapital intelektualny i jego ochrona. Instytut Wiedzy i Innowacji. Warszawa (2009)
6. Polonka, J., Pankowska, M., Zytniewski, M.: Modele techniczno-społeczne wirtualizacji i udostępniania na żądanie zasobów IT. Wydawnictwo Uniwersytetu Ekonomicznego, Katowice (2016)
7. Report of the Gdansk Institute for Market Economics. http://www.ibngr.pl/Publikacje/RaportyIBnGR/Cloud-Computing-elastycznosc-efektywnosc-bezpieczenstwo. Accessed 28 Jan 2019
8. Juszczyk, M., Wita, B.: Technologie mobilne, przetwarzanie w chmurze obliczeniowej: nowe narzędzia, nowe możliwości. Polskie Towarzystwo Informatyczne, Lublin (2012)
9. Website information: gatner.com. Accessed 27 Jan 2019
10. Listwan, T.: Rozwój badań nad zarządzaniem zasobami ludzkimi w Polsce. In: Osiągnięcia i perspektywy nauk o zarządzaniu, Oficyna a Wolters Kluwer business. Warszawa (2010)
11. Agile Manifesto: agilemanifesto.org/iso/pl/. Accessed 28 Jan 2019
12. Bienkowska, J.: Problemy kształtowania sytuacji pracy w warunkach niepewności. Humanizacja Pracy, Warszawa (2014)
13. Wiewiórowski, W.R., Wierczyński, G.: Informatyka prawnicza: nowoczesne technologie informacyjne w pracy prawników i administracji publicznej. Wolters Kluwer, Warszawa (2016)

14. Czerwonka, P.: Zastosowanie chmury obliczeniowej w polskich organizacjach. Wydawnictwo Biblioteka, Łódź (2016)
15. Szewczyk, A., Wojarnik, G.: Chmury nad e-biznesem, Wydawnictwo Naukowe Uniwersytetu Szczecińskiego. Szczecin (2014)
16. Oleński, J.: Ekonomika informacji. Wydawnictwo PWE, Warszawa (2003)
17. Allen, D.: Sztuka efektywnosci. Skuteczna realizacja zadan. Helion, Warszawa (2006)
18. Allen, D.: Getting things Done, czyli sztuka bezstresowej efektywnosci. Helion, Warszawa (2008)
19. Trziszka, M.: Narzedzia komunikacji wykorzystywane w modelu pracy zdalnej w firmach rodzinnych. In: Firmy Rodzinne - rowoj teorii i praktyki zarzadzania, Przedsiebiorczosc I Zarzadzane, Lodz-Warszawa (2017)
20. Siergiejczyk, M.: Problematyka bezpieczeństwa informacyjnego w chmurze obliczeniowej. Logistyka (2015)
21. Predicts 2019: Increasing Reliance on Cloud Computing Transforms IT and Business Practices. https://www.gartner.com/doc/3895580?ref=mrktg-srch. Accessed 27 Jan 2019

Competencies in Work Environment

Competency Model for Logistics Employees in Smart Factories

Markus Kohl[1](✉), Carina Heimeldinger[2], Michael Brieke[2], and Johannes Fottner[1]

[1] Chair of Materials Handling, Material Flow, Logistics, Technical University of Munich, Boltzmannstr. 15, 85748 Garching, Germany
{Markus.Kohl,J.Fottner}@tum.de
[2] MAN Truck & Bus SE, Dachauer Str. 667, 80995 Munich, Germany

Abstract. New technologies are changing today's logistics processes. The role of humans and the tasks they perform will therefore undergo decisive changes in the future. Companies will need to develop qualification strategies to cope with changing tasks, for which competency models are essential. Current research has dealt little with competency changes caused by the digitalization of automotive logistics. This leads to a lack of future-oriented competency models. Thus, this paper designs a competency model for logistics employees that integrates future competencies for operational and planning activities. General and logistics-specific competency models and studies on competency changes by digitalization are analyzed. The competency model created consists of professional, methodological, social, and personal competencies. It also considers the process for implementation and future adjustments. The model is applied and evaluated in selected logistics processes and identifies process-specific competencies and future changes. In addition, the competency model can conduce to strategic personal planning.

Keywords: Competency model · Competency development · Future logistics · Smart factory · Human resource management

1 Introduction

Today's production and logistics processes are changing due to the introduction of new technologies, such as autonomous robots, big data, and artificial intelligence [1]. Digitalization enables the communication between humans, machines, and products [2]. This leads to major changes, not only in production but also in external and internal logistics processes. These changes occur, for example, whenever automated guided vehicles (AGVs) are implemented. Previously, for example, forklifts were used for material transport, whereas AGVs could be applied to automate material transport [3]. These process-related changes often imply adjustments in work content and thus in required competencies. It should not be underestimated that, besides technical challenges, people also have to adapt to these changes [4]. In addition, to getting used to the adapted situation, new skills and abilities are particularly necessary. To prepare

© Springer Nature Switzerland AG 2020
W. Karwowski et al. (Eds.): AHFE 2019, AISC 971, pp. 133–145, 2020.
https://doi.org/10.1007/978-3-030-20494-5_13

employees for new tasks such as monitoring and maintaining the AGVs [3] new qualifications are needed.

This requires a framework to analyze competencies systematically and which should be an integral part of strategic personnel management. Within such a framework, the process changes due to the introduction of new technologies must be determined. This requires a detailed understanding of the process as well as a technology roadmap that sequences future technologies for their implementation.

The basic building block for the whole framework is an industry-specific competency model, in this case for logistics. The aim is to analyze current and future competencies for automotive logistics tasks in the operational and planning area. Even though competency models already exist, not enough research analyzes changes in planning and operative logistics tasks caused by digitalization. Therefore, this paper deals with the question of which existing competency models can serve as a basis in logistics. It is also necessary to clarify, which current and future competencies need to be added to the model.

To answer the relevant research questions, first a short overview of existing competency models is given. Afterwards the theoretical background of changes in competencies due to digitalization will be given and the research gap will be described. The research methodology of creating a competency model will be introduced in the third section followed by the logistics-specific competency model. To validate the competency model, the latter was used in a company producing commercial vehicles. This paper concludes with a short discussion and an account of further steps.

2   Background

2.1   Competency Models

In order to cope with changing processes and tasks in the future, companies must develop qualification strategies and for these, competency models can be considered as essential. MANSFIELD defines a competency model as "a detailed, behaviorally specific description of the skills and traits that employees need to be effective in a job" [5]. Researchers most often divide competencies in the four following categories: social, self/personal, methodological/action-related and professional/technical/domain-related [4, 6–9]. This paper also uses these four categories, since this classification has proved its worth and appeared to be reasonable for this work.

To obtain an overview of existing competency models, a review of the literature was conducted and 16 competency models were compared. During this process, the focus was to identify specialized, logistics models or else to consider the four competency categories. Most of the competency models are not applicable for logistics-specific tasks. Some models only consider general leadership [10, 11], while other focus on professional and methodological competencies [12], or on social competencies [13]. Other models pursue a different goal such as competency-oriented program planning [14], support the selection of employees [15], or mainly track current competencies [16]. Competency models specific to group work [17] or focusing on competency measurement and not using their own model [18] are not considered. After the pre-selection, seven competency models, which can be divided into more general and division-specific, were examined in Sect. 3.2.

2.2 Change to Competencies Caused by Digitalization

The decisive question for the relevance of this work is whether and to what extent the necessary competencies change due to digitalization and automatization/autonomization. Studies concur that there will be significant changes [19, 20]. One study by ACATECH compares competency changes in companies. Small and medium-sized companies predict that social and communication competencies will become more important while large companies put emphasis on interdisciplinarity [19]. This development of competencies confirms a study by INGENICS. Lifelong learning and IT-competencies are expected to gain in importance additionally [20].

Another method to identify change in competencies is the use of competency models. BUTSCHAN NOOTLE ET AL. [21] consider the competency model by NORTH [22] to analyze competency changes in production. The authors note that employees in production will need a broader and deeper understanding of the interrelationships between production, logistics and procurement. In addition, skills such as openness to change, adaptability and willingness to learn will play an increasing role in smart factories.

The review of the literature implies that there are different statements about which exact competencies will become necessary in the future. In summary, however, it can be seen that today's competencies will differ in their significance in the future and new competencies emerge.

2.3 Summary of the State of the Art and of the Research Requirement

The literature review clarifies that logistics processes are changing and will change in the future due to digitalization and automatization/autonomization. Therefore, the tasks of personnel and thus the necessary competencies are also subject to change [23]. In order to overcome the challenges, new qualifications need to be implemented in the industry [24].

A lack of personnel resources, non-existent internal training concepts, and the unknown nature of future requirements are among the internal hurdles and challenges in the further qualification of employees [25]. Therefore, strategic development, planning and implementation of competencies will become more important [4]. A solution to counteract the lack of competencies can be the use of competency models.

Researchers criticize that competency models often only consider the current competencies that are needed in a job, but do not address the required future skills [26]. For example, the transportation, distribution, and logistics competency model depicts the logistics tasks, but future competencies are not analyzed in detail [27]. The competency model by HECKLAU ET AL. concentrates on the competency change only in production, but unconsidered logistics [4]. The same applies to DEMEL's competency wheel that takes future competencies into account, but has not yet been applied in logistics [28]. Most of the other competency models are not adapted to logistics or have not been applied there [21, 22]. This is not the case with the ABEKO competency model that incorporates operative logistics, but planning tasks are disregarded [29].

Accordingly, none of the identified competency models is logistics-specific and takes current and future competencies into account. Hence, there is a need for a logistics-specific competency model that considers digitalization factors.

3 Research Methodology

In order to develop a competency model that incorporates logistics activities and displays competencies for future work, as demanded in Sect. 2.3, a structured procedure is followed. Even if none of the existing competency models meets the requirements sufficiently, established models can nevertheless provide a basis for adjustments.

In order to design a logistics-specific competency model, the methodology of SAUTER AND STAUDT [30] was used. This consists of three phases:

The *modelling phase* analyzes and examines existing competency models, in order to determine whether a suitable competency model can serve as the basis for the new model [30]. After defining the requirements for a logistics-specific competency model, the preselected competency models (see Sect. 2.3) were assessed. Since even the best-rated model could not meet all of the requirements sufficiently, it was only taken as a foundation and had to be adapted accordingly.

In the *identification phase*, the competencies were selected and their definition compiled. The consideration of different competency models and studies on competency changes through digitalization helped to gather future competencies. The competencies identified from literature were classified and allocated to the four main categories. Behavioral anchors were defined, in order to measure the respective competencies [31].

The *validation phase* consists of the practical implementation of the competency model created. In this phase, the existing competencies are recorded [30].

3.1 Requirements for the Competency Model

To create a competency model, requirements must be defined and fulfilled, which are based on the objectives and expert discussions. The following requirements were devised for the logistics-specific competency model for smart factories:

Since digitalization is expected to bring major changes in all logistics processes, the **logistics-specific** competency model should be applicable throughout the supply chain and along the whole process chain. In addition to operational tasks, this includes planning tasks to consider all activities as networking increases. The main focus in this case is on automotive logistics, whereas the model should be transferable to other areas. Leadership competencies are not considered in more detail in this competency model, as they are mostly process and area unspecific. Due to digitalization, **future competencies** must be included in the model. In addition, the model should be **process-based**, i.e. the model should show competencies as a function of the process under consideration and can therefore be adapted to the situation. To guarantee clarity and ease of use, the model should consist of a **manageable number of competencies**. Moreover,

competency-specific behavioral anchors, also known as competency levels, need to be verbalized, in order to ensure a clear classification of the respective competencies.

3.2 Assessment of Existing Competency Models

The seven preselected competency models were evaluated regarding the defined requirements arising from Sect. 3.1, in order to check whether suitable competency model exists. To rank the competency models, an assessment method featuring simple scoring was used. Thus, a ranking system starting from zero (does not fulfill requirement) to four (completely fulfills requirement) was devised. The models, their respective advantages and limitations were analyzed, in order to assess them conscientiously. Table 1 shows the result of the assessment.

Table 1. Assessment of the preselected competency models.

Competency model	Logistics-specific	Future competencies	Process-based	Number of competencies	Specific behavioral anchors	Score
Competency Wheel and Competency Matrix [22]	2	2	2	3	3	12
Competency Profile for Leaders in Construction Management [7]	1	2	4	4	1	12
ABEKO [29]	3	3	3	1	1	11
Competency Wheel [28]	2	3	1	3	1	10
Competency Model [4]	2	2	2	2	1	9
Transportation, Distribution, and Logistics Competency Model [27]	3	1	1	2	0	7
Competency Atlas [32]	2	1	2	1	0	6

The competency wheel by NORTH [22] is industry-independent and pursues the goal of analyzing and illustrating existing competencies in comparison to the competencies required in the future. DEMEL's competency wheel [28] is similar to this, but is mainly used for leaders. the competency atlas is also an industry-independent competency model containing 64 competencies clustered in the four main competencies [32].

The competency model by HECKLAU ET AL. [4] is more division-specific and displays the change in competencies caused by digitalization in production. The ABEKO competency model contains competencies for operative tasks in logistics and concentrates on process-orientation [29]. The transportation, distribution, and logistics competency model is a logistics-specific model divided into seven areas [27]. The competency model for the construction sector is specified for leading employees and offers the opportunity to illustrate competencies depending on the process being considered [7].

The competency wheel [22] achieved the highest result. However, it has not yet been used in logistics and future competencies are mostly missing. The visualization in form of a wheel and a matrix display the competencies superficially. This illustration can also be applied when competencies are adapted and is therefore used for the logistics-specific competency model, as well as the behavioral anchors. LIESERT's competency profile [7] is in second place due to its process orientation. Although the model was originally developed for the construction industry, it shows a process-oriented application that can be transferred to logistics.

The two best-placed models form the basis for the new competency model, since the other competency models provide only marginal advantages. They offer sufficient adaptability to focus on the logistics of the future and this ensures practical applicability. These adjustments, which are necessary, are explained in the following section.

3.3 Adaptation of the Competency Model

The next step is the implication phase advocated by SAUTER AND STAUDT [30]. To fulfill the requirement of taking future competencies into account, studies [20, 21, 33] were used that consider changes in competencies caused by digitalization. In addition, all the competencies of the competency models selected were collected and clustered into the four groups: personal, social, methodological and professional competencies. This information gathering resulted in a total number of over 200 competencies.

Furthermore, logistics-specific studies were analyzed [19, 20, 25] and future competencies like lifelong learning or IT knowledge could be identified [20]. According to a study by ACATECH [19] competencies, such as problem-solving skills and interdisciplinarity, will also become more important in the future.

All of the competencies identified in both ways were clustered with the support of experts in logistics and, as far as possible, combined and reduced in order to create a logistics-specific competency model.

Using the methodology of LEINWEBER [31] to specify the competencies and to ensure their practical application, behavioral anchors were verbalized with experts in employee-training. In addition, the validation phase advocated by SAUTER AND STAUDT [30] was implemented within the commercial vehicle sector. In the following section, the resulting logistics-specific competency model is presented.

4 Logistics-Specific Competency Model for Smart Factories

The logistics-specific competency model consists of four different categories: self, social, methodological and professional competencies as shown in Fig. 1.

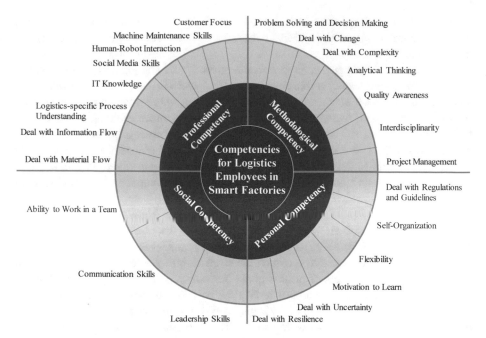

Fig. 1. Main competencies for logistics employees in smart factories.

These categories are divided into 24 sub-competencies in order to detail the main groups without a prioritization or weighting between the sub-competencies. For practical application, in eight cases additional competencies are added on a lower level. This allocation helps to keep track of the competencies at the second tier, but also to explain others more specifically at the third tier. In order to simplify the application, behavioral anchors were added for the user. A behavioral anchor helps to classify the skills defined at a particular competency level [31].

The following sections give a detailed explanation of the logistics-specific model. The competencies, behavioral anchors, applications, and the review process are described.

4.1   Competency Categories

Social competencies refer to the fact that an individual embedded in a coal context requires the ability to communicate, cooperate and to establish social connections [8].

In the model, social competencies are defined by the *ability to work in a team*, *communication skills* and *leadership skills*.

Personal competencies are the ability of a person to act in a reflective and autonomous way, the ability to learn and to develop an own attitude and ethic value system [8]. This category contains *dealing with regulations and guidelines, self-organization, flexibility, motivation to learn, dealing with uncertainty*, and *dealing with resilience*. *Flexibility* was added, in order to be able to adapt to rapidly changing tasks and to map work forms, such as job rotation. Lifelong learning is part of *motivation to learn* and has been added as a future competency, based on studies. *Dealing with resilience* is also frequently considered in studies. In order to check whether an employee must be physically and mentally more or less resilient, this competency was included [16, 32].

Methodological competencies define the approach of solving tasks and problems [4]. Therefore, competencies like *problem-solving and decision-making, dealing with change, dealing with complexity, analytical thinking, quality awareness, interdisciplinarity,* and *project management* form part of the competency model. The competency *interdisciplinarity* was added, because logistics processes relate to process partners along the entire value chain and this is likely to be needed even more in the future [20]. Another future competency that has been considered is *dealing with complexity*. Increasing data volumes contribute to this competency becoming more important, as does *analytical thinking* [33].

Professional competencies comprise all job-related knowledge and skills [4]. Therefore, *customer focus, machine maintenance skills, human-robot interaction, social media skills, IT knowledge, logistics-specific process understanding* and *dealing with information* and *material flow* have been taken into account. Figure 2 shows an excerpt of the different tiers of professional competencies. The logistical processes are defined based on information and material flow between the sub-processes. To enable use of the competency model in every logistics process, the competencies *dealing with information* and *material flow* were added. Studies predict that *logistics-specific process understanding* will become increasingly important [19, 20]. Brühl classifies social media as an important form of communication within a company, as its use enables more transparent coordination of tasks and projects [34]. Therefore, *social media* was included as competency in the model.

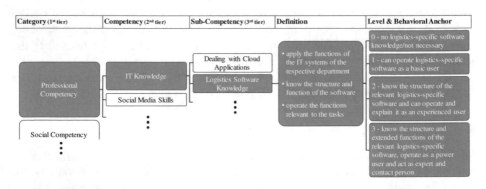

Fig. 2. Excerpt of the structure of the logistics-specific competency model.

4.2   Behavioral Anchors

In order to simplify use, definitions are given for each competency and a rating scale to measure the competency requirement of the task. The scale relies on the model by North et al. [35] that features the following ranking: novice-proficient-expert.

The first level means that the competency is recognizable, but the employee only acts as per advice. The next level induces independent acting as well as the transfer of a task to different situations. To reach the highest level of competency, an employee operates completely independently, can use the required competency in complex situations, and is aware of his or her own actions.

According to MANSFIELD, it is important to define consistent levels for the competency model. Nevertheless, the levels should not rely on the specific job as the model mainly is used across the whole organization [5]. The competency model is logistics specific. Therefore, each competency level is defined from the logistics point of view.

These common verbalizations provide the basis for competency-specific behavioral anchors. Following this, a specific anchor was formulated for each competency to enable easy practical use. An example for *ability to communicate* is shown in Fig. 2.

4.3 Possible Applications of the Logistics-Specific Competency Model

For the logistics-specific competency model, there are various possibilities for application. In addition to evaluating the necessary competencies for the current processes or activities, the model can also be used to measure employee competencies based on the specific behavioral anchors. By comparing the two assessments, the current situation can be tracked and a competency surplus or shortage can be recognized.

A similar approach can be used for the introduction of new technologies. Analyzing the process changes due to the implementation and deriving the competencies is necessary. Here, the comparison can be done regarding current processes to identify processual competency changes or regarding existing employee competencies to identify possible qualification requirements. Examples are described in Sect. 5.

In addition to the use for implementing a new technology, competencies for a possible future status can also be assessed. Scenarios are used to obtain a holistic assessment of competencies. By identifying changes and deriving trends, it is possible to prepare early for transformations such as digitalization.

4.4 Review the Logistics-Specific Competency Model

Logistics processes are subject to constant change due to new technologies. Therefore, there is a need to keep the logistics-specific competency model up to date. Regular reviews can ensure that the required competencies for future changes are systematically identified.

Indicators for this can be, for example, extensive process changes. As a result, the activities might no longer be represented by the previous competencies. A further indicator for a revision are new technologies, which were not known up to now or which were not taken into account when designing the competency model. Moreover, organization-specific adjustments can occur in the company, which can also be an indicator for a revision of the competency model.

In the case of small adjustments, updates can be done by including newly identified competencies or assigning or redefining existing ones. In this case, the added or adapted competencies must be re-evaluated with regard to the processes or activities. Nevertheless, it is necessary to apply the process for creating the competency model iteratively, approximately annually. This is necessary to continuously identify the competencies required in the future, to implement them in the competency model and to recognize changes at an early stage. New competency models, studies on competency changes, and process as well as technology changes can serve as sources for this. The revision should also critically examine whether the existing competencies are still relevant or whether they can be removed from the model.

5   Case Studies and Evaluation

To ensure practical applicability, the competency model was evaluated in three different processes at the commercial vehicle manufacturer MAN Truck & Bus SE. The model was applied in processes along the supply chain to assess competencies for the introduction of new technologies, as described in Sect. 4.3. The aim of these examples is to validate the model in various processes and to prove whether the competencies adequately reflect current and future activities. The material flow and goods receipt processes represent operative tasks, while the disposition process was chosen as a planning task. For incoming goods, automation as well as tracking and tracing technologies, such as RFID, will require IT competencies and lifelong learning, while physical resilience will decrease due to assumed higher degree of automation. In the future, disposition will be able to focus on complex tasks that cannot be performed by algorithms. This leads to an increase of the competencies problem-solving and decision-making. The material flow including the application of the competency model and its results are described in more detail below.

Material flow is one of the in-house logistics processes. Forklifts are the most widely used technologies today to provide material for the tugger trains. The application of the logistics-specific competency model to the current process shows that deal with material flow and physical resilience are among the most important competencies today. An evaluation of the current competencies was carried out before the effects of process automation were examined with experts from the department. As a future scenario, the provision of materials using AGVs instead of forklifts was set up. After assessing the competencies of the scenario, increased requirements on machine maintenance skills could be determined, since the fast repair of AGVs is crucial for ensuring short-term material availability in the event of a failure. In addition, basic human-robot capabilities are required for the interfaces between the systems, as can be seen in Fig. 3. Due to automation, physical resilience decreases, while control and monitoring require increased IT competencies and decision-making and problem-solving competency.

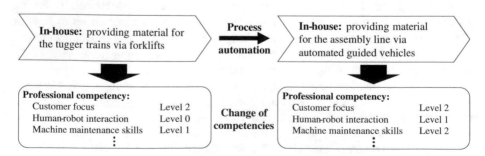

Fig. 3. Process changes and their impact on the necessary competencies.

Summing up, the competency model is applicable in different processes and fully depicts the logistical activities. The competency model obtained high acceptance among employees and managers of the evaluated departments and in the human resources department. With a manageable number of competencies, it simplifies implementation for the user. The competencies were sufficient to map future scenarios in the processes. The possibility of determining the logistics-specific competencies clarified and specified upcoming changes and induced an early engagement with coming challenges. Due to competency-specific behavioral anchors, the user can estimate future changes in competencies.

6 Conclusions

This paper describes the development and evaluation of a competency model for logistics employees in smart factories. It is derived and adapted from existing competency models. The focus of the adaption is the inclusion of future competencies while focusing on logistics. General and logistics-specific models and studies on competency changes were analyzed for creating a competency model consisting of professional, methodological, social and personal competencies. The focus is on automotive logistics tasks, with the competency model being transferable to other areas. In addition, it can serve as a basis for strategic personnel planning in the course of digitalization.

However, the model has so far only been evaluated in regard to three corporate processes. Accordingly, it cannot be ensured whether the competency model fully reflects the activities involved in all logistics processes. Further applications would be necessary, especially in the area of planning activities. Nevertheless, the case studies and the feedback from different fields have already shown that the competency model was able to provide important insights with regard to current and future competencies. These are used for the development of qualification strategies to deal with changed processes and tasks in the future.

Ultimately, it would be beneficial to integrate the logistics-specific competency model into a framework for strategic personnel planning. A corresponding procedure should take into account the determination of qualitative and quantitative employee competencies for automotive logistics in the future. In particular, the formulation of quantitative competency requirements, using a systematic framework and taking into account future scenarios, would support strategic personnel planning. As such, it represents an area meriting further research.

Acknowledgements. This research was funded in part by the MAN Truck & Bus SE and in part by the Chair of Materials Handling, Material Flow, Logistics (fml) at the Technical University of Munich. All support is gratefully acknowledged. Any opinions, findings, conclusions, or recommendations expressed in this paper are those of the writers and do not necessarily reflect the views of the MAN Truck & Bus SE or the fml.

References

1. Lorenz, M., Ruessmann, M., Strack, R., Lueth, K.L., Bolle, M.: Man and machine in industry 4.0: how will technology transform the industrial workforce through 2025. In: The Boston Consulting Group (2015)
2. Dillmann, R.: Digital Transformation in Supply Chain Management. BearingPoint (2016)
3. Ullrich, G.: Automated Guided Vehicle Systems. Springer, Heidelberg (2015)
4. Hecklau, F., Galeitzke, M., Flachs, S., Kohl, H.: Holistic approach for human resource management in Industry 4.0. Procedia CIRP, 1–6 (2016)
5. Mansfield, R.S.: Building competency models: approaches for HR professionals. Hum. Resour. Manage. 1, 7–18 (1996)
6. Le Deist, F.D., Winterton, J.: What is competence? Hum. Resour. Dev. Int. 1, 27–46 (2005)
7. Liesert, A.: Prozessorientierte Qualifikation von Führungskräften im Baubetrieb: Ein Kompetenzmodell. Springer, Wiesbaden (2015)
8. Erol, S., Jäger, A., Hold, P., Ott, K., Sihn, W.: Tangible industry 4.0: a scenario-based approach to learning for the future of production. Procedia CIRP, 13–18 (2016)
9. Erpenbeck, J., von Rosenstiel, L., Grote, S., Sauter, W. (eds.): Handbuch Kompetenzmessung: Erkennen, verstehen und bewerten von Kompetenzen in der betrieblichen, pädagogischen und psychologischen Praxis. Schäffer-Poeschel Verlag, Stuttgart (2017)
10. Dörr, S.L., Schmidt-Huber, M., Maier, G.W.: Messung von Führungskompetenzen - Leadership effectiveness and development (LEaD). In: Erpenbeck, J., et al. (eds.) Handbuch Kompetenzmessung, pp. 113–135. Schäffer-Poeschel Verlag, Stuttgart (2017)
11. APICS: Distribution and Logistics managers Competency Model. APICS (2014)
12. Supply Chain Council: Supply Chain Operations Reference Model. http://docs.huihoo.com/scm/supply-chain-operations-reference-model-r11.0.pdf
13. Kanning, U.P.: Inventar sozialer Kompetenzen (ISK/ISK-360°). In: Erpenbeck, J., et al. (eds.) Handbuch Kompetenzmessung, pp. 318–325. Schäffer-Poeschel Verlag, Stuttgart (2017)
14. Dekena, B., Nyhuis, P., Charlin, F., Meyer, G., Winter, F.: Kompetenzorientierte Produktionsplanung: Simulationsbasierte Produktionsplanung unter Berücksichtigung von Mitarbeiterkompetenzen. In: Werkstatttechnik online, vol. 3, pp. 216–220 (2013)
15. Montel, C., Hiltmann, M., Mette, C., Zimmer, B.: Das PERLS-system. In: Erpenbeck, J., et al. (eds.) Handbuch Kompetenzmessung, pp. 441–451. Schäffer-Poeschel Verlag, Stuttgart (2017)
16. Hossiep, R., Paschen, M.: Das Bochumer Inventar zur berufsbezogenen Persönlichkeitsbeschreibung: BIP. Hogrefe, Verlag für Psychologie (2003)
17. Kauffeld, S., Grote, S., Frieling, E.: Das Kasseler-Kompetenz-Raster (KKR, act4teams). In: Erpenbeck, J., et al. (eds.) Handbuch Kompetenzmessung, pp. 326–345. Schäffer-Poeschel Verlag, Stuttgart (2017)
18. Mollet, A.: COMPRO + Competence Profiling. In: Erpenbeck, J., et al. (eds.) Handbuch Kompetenzmessung, pp. 430–441. Schäffer-Poeschel Verlag, Stuttgart (2017)
19. ten Hompel, M., Anderl, R., Gausemeier, J., Meinel, C., Schildhauer, T., Beck, M.: Kompetenzentwicklungsstudie Industrie 4.0 - Erste Ergebnisse und Schlussfolgerungen (2016)
20. Schlund, S., Hämmerle, M., Strölin, T.: Industrie 4.0 - Eine Revolution der Arbeitsgestaltung: Wie Automatisierung und Digitalisierung unsere Produktion verändern werden (2014)
21. Butschan, J., Nestle, V., Munck, J.C., Gleich, R.: Kompetenzaufbau zur Umsetzung von Industrie 4.0 in der Produktion. In: Seiter, M., Grünert, L., Berlin, S. (eds.) Betriebswirtschaftliche Aspekte von Industrie 4.0, pp. 75–110. Springer Gabler, Wiesbaden (2017)

22. North, K.: Kompetenzrad und Kompetenzmatrix. In: Erpenbeck, J., et al. (eds.) Handbuch Kompetenzmessung, pp. 465–477. Schäffer-Poeschel Verlag, Stuttgart (2017)
23. McKinsey Global Institute: Skill Shift: Automation and the Future of the Workforce (2018)
24. Vuorikari, R., Punie, Y., Carretero Gomez, S., van den Brande, L.: DigComp 2.0: The Digital Competence Framework for Citizens. Publications Office, Luxembourg (2016)
25. Kersten, W., Seiter, M., von See, B., Hackius, N.; Maurer, T.: Trends and Strategies in Logistics and Supply Chain Management–Digital Transformation Opportunities. BVL International (2017)
26. Campion, M.A., Fink, A.A., Bruggeberg, B.J., Carr, L., Philipps, G.M., Odman, R.B.: Doing competencies well: best practices in competency modeling. Pers. Psychol. **64**, 225–263 (2011)
27. U.S. Department of Labor: Transportation, Distribution, and Logistics Competency Model (2018)
28. Demel, B.: Strategische Kompetenzentwicklung - Kompetenzen maßgeschneidert erarbeiten, einschätzen, messen und entwickeln. In: Erpenbeck, J., et al. (eds.) Handbuch Kompetenzmessung, pp. 99–111. Schäffer-Poeschel Verlag, Stuttgart (2017)
29. Henke, M., Hegmanns, T., Straub, N., Kaczmarek, S., May, D., Härtel, T., Rudolph, B., Sobiech, D., Müller, S., Dehler, J., Möllmann, A., Zaremba, B.: Assistenzsystem zum demografiesensiblen betriebsspezifischen. Kompetenzmanagement für Produktions- und Logistiksysteme der Zukunft (ABEKO) (2017)
30. Sauter, W., Staudt, F.-P.: Strategisches Kompetenzmanagement 2.0: Potentiale nutzen - Performance steigern. Springer Gabler, Wiesbaden (2016)
31. Leinweber, S.: Etappe 3: Kompetenzmanagement. In: Meifert, M.T. (ed.) Strategische Personalentwicklung, pp. 145–178. Springer Gabler, Wiesbaden (2013)
32. Heyse, V.: KODE und KODEX - Kompetenzen erkennen, um Kompetenzen zu entwickeln und zu bestärken. In: Erpenbeck, J., et al. (eds.) Handbuch Kompetenzmessung, pp. 245–273. Schäffer-Poeschel Verlag, Stuttgart (2017)
33. Weiß, Y.M.-Y., Wagner, D.J.: Die Zukunft der Arbeitswelt: Arbeitswelt 4.0. In: Jochmann, W., Böckenholt, I., Diestel, S. (eds.) HR-Exzellenz, pp. 203–217. Springer Gabler, Wiesbaden (2017)
34. Brühl, V.: Wirtschaft des 21. Jahrhunderts: Herausforderungen in der Hightech-Ökonomie. Springer Gabler, Wiesbaden (2015)
35. North, K., Reinhardt, K., Sieber-Suter, B. (eds.): Kompetenzmanagement in der Praxis: Mitarbeiterkompetenzen systematisch identifizieren, nutzen und entwickeln. Springer Gabler, Wiesbaden (2013)

Competency Profiles as a Means of Employee Advancement for a Resource-Efficient Chipping Production

Leif Goldhahn[⊠], Robert Eckardt, Christina Pietschmann,
and Sebastian Roch

Faculty Engineering Sciences, University of Applied Sciences Mittweida,
InnArbeit – Centre of Innovative Process Planning and Ergonomics,
Technikumplatz 17, 09648 Mittweida, Germany
{Goldhahn, Eckardt, Pietschm, Roch}@hs-mittweida.de

Abstract. Target-oriented qualification concepts that train and sensitise the process planners and other actors involved for resource-oriented manufacturing are essential in the field of chipping production. Concepts supporting the conscious and sustainable handling of resources such as material and energy requirements are currently not available for employees.

The specification of a work task occurs through process planning. Therefore, the essential knowledge, skills and proficiency ought to be determined for the respective work task. Derived from this is the competency profile. By means of this profile, an employee-related target-performance comparison is possible. Subject of this comparison are the competencies required for the considered task. Furthermore, a presentation of the qualification demand is feasible with the help of this profile.

A method has been developed by the junior research group *MoQuaRT* and allows the identification and development of competencies with the aid of a partially standardised competency profile for the resource-efficient chipping production.

Keywords: Qualification concepts · Chipping production ·
Resource-oriented production · Manufacturing process ·
Knowledge management · Organizational learning

1 Introduction

Highly complex production processes as well as strategies are determined by two major points. These points are the demands on the product diversity and variability of globally operating chipping production companies but also their flexible true-to-quality production.

Furthermore, globally progressively growing energy and material needs become a central focus of process planning within the company. Prospectively, these needs are in particular a scarcity of resources [1, 2].

Research [3, 4] together with the aviation supply industry show that up to 97% of difficult to machine materials must be resupplied to the recycling process after the

© Springer Nature Switzerland AG 2020
W. Karwowski et al. (Eds.): AHFE 2019, AISC 971, pp. 146–157, 2020.
https://doi.org/10.1007/978-3-030-20494-5_14

manufacturing process. This is the case due to unfavourable raw material geometry. Besides a high tool wear, production times of more than 24 h are possible.

To counter this inefficient use of valuable resources, solutions must be found to create sustainable production processes. Hereby, the contestability of the company should not be negatively affected [3, 5, 6].

A possible solution approach represents the modelling, selection and application of alternative production processes. The resource-oriented specification concerning geometry as well as dimensions of necessary raw materials also are such a solution approach.

These are specified by the process planning as a part of the order processing process (see Fig. 1, Sect. 2). Here, the process planning contributes to reduce the needed process-related resources [7, 8].

The employee assumes a control role, for example, through the specification of production processes and strategies or the selection of a fundamentally suitable initial material geometry [9]. According to [2] and [10], humans cause an inefficient use of resources because of:

- missing awareness of resource needs
- lacking knowledge of possible savings
- insufficient process monitoring or
- lacking knowledge of energy-oriented approaches to question the production process against the background of sustainability.

On the one hand, the mentioned aspects illustrate the necessity of the sensitizing of employees concerning the resource-oriented chipping production. On the other hand, they underline the importance of mediating the links between the choice of suitable production processes and the accompanying material, energy and time savings. The mediation of these links, but especially the sensitizing of the employees, demand resource-oriented qualification concepts. For these concepts, the systematic analysis and conditioning of the skills, knowledge and qualification of the employees for their respective fields of activity in the form of a competency profile are of significance. The competency profile reveals the employee's necessary qualification need. In this article, the procedure to create competency profiles will be examined by means of specific examples.

2 Background

Especially the topic of an efficient use of internal resources acquire new relevance with the responsible employees. The employee assumes an elementary role in the sustainable alignment of production processes (method and strategy oriented), e.g. process planning [2]. Therefore, the employees' self-conception and the responsibility ought to be adapted and modified accordingly [11].

In the ESF junior researcher project *Entwicklung innovativer Verfahrens- und Betriebsmittelmodelle sowie Qualifizierungskonzepte für die ressourceneffiziente Fertigung hochbeanspruchter Bauteile – MoQuaRT*, qualification concepts have been developed. These concepts contribute generally to save resources and especially touch upon energy, material and time savings during the process planning (see Fig. 1). The process planning subtask work plan creation includes four tasks: initial part determination, process sequence calculation, selection of manufacturing means and standard time determination (highlighted in green in Fig. 1). In this article, the development of a competency profile for the process planner as well as another competency profile for employees in the chipping production will be presented in extracts.

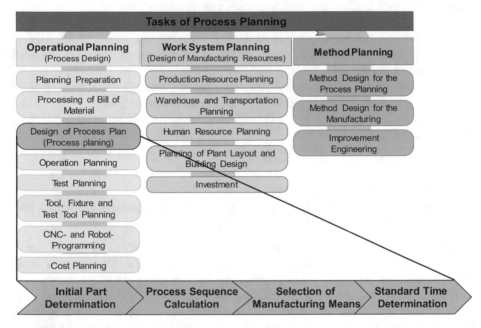

Fig. 1 Tasks of the process planning, incl. Subtask *Design of Process Plan* [12]

A qualification concept is a unitized knowledge-based modular system to methodically didactically further develop individual resources by employees for work systems. Hereby, organizational and technical framework conditions must be taken into account. With this, a specific group should be sensitized and motivated goal-orientedly.

The participants (employees in companies with variable activities) gain knowledge on the tasks of the resource-oriented work plan creation with the help of the qualification concept. This could be the panning of unmachined part measuring or types, production processes, selection of machinery and tools or cutting parameters. Furthermore, knowledge and skills to compare technological variants are developed. These concern material and energy on the basis of technological planning algorithms and energetic machine models. With the help of the developed qualification concept, systematic, resource-oriented and analytic thinking (professional skills) as well as creative, resource-oriented work (methodical expertise) are promoted. The didactic goal is the reorientation away from a mere transfer of knowledge through the instructor towards a knowledge transfer with the aid of online learning methods, so called Blended-

Learning-Concepts (pedagogical didactical) [13]. Therefore, the learner is able to autonomously determine their learning speed and learning need [14].

The focus of the developed qualification concept makes up the resource-oriented process planning, incl. its subtasks (cf. Fig. 1) and the subsequent production of the sample part *Koordinatensystem* (see Fig. 2) [15].

Fig. 2. Reference part *Koordinatensystem*

In the first step the sample part and the task is presented to the participants. Based on this, they create the work plan and the NC programme for a sample solution under the guidance of the instructor. The participants also set up the machinery, measure important consumption parameters and evaluate these. After that, the participants independently work out an alternative solution and realize it. They then evaluate their alternative solution, especially in comparison with the sample solution. The process planning and production takes places with the resources (machinery, tools, etc.) provided by the University of Applied Scienes Mittweida.

The reference part *Koordinatensystem* is provided for the participants in two CAD model variants (Fig. 3).

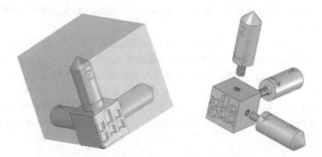

Fig. 3. CAD model of the sample (V1) and the alternative solution (V2) of the reference part

Version one (V1) is the sample solution. It is designed as a whole component and as such produced in the production process milling.

Version two (V2), as an alternative solution, includes an assembly made up of four components. These components are the basic body and three individual axes. The four subtasks of the work plan creation are: *planning preparation, design of process plan, operation planning* and *NC programming* are part of both (sample and alternative solution) (Fig. 1).

Figure 4 shows the fundamental approach when creating a qualification concept. An elementary step (development step 2: analysis employee competence, highlighted in red) is to analyse and structure the employee's competences, incl. their already acquired knowledge, their skills and qualifications (cf. [16]). The competency profile serves as a perimeter of the qualification needs of the employee with the goal to support problem solving.

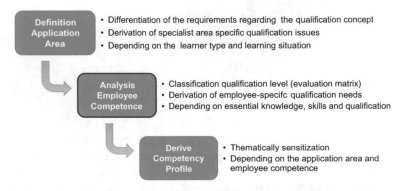

Fig. 4. Classification of the employee competence development in the qualification development process

The already existing knowledge and the qualification serve as the precondition of a competency development.

Sauter and Sauter define competences as skills to operate creative and self-organized in hardly comprehensible and understandable, complex, dynamic and also chaotic situations. Competences show in actions during the execution of the person's work contents [17].

According to [12], certain work contents explicitly determine the requirements of work concerning the executing person. Modified demands are caused by the consequences of work contents (activities) that are changed due to organizational design measures.

Based on the sum of requirements on the respective employee (necessary competences) concerning the workable action, an appropriate competency profile can be derived.

3   Competency Profiles

The sequential steps shown in Fig. 5 are necessary to create a competency profile. These steps are explained subsequently.

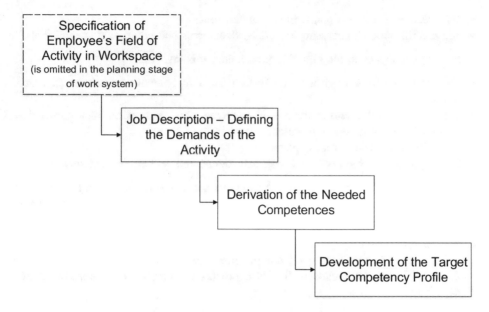

Fig. 5. Approach to developing competency profiles

Specification of the Employee's Field of Activity (in the Workspace)
The specification of the field of activity includes the determination in which operational activities the considered employee is active in. Operational activities are, for example, construction, work preparation (production planning and control activities), production or logistics [18]. The employees within the process chain (constructors, production planners, masters, machine setters, operators, etc.), who need to be qualified, vary in their activities, skills, level of knowledge, qualifications and their willingness to support processes of change [14].

The present article exemplarily focusses on the operational activities production planning and its subtasks as well as the skills needed in the chipping production.

This partial step of specifying the employee's field of activity is omitted in the planning stage of work systems.

Job Description – Defining the Demands on the Activity
A job description comprises the written description of a work task [19]. This provides insight into work processes, following duties, experiences, necessary expertise and its scope and depth.

For the employees in the mentioned example of production planning, this involves:

- the development of new techniques in the field of production organization with a high level of responsibility and autonomy
- the expertise in the field of production planning, resource-efficient parts production, incl. the knowledge about alternative production variants and raw material geometry
- the determination of detailed work sequences for production and process testing

- the creation of process plans, incl. standard times
- the identification of necessary materials, documents and equipment special tools.

For the employees in the chipping production, this involves:

- the implementation of production tasks with a high level of responsibility and autonomy
- the expertise in the field of the resource-efficient parts production, incl. knowledge about alternative production variants
- the skills to interpret process plans
- the acquisition of the necessary materials, documents and equipment special tools.

Based on the following duties of a process planner and an employee in the chipping production as well as their necessary expertise, the needed competences can be derived.

Derivation of the Needed Competences

In the literature, manifold approaches to the classification of competences can be found (cf. [14, 20]). For the developed competence, the approach based on Risch and Wadewitz [21] has been chosen. In this approach, the employee's competences are divided into:

- technical and fundamental competences
- organizational competences
- social competences

Goldhahn uses this approach to define and structure these competences for the demands of process planning tasks. He thus describes the necessary competences for every process planning task. In part, this is presented in the following table (see Table 1) [12].

Table 1. Excerpt of necessary competences in process planning cf. [12]

Task	Technical and fundamental competences	Organizational competences	Personal and social competences
Determination of the initial part	• read engineering drawing • expertise of possible of kinds and dimensions of raw parts • ability to evaluate efficiency • …	• consultation with construction, material management and process planning • …	• spatial imagination • systematic thinking • …
…	• …	• …	• …

As a result of the characterisation of process planning tasks concerning their necessary competences, the support of a resource-oriented process planning is possible.

Aspects such as:

- the promotion of systematic, resource-oriented and analytic thinking as well as the employee's creative, resource-oriented work
- the knowledge about the resource-oriented tasks of process planning, e.g. the planning of kinds and dimensions of raw material, production processes, selection of machinery and tools and the determination of cutting parameters must be supplemented.

In Table 2 the necessary competences of an employee in the chipping production are deposited in excerpts.

Table 2. Excerpt of necessary competences for employees in the chipping production cf. [12]

Task	Technical and fundamental competences	Organizational competences	Personal and social competences
determination of the initial part	• read engineering drawing • expertise of the dependence of cutting parameters on surface quality as well as the technologically-dependent energy demand (e.g. cutting parameters)	• consultation with masters and technologists • ...	• spatial imagination • systematic thinking • ...
...	• ...	• ...	• ...

To characterize the tasks of the resource-oriented chipping production, the following aspects must attract attention:

- resource-oriented thinking as well as the support of the employees' creative, resource-oriented work
- knowledge of resource-oriented planning of types and dimensions of the raw part
- essential knowledge, skills and proficiency of a machine tool in resource-oriented chipping production (e.g. Stand-By-Regime).

Furthermore, a tool in the form of a checklist must be created to describe and evaluate the employees' competence development with central questions. These central questions should, for example, discuss and criticize the targeted achievements. Possible questions could be:

- After the implementation of the qualification concept, what should the employees be able to do?
- Can the targeted learning outcomes be checked?

Development Target Competency Profile
The competences needed for a resource-oriented process planning worked out in Derivation of needed components, are hereafter transferred in a competency profile.

The competency profile in Fig. 6 exemplarily includes the technical and fundamental competency demands on the process planner. Furthermore, it shows how pronounced each necessary competence should be (highlighted in green in Fig. 6).

			Specificity					
			0	1	2	3	4	5
technical-specialist competencies								
Base Part Determination								
		practical exerience of base part determination			x			
		knowledge about interpretation of a technical drawing				x		
		skills concerning technically relevant of material properties				x		
		skills concerning production processes			x			
		knowledge and skills for use of an CAP- or ERP-System						x
		skills about ressource-oriented scheduling of the dimension of raw material and type of raw material		x				
		...						
Process Sequence Calculation								
		practical exerience of process seqeunce calculation			x			
		interpretation of a technical drawing				x		
		skills about resource-oriented production processes		x				
		skills concerning tooling		x				
		skills concerning the possibles of forced sequence				x		
		konwledge of comparison of technological variants			x			
		knowledge and skills for use of an CAP- or ERP-System						x
		...						
Specificity: 0 ... does not apply at all / 5 ... fully correct								
minimum specificity								
x =self-assessment employee								

Fig. 6. Excerpt of the target competency profile: technical and fundamental competencies of the process planner

Every qualifiable employee ought to determine their personal actual competence situation via self-assessment. Subsequently, the employee should define their individual competence goals with those of their immediate supervisor (cf. "x" in Fig. 6) [17].

Hereinafter, the actual comparison of the competency profile is required. The demanded level of qualification is compared to the employee's actual level of qualification. The extent of the qualification need becomes apparent this way. Furthermore, the respective qualification concepts are defined.

The competency profile shown in part in Fig. 7 exemplarily shows the technical and fundamental competency demands on an employee in the chipping production. Moreover, the competences demanded by the company as well as the employee's actual competences are outlined (highlighted in green in Fig. 7).

		Specificity					
		0	1	2	3	4	5
technical-specialist competencies							
scheduling of production flow							
	skills for systemic limitation of the maufacturing (back taper (feature), component size, technology)			x			
	skills of the depent of cutting parameters and the quality of surface (Rz-value) as well as the technology-denpendent energy demand (tool- and cutting		x				
	skills about the used supplies (e.g. cutting fluid) as well as the effects thereof surface of the component			x			
	skills about resource-oriented scheduling of the dimension of raw material and type of raw material	x					
	…						
Setup of machines							
	Skills concerning the determination of manufacturing processes			x			
	…						
Specificity: 0 … does not apply at all / 5 … fully correct							
minimum specificity							
x =self-assessment employee							

Fig. 7. Excerpt of the target competency profile of a chipping production employee's technical and fundamental competences

Here too, a variance analysis of the competency profile is conducted. The demanded level of qualification is compared to the employee's actual level of qualification. This way, the extent of the qualification need and its respective contents become apparent. These targeted profiles determine the necessary range of a competence spread for a defined field of activity in the company. Therefore, a company-specific adjustment through specialists and executives is possible here.

4   Summary

The competency profile in its respectively presented variant for the evaluation of an employee in process planning and chipping production contributes to the fundamental analysis of the employee's qualification demand.

Dependent on the differentiated scope of application, area specific qualification topics are worked out. These define the demands on the work task. This makes up the basis for the determination of the necessary competences of the respective process planning and production tasks. It also enables the designation of the necessary competences to implement them.

The developed competency profile exemplarily takes up the necessary competences for the process planning. It also supplements them with the competences needed for a resource-efficient production. This way, the target/actual comparison of the employee's actual and demanded competences becomes apparent.

The competency profile thus creates a basis for the systematic, resource-oriented, creative and analytic work which is implemented with the help of the qualification concept. Consequently, the employee receives all information needed for the development process of competences.

Acknowledgements. The authors thank the European Social Fund (ESF) and the Federal State of Saxony for supporting the junior research group „Entwicklung innovativer Verfahrens- und Betriebsmittelmodelle sowie Qualifizierungs-konzepte für die ressourceneffiziente Fertigung hochbeanspruchter Bauteile"– MoQuaRT as well as the project executing organisation Sächsische Aufbaubank – Förderbank.

References

1. Bundesministerium für Bildung und Forschung: Umwelt und Klima. https://www.bmbf.de/de/umwelt-und-klima-145.html. Accessed 10 Jan 2019
2. Neugebauer, R.: Handbuch ressourcenorientierte Produktion [Elektronische Ressource]. Hanser (Hanser eLibrary), München (2014). http://www.hanser-elibrary.com/action/showBook?doi=10.3139/9783446436237
3. Goldhahn, L., Pietschmann, C., Eckardt, R.: Process for the machine specific analysis and modeling of the technology based energetical demand forecasts. Procedia CIRP **77**, 405–408 (2018). https://doi.org/10.1016/j.procir.2018.08.298
4. Goldhahn, L., Bock, D., Eckardt, R., Pietschmann, C., Weber, H., Loll, J.: Ressourceneffiziente technologische Planung. ZWF Zeitschrift Für Wirtschaftlichen Fabrikbetrieb **112**(5), 332–336 (2017)
5. IHK Berlin: Wettbewerbsvorteil Energieeffizienz. www.upl-lichtenberg.de/fileadmin/files/upl/ver-anstaltugen/Energieeffizenz/IHK.pdf (2018)
6. Müller, E., Engelmann, J., Löffler, T., Strauch, J.: Energieeffiziente Fabriken planen und betreiben. Heidelberg. Springer, Berlin (2013)
7. Abele, E., Schrems, S., Schraml, P.: Energieffizienz in der Fabrikplanung. Werkstatttechnik online, vol. 102, no. 1/2, pp. 38–42 (2012)
8. Goldhahn, L., Eckardt, R.: Sustainable process planning of manufacturing variants for high-precision parts. Procedia CIRP **46**, 344–347 (2016). https://doi.org/10.1016/j.procir.2016.04.127
9. Eversheim, W.: Organisation in der Produktionstechnik 3. Arbeitsvorbereitung. 4., bearbeitete und korrigierte Auflage. Springer, Heidelberg (VDI-Buch) (2002)
10. Agricola, A., et al.: Steigerung der Energieeffizienz mit Hilfe von Energieeffizienz-Verpflichtungssystemen. Kurz: Energieeffizienz-Verpflichtungssysteme. Hg. v. Deutsche Energie-Agentur GmbH (dena). Köln. https://shop.dena.de/fileadmin/denashop/media/Downloads_Dateien/esd/9099_Studie_Energieeffizienz-Verpflichtungssysteme_EnEffVSys.pdf. Accessed 10 Jan 2019
11. Rohs, M. (Hg.): Handbuch Informelles Lernen. Springer VS (Springer Reference Sozialwissenschaften), Wiesbaden (2016). http://dx.doi.org/10.1007/978-3-658-05953-8
12. Goldhahn, L.: Gestaltung des arbeitsteiligen Prozesses zwischen zentraler Arbeitsplanung und Werkstattpersonal. Dissertation. TU Chemnitz. Institut für Betriebswissenschaften und Fabriksysteme. Chemnitz (2000)
13. Goldhahn, L., Pietschmann, C., Eckardt, R., Roch, S.: Concepts for improving employee quali fications for resource-efficient chipping production. In: 13th International Scientific Conference on Sustainable, Modern and Safe Transport (TRANSCOM) (2019, unpublished)
14. Bremer, C.: Bremer_Szenarien. (2018). www.bremer.cx/material/Bremer_Szenarien.pdf. Accessed 14 Dec 2018

15. Goldhahn, L., Roch, S., Eckardt, R., Pietschmann, C.: Mitarbeiterorientiertes Quali-fizierungskonzept für die ressourcenorientierte spanende Fertigung. 65. GfA- Frühjahrskongress (2019, unpublished)
16. Goldhahn, L., Eckardt, R., Pietschmann, C., Roch, S.: Qualifizierungskonzept für die ressourcenorientierte Teilefertigung. In: Wissenschaftliche Berichte (Scientific Reports) der Hochschule Mittweida. Mittweida (2018)
17. Sauter, W., Sauter, S.: Workplace Learning. Integrierte Kompetenzentwicklung mit kooperativen und kollaborativen Lernsystemen. Springer Gabler, Berlin (2013). http://dx.doi.org/10.1007/978-3-642-41418-3
18. Schenk, M., Wirth, S., Müller, E.: Fabrikplanung und Fabrikbetrieb. Methoden für die wandlungsfähige, vernetzte und ressourceneffiziente Fabrik. 2., vollständig überarbeitete und erweiterte Auflage 2014. Springer Vieweg, Berlin (2014)
19. https://www.uni-heidelberg.de/md/zuv/personal/aktuelles/leitfaden_td__veroffentlicht_-_stand_august_2014_.pdf
20. Schlick, C., Bruder, R., Luczak, H.: Arbeitswissenschaft. 4. Auflage. Springer Vieweg, Berlin (2018). http://dx.doi.org/10.1007/978-3-662-56037-2
21. Risch, W., Wadewitz, M.: Analyse des Ist-Standes. In: Wiebach, H. (ed.) Facharbeiterorientierte Betriebsmittel- und Arbeitsplanung in KMU. Handlungshilfe zur betrieblichen Umsetzung, pp. 65–72. RKW, Eschborn (1997)

Human Aspects in Industrial and Work Environment

Cognitive and Organizational Criteria for Workstation Design

Salvador Ávila[1(✉)], Beata Mrugalska[2(✉)],
Magdalena K. Wyrwicka[2(✉)], Maraisa Souza[1(✉)], Jade Ávila[1(✉)],
Érica Cayres[1(✉)], and Júlia Ávila[1(✉)]

[1] Polytechnic Institute, Federal University of Bahia,
Rua Professor Aristides Novis, 2, 40210630 Salvador, Brazil
avilasalva@gmail.com, jade.engavila@gmail.com,
julia.savila@gmail.com, mel.mara@hotmail.com,
ericaayros@hotmail.com
[2] Faculty of Engineering Management, Poznan University of Technology,
ul. Strzelecka 11, 60-965 Poznań, Poland
{beata.mrugalska, magdalena.wyrwicka}@put.poznan.pl

Abstract. The workstation becomes comfortable environment when adopting appropriate criteria for its design and adjustment in the accomplishment of the routine tasks. Organizational and cognitive design deserves attention of entrepreneurs to avoid human error and failure. This design is based on discussions about cognitive processing, standards with risk analysis and guidelines for good practice. The writing and the execution of the procedures demand revisions to the control of the tasks and their conformity in the multidisciplinary aspects. The failed on the task analysis due to human and organizational factors investigates the chronology, logic and materialization of the failure indicating the need of revision the culture design and the process. This research discusses the workstation to achieve the best performance of the task with criteria that decrease or nullify the human error. This activity results in increased productivity, profit and quality of life in addition to the reduction of human error, fatigue and cost of production reaching industrial sustainability.

Keywords: Workstation criteria · Risk · Human error · Cognitive factors · Organizational factors

1 Introduction

In the last decades, technology has expanded enormously and contributed to more advanced human-machine systems. In such configurations the design of automatic systems and the control of the interaction with human operators have become much more complex. The roles and duties of operators have undergone changes as they often became supervisors and monitors of automatic procedures [1]. However, in spite of all these facts, the activities performed by a human being still generate an opportunity for error. The practice of human error is equivalent to any human action that exceeds the tolerances defined by the system [2], being considered the natural and inevitable result

© Springer Nature Switzerland AG 2020
W. Karwowski et al. (Eds.): AHFE 2019, AISC 971, pp. 161–173, 2020.
https://doi.org/10.1007/978-3-030-20494-5_15

of human variability in relation to the interactions with the system, such as the workstation. It is stated that they are perceived as a cause or contributing factor in approximately 90% of accidents [3]. This is particularly visible in the case of accidents in construction and manufacturing, where errors are viewed as inappropriate human decisions that have a negative impact on the operation of technical systems. They lead to problems of quality, production or accident in the installations, having as main causes insufficient knowledge and errors of design, procedure and operator, factors that are related to the qualities and characteristics of the workstation [2]. The workstation project, therefore, must meet the demands of the activity in relation to product handling, process control, and physical or cognitive load limitations. These characteristics of the activity make up the risk analysis based on the classification of events regarding frequency, impact and complexity. The complexity can be technical [4], task operations and social relations [5]. The risk can be calculated and classified according to the complexity of the task environment in which the operational culture is established.

In the paper the issues of workstation design are discussed in reference to its best performance of the task with recognition of cognitive and organizational criteria. The investigation is carried out in order to verify them in industrial practice. The results show that among cognitive and organizational factors the following: procedures; stress; work process; available time; training/ experience, leadership and management; ability to work; communication should be undertaken in workstation design.

2   The Workstation and Its Design Criteria

Generally, ergonomics is defined as a scientific discipline that studies work, its processes, products and systems in relation to a man who operates or uses them. Its primary goal is to improve the worker performance and safety through the study and development of the following issues [6, 7]:

1. adjustment of work to human physical abilities and capabilities,
2. balance between human cognitive abilities and limitations, and the work system,
3. correct labour division, choice of technologies and production processes,

what reflects three domains of ergonomics such as: physical, cognitive and organizational. Although the former is the most widely known, cognitive and organizational ergonomics have gained much interest in the last decades as work environments is getting more and more under time restrictions and events are unpredictable [8].

When dealing with physical ergonomics, the work environment is analyzed according to the facilities and layout. Its goal is to guarantee both the physical capacity of the operator and the device, from checking systems, to equipment and the perception of stress and fatigue of the team and the operator [7]. The physical capacity of the operator relies on conditions of the work station that maintains the acoustic, thermal and luminous comfort of the environment and the adequacy of safety measures such as the use of personal protective equipment [9]. Furthermore, it should be underlined that decisions and interventions in the control room, while involving the same physical environment, do not follow the same pattern. These decisions and interventions depend on physical, social, cognitive and organizational factors [10]. The human–machine interface considers physical and logical screen adjustments as part of the control room design review. The configuration screens must meet the design criteria: distribution, color, alarm types, switches, variable display, distance between man and equipment, and security aspects [9].

According to Stenberg [11] cognitive psychology represents structures of human behavior in which human errors are defined. In this context, human errors can be omission or decision making [12], both involve cognitive aspects of the operators and are related to the mode of performance based on knowledge [13]. When analyzing cognitive ergonomics, mental processes that allow the operator to reason and make decisions at the workstation, are considered [14]. The work environment is analyzed in such domains as: mental workload; skilled performance; human-computer interaction; human reliability; work stress; mental training [8].

When dealing with social factors, archetypes stand out that are representations of patterns of behavior and memory rescue over time, such as rituals and cultural aspects [15]. Therefore, the archetype represents a character that has a preformatted behavior that may have an impact on industrial security [9].

To complete the ergonomic analysis of the work, the organizational ergonomics of the workstation, which involves factors such as division of work, supervisory and control systems, communication and cooperation, and recruitment and selection of people for work, including methods of training and job training [7], should be also discussed. The formation of strategic teams, communication tools and shift book use are important organizational features for the job design. These procedures are capable of interfering in interpersonal and intrapersonal variability for the construction of the essential skills necessary for the control and execution of the task. When there is good management and good communication it is possible to maximize the operator's knowledge linkage in the operation, favoring cooperation, emotional balance, team health, and feedback. This factor can guarantee the increase of the group's operational competence in the industry, aiding in decision making in situations of risk and identification of the root cause of the problems [9].

2.1 Factors of Planning and Execution of Task

The analysis of the cause of the human error in the task is complex due to the influence of the environments. The operator, who performs a certain task, has a variable behavior influenced by several factors such as: unavailability of tools, inadequate socio-organizational environments that cause low commitment and issues related to the preparation of competences for work [16].

According to PADOP methodology (Standards and Procedures in Operation) [17], technical proposals are proposed (IAT, PCET and AEET) to analyze the task from the influence of the environments, the type of processing and verification of the efficiency of the activity. In the Technique Identification of the Work Environment (IAT) technique the technical and communication requirements of the task, the activity schedule involved and the flow of information data to control normality are analyzed. The Cognitive Procedure and Work Execution (PCET) involves the design of the physical and cognitive workstation from the analysis of task standardization, procedure verification, memory retrieval and training. And, finally, The Efficiency and effectiveness

Analysis (AEET), this technique involves task control from the task review to meet the new requirements. From the application and concepts of these techniques, together with the quantification and analysis of the human factors at the workstation it is possible to design it efficiently.

2.2   Seeking the Good Performance of the Task – Quantification

According to Embrey [18], human error happens because of unsuitable work because human limitations. The adequacy analysis of the work requires the identification and classification of human errors. This research requires an earlier study on the cultural characteristics prevailing in the work environment and that create bad habits impacting on the work behavior. The technical and social characteristics also influence the cognitive functions used to perform the task.

The analysis of human reliability is a tool adopted to improve the performance from the analysis of qualitative information on the factors of performance that cause failures and from calculations of the probability of human error in the task [9]. In this context, the performance of the task depends on requirements that include several aspects such as: (a) organizational climate (b) management structure for routine and emergency operational communication; (c) structuring of indicators for decision-making in the production environment; (d) adequate knowledge of the team; (e) communication tools for change shifting and team feedback for deviations and occurrences in the routine; (f) commitment, cooperation, understanding of risks; (g) standards of performance indicators [2, 19, 20, 21, 22, 23].

The SPARH method calculates the human reliability in the risk activity considering the routine situation in the workstation. This method considers that the input of organizational and technical characteristics to the activity has an answer cognitive and behavioral, thus causing success or failure in the task. These characteristics are called human performance factors, considered basic elements for information processing resulting in the probability of human error. One of the major models used to estimate the impact of human factors on information processing includes the perception of environmental cues, quality of working memory, memory retrieval strategy, long term memory quality and characteristics in the decision-making process [19].

Performance Modeling Factors The operator is considered the most important element in any process system because they are the performers of the actions prescribed by the operational staff. Leaders at the front line should ensure that the operator-machine interface is compatible with the capabilities and limitations of the worker to avoid or minimize human error.

The characteristics that affect the performance of the worker in the task are considered human-organizational-managerial factors, which in [19] is dubbed as a factor of performance modeling. These human factors that can interfere with the performance of the task are internal, external or even stress-related factors that relate to the previous ones.

According to Lorenzo [2], the internal factors are: training (which defines ability); practice that promotes experience; knowledge of required performance standards; stress levels, triggers and resulting environments; level of motivation; emotional state of the team and that can impact on the cognitive apparatus; physical condition of the worker that can alter the level of perception; and the cultural characteristics that promote bad habits. On the other hand, the external factors that cause human error are: architecture of the work place within the industrial environment; characteristics of this environment that can generate stress (temperature, noise, vibration, cleaning); workload; tools to support the work activity; organizational structure with the respective definition of guidelines; procedures, their interpretation, review and efficiency; type and quality of written or oral communications; structure of the team with their respective functions; needs of anticipation of action from the perception of deviations; interpretation of scenarios for decision making; complexity of task and process; amount of memory (short and long term) demand for the task; quality and quantity of feedback in the routine; interfaces between the worker and the machine-processes; among others. The interaction or non-combination between these factors (internal and external) can promote stress in the individual who is developing the task, thus causing a low performance.

The factors that cause stress are high speed and intensity in the execution of the task; risk of physical exposure of the worker; physical and functional threats that destabilize the worker's function; monotonous work and distractions that alter perception and the mental map; lack of rewards or recognition for good results; and fatigue from repetitive movements, pain or discomfort.

The tasks required by the operator require several functions discussed in engineering and cognitive psychology such as perception, attention, memory, interpretation, decision [11]. If the operator does not perform his function with a good performance, these factors can be accentuated, since the information does not transform into the operator's action. Ávila et al. [24] states that the information cycle must remain current to avoid human errors in the decision, leading to inappropriate actions in critical activities. Ávila et al. [9] also notice that the executive function in the task depends on the operator's preparation (knowledge and skills), commitment, task feedback, environmental perception in relation to the process state and the activity schedule required by the organization.

It is known that the vast majority of human errors occur due to improper design of the workstation. Therefore, by providing the resources needed to identify and eliminate situations of possible error, managers can improve performance modeling factors and reduce the frequency of human error.

2.3 Perception of the Risk and Human Error

According to Embrey [18], human error can be initiated from work situations not adequately designed for human limitations. They are influenced by individual, organizational and technological factors [25]. Moreover, Lorenzo [2] and Embrey [18] have developed a method of human reliability analysis in which human error is a function of factors that affect workers performance.

The reliability analysis of the system should be prioritized for the development of technical and organizational standards, in the design phase and workstation details.

Language and behavior are human factors that change the routine of work and can direct operators to appropriate action or not. Thus, certain population stereotypes [2], organizational resistance derived from the conflict between politics and practice, and managerial conflicts can create an environment that induces human error [9].

In the flow of data and information of the products and processes to control the activity normality are indicated the critical points to keep the operation working and to allow the cognitive processing of the people. This analysis helps to avoid making wrong decisions in critical situations because of the amount of information that circulates and keeps the operator's attention. Therefore, the current data should be classified, and the priority established for decision making.

From the operator's discourse, leaders' perception and routine analysis it is possible to verify the actual demand for knowledge, skill, communication and construction of mental map to mitigate human errors, bad habits and cognitive gaps capable of implementing good engineering of human factors in control systems, process equipment and the work environment.

3   Methodology

The general scenario of the evaluated workstation is represented by Fig. 1. In this figure, it is verified that there is impact of social, human, organizational and physical factors in the work environment. These factors directly influence the constant flow of information for process and product control within the workplace. Loading important information between the operators (and the leader) and the machine (equipment) through a man-machine interface to ensure process control and minimize risks. However, in order to ensure good decision-making, avoiding risks and losses and optimizing the process, one should analyze the set of actions of the man who plans, perceives and executes and diagnoses the occurrence of imminent risk in the execution of the task. However, the methodology of the article intends to evaluate human and organizational factors.

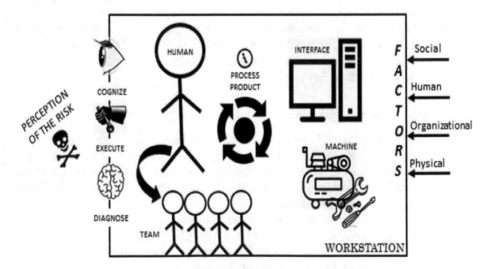

Fig. 1. The scenario of workstation

The workstation involves the flow of product and process information, the man-machine interface for process control, and the relationship between equipment (maintenance and capacity) and man. This set of information and the structure of the workstation guide the leader to make the best decisions in the execution of the task. These decisions are of great impact on the performance of the task and must be exercised from the set of skills involving the operator's cognitive, task-planning ability, and execution. This set of characteristics of the factors demand the establishment of criteria for the design of the workstation. In this way, with the established criteria the perception of the risks of human error and the equipment failure is improved.

On the basis of the losses generated by human errors, this article aims to develop and apply a methodology able to minimize the risks of human errors, through the creation of criteria to improve the design of the workstation (Fig. 2).

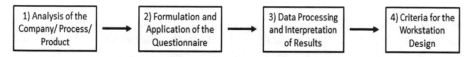

Fig. 2. Step by step of methodology

Initially a study is done on the company where the workstation will be designed. The data of processes, products, routine and management are used in this work. After establishing the overall picture of the company, with the help of a manager, questions are formulated based on external and internal factors that interfere with the execution of the task. Subsequently the quiz is applied to the operators of the job. Based on the data processing and interpretation of the results of the quiz, it is diagnosed which organizational and cognitive factors deserve to be highlighted for the establishment of these criteria that must be followed for this design. After the application of the quiz a mapping of the work station is done and the deficiencies of the task that can promote the process losses are verified.

From the results of the investigation and mapping of the cognitive and organizational factors of interest, with the follow-up of a manager, the criteria for the design of the work station are formulated. It establishes a set of procedures capable of improving the routine of the task and minimizing the human error in the process.

4   Results and Discussions

The methodology was applied in a Temperate Glass Factory located in Bahia-BR. This company aims to deliver the product with quality, with the lowest loss possible in production, remaining competitive in the market and financially profitable.

4.1   Characterization of the Scenario

The factory has been operating in the industrialization, processing and distribution of glass since 2005. Serving the civil construction, locksmith, furniture, architecture and decoration sectors. Investments in state-of-the-art technology and equipment to meet the highest quality standards have been highlighting the company in the market.

Strategically located in the interior of Bahia, a link between several regions, the company serves several states as: Bahia, Sergipe, Alagoas, Pernambuco. Serving the sectors of construction, locksmithing, furniture, architecture and decoration. Investments in state-of-the-art technology and equipment to meet the highest quality standards have been highlighting the industry in the market.

The Productive Process and the main characteristics of this industry are discussed to indicate the probable losses based on the experience of the managers.

Step 1: Cutting - Start of all productive process; Step that receives the plane of cut of the pieces; There is a probability of loss due to improper handling of the parts highlight and change of cutting plane. Step 2: Automatic Lapidation and COPO; In the automatic step the loss is accentuated due to inadequate command in the machine, preventive procedures not executed correctly and lack of observation. Step 3: Drilling, scarification and washing - The loss is considered high, with the same reasons as the automatic lapidation line. Step 4: Quality Inspection - Product can only be tempered if it is totally clean and with the drilling specified by the customer. If there are records of losses due to incorrect manufacturing, they are assigned to the quality inspector. This step is fully performed by the operator, thus being the focus of many human errors. Step 5: Tempering - It is the procedure where the glass is tempered. The glass becomes five times stronger, being an irreversible process, that is, manufactured glass wrong, is scrapped.

4.2   The Questionnaire

The questionnaire, which the operators responded, was based on organizational and cognitive factors that can interfere in the performance of the task. The main factors were determined through a survey made to the managers of the company and the types of activity developed.

(1) What are the main steps in the manufacturing process? (2) Which areas of the plant have bigger losses? (3) What actions within the workplace lead to losses in these areas?

Within this manufacture, the specialist indicated the areas of Automatic Drilling, Quality Inspection and Reheating Furnace as the main sources of loss. According to managers, in these areas there are actions that can be considered the main causes of production losses.

Automatic Drilling - Incorrect operating command in machine execution; Procedures determined, but not obeyed by the operator; Lack of observation in the execution of the machine; Misaligned drilling execution. Quality Inspection - bonding the labels on the wrong part; Error in the conference of the drillings in the piece; Incorrect handling causing the part to break, the operator does not obey the conference checklist leading to the error or forgetting of a certain item. Quenching Furnace - Execution of inadequate glass tempering recipe; Poor oven maintenance; Change of identification tags of the pieces after the tempering; Incorrect handling resulting in breakage or worsening product function; Communication failure between operators.

The indicated causes are mostly related to human factors and the conditions of the work. From this information, the questionnaire was formulated with a sample of 20% of the factory operators, obtaining the results necessary to determine the criteria for the design of the workstation.

From the collection of information regarding the productive process of the plant, discourse with managers and the factors that can affect the performance of the task, the following questionnaire is proposed to the operators (Fig. 3).

Questionnaire	Y	N
1. The procedures that the company proposes contribute to the good performance of its tasks?		
2. Are the equipment manuals easy to access?		
3. Are the goals reported by the leaders clear and objective so that they can assist in the progress of the tasks?		
4. How does the leadership communicate with the team collaborate for the good performance of the task?		
5. Does the leadership use tools to inform the progress and evolution of the established goals?		
6. Do you believe it is necessary to exert pressure to achieve productivity?		
7. Do you agree that the company provides the necessary training for your work?		
8. Do the training provided by the company match the work performed?		
9. Are the instruments available for carrying out its activities adequate to achieve the result that the company expects?		
10. Do the workers receive feedback on the results achieved?		

Questionnaire	Y	N
11. Do you believe that feedback is essential, or is it encouraging the periodic dissemination of the results of the goals?		
12. Do you assume that the time available to perform your tasks matches the goal set to ensure the desired productivity?		
13. Do you feel your work is threatened when leaders talk about achieving goals?		
14. Do you feel unnatural pain at the end of a work day?		
15. Do you consider your work environment healthy (Noise, Temperature ...)?		
16. Does the company usually recognize the good performance of the team?		
17. Do you agree that the company has good order logistics and product delivery?		
18. Is there good maintenance of the equipment involved in your task?		
19. Do you think the company's automated processes well calibrated?		
20. Do you agree with the deadlines set by the company for the delivery of the products?		

Fig. 3. The questionnaire

4.3 Evaluation of the Results

After the questionnaire was applied to the operators, it is evidenced that the human factors are decisive items for the performance of the task and reduction of the losses. In relation to the training provided by the company, it was possible to verify that these collaborate for the performance of the task, however it is observed that the operator does not have easy access to the procedures when there is a need to consult or fix the learning of the training.

It was noticed that the leadership has good communication with the team. However, the company's feedback to the operator in relation to the target set for the period still needs to improve. This communication motivates operators to do the work effectively. The responses regarding the equipment were favorable due to modern technology in relation to the controls and the man–machine interface. The physical environment does not help due to high temperature and noise requiring the use of equipment for individual safety that harm the operator's attention and can cause losses. Thus, the monitoring of the manufacturing must be constant to avoid human error.

Some operators have commented that pressure by results is needed to achieve productivity. They also stated that the time available does not fully meet the needs to achieve the goals set. It can lead to loss and demotivation of the team. The continuous monitoring of the effectiveness of the training allows checking the level of learning of the team. The adoption of feedback behavior of the results and the comparison with the established goals is essential for the search of the best performance in the operation team (Fig. 4).

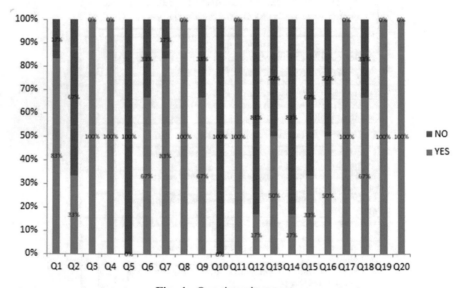

Fig. 4. Questionnaire answers

Factors and Criteria Selection According to the data treatment of the survey results, the main verified factors in workstation in factory are: (1) cognitive - procedures; stress; work process; available time; training/ experience; (2) organizational - leadership and management; ability to work; & communication. The factors identified with the characteristics of the company are presented in Table 1. It helps to construct the workstation design criteria.

Table 1. Workstation design criteria

Cognitive factor	Criteria description	Organizational factor	Criteria description
Procedures	Facilitate operators' access to operating manuals	Leadership/Management	Promote courses for managers as to the motivation of the team
	Periodic review - procedures		Create organizational standards with specific management goals
Stress	Use of personal protective equipment that reduces noise and is easy to handle	Work capacity	Accompany worker's health with routine tests for visual and auditory acuity
	Create a thermal rest environment		Analyze the difficulties of performing the work due to lack of tools or procedures
Available time	Re-evaluate the time of each step of the process and correct constraints	Comunication	Create clearer communication between leaders and employees
	Keep the operators up to date on the targets of time to be met		Create dual redundant communication
Work process	Carry out periodic follow-ups for chrono-analysis		Practice operator feedback exercises for the manager and vice versa
	Review layout and procedures		
Training/Experience	Keep the operators up-to-date on the goals to be met		
	Analyze operational deviations and transform into training		

5 Conclusions

From the application of the questionnaire it was possible to map the areas, activities and factors (cognitive and organizational) that cause greater loss of materials in the workstations in the glass factory. To reduce the risks of human error in the execution of the task, criteria were suggested to be applied to this work. The intesa operating routine often does not allow visualizing incorrect work patterns.

In order to confirm the given criteria, it is proposed to establish indicators resulting from the task and to monitor the improvements. Following the ergonomic analysis of the task, one should compare the adopted pattern with the found scenario. If the results are not satisfactory, the indications after the response of other operators to new questionnaires should be discussed.

Cultural movements with emerging demands, that are not visible, may not be recognized by requiring the assistance of other techniques to complete the analysis of the work. The operator's speech helps this detection as well as the observation of the workstation and the comparison of similar plants.

Engineering and Ergonomics need to develop in more detail the analysis of cognitive requirements and care for organizational factors that cause stress.

References

1. Mrugalska, B.: Health-aware model-predictive control of a cooperative AGV-based production system. Sensors **19**(3), 532 (2019)
2. Lorenzo, D.K: API 770 – A Manager's Guide to Reducing Human Errors: Improving Human Performance in the Process Industries. API Publishing Services, Washington (2001)
3. Mrugalska, B., Nazir, S., Tytyk, E., Øvergård, K.I.: Human error and response to alarms in process safety. DYNA **83**(197), 81–86 (2016)
4. Perrow, C.: Normal Accidents: Living with High-Risk Technologies. Basic Books, New York (1984)
5. Ávila, S.: Failure analysis in complex processes. In: Proceedings of 19th Brazilian Chemical Engineering Congress. COBEQ, Búzios (2012)
6. Kawecka-Endler, A., Mrugalska, B.: Analysis of changes in work processes, In: Vink, P. (ed.) Advances in Social and Organizational Factors, pp. 672–681. CRC Press, Boca Raton (2012)
7. Vidal, M.C.: Guide to Ergonomic Work Analysis (AET) in the Company. Virtual Scientific, Rio de Janeiro (2008)
8. Badiru, A.B, Bommer, S.C.: Work Design: A Systematic Approach. CRC Press. Taylor & Francis Group (2017)
9. Ávila, F.S., Menezes, M.L.A.: Influence of local archetypes on the operability & usability of instruments in control rooms. In: Proceedings of European Safety and Reliability Conference – ESREL, Zurich (2015)
10. Chavienato, I.: Gestão de Pessoas. Manole, São Paulo (2014)
11. Stenberg Junior, R. Cognitive Psychology. Artmed, Porto Alegre (2008)
12. Swain A.D., Gutmann, H.E.: Handbook of Human Reliability Analysis with Emphasis on Nuclear Power Plant Applications (NUREG/CR-1278, SAND800 200, RX, AN) Sandia National Laboratories, Albuquerque (1983)

13. Rasmussen, J.: Risk management in a dynamic society: a modelling problem. Saf. Sci. **27** (2/3) (1997)
14. Vidal, M. C., Carvalho, P. V. R. de.: Cognitive Ergonomcs. Virtual Scientific, Rio de Janeiro (2008)
15. Marais, K., Leveson, N.G.: Archetypes for organizational safety. Saf. Sci., August (2006)
16. Ávila, S.: Etiology of operational abnormalities at industry, a model of learning. Doc-torate Thesis at Federal University of Rio de Janeiro, Chemistry School, Rio de Janeiro (2010)
17. Ávila S.F., Pessoa F.L.P.: Proposition of review in EEMUA 201 & ISO Standard 11064 based on cultural aspects in labor team. Proc. Manuf. **3**, 6101–6108 (2015)
18. Embrey, D.: Preventing Human Error: Developing a Consensus Led Safety Culture Based on Best Practice. Human Reliability Associates Ltd., London (2000)
19. Boring R, Blackman H. The Origins of SPAR-H Method's Performance Shaping Factors Multipliers. EUA, Idaho. August (2007)
20. Ávila, S., Cerqueira, I., Drigo, E.: Cognitive, intuitive and educational intervention strategies for behavior change in high-risk activities - SARS. In: Karwowski, W., Trzcielinski, S., Mrugalska, B., Di Nicolantonio, M., Rossi, E. (eds.) Advances in Manufacturing, Production Management and Process Control. AHFE 2018. Advances in Intelligent Systems and Computing, vol. 793, pp. 367–377, Springer, Cham (2019)
21. Cerqueira, I.; Drigo, E.; Ávila, S.; Gagliano, M.: C4t: safe behaviour performance tool. In: Karwowski, W., Trzcielinski, S., Mrugalska, B., Di Nicolantonio, M., Rossi, E. (eds.) Advances in Manufacturing, Production Management and Process Control. AHFE 2018. Advances in Intelligent Systems and Computing, vol. 793, pp. 343–353, Springer, Cham (2019)
22. Drigo, E.S., Ávila Filho, S., Sousa C.A.O. Operator discourse analysis as a tool for risk management. In: Proceedings of European Safety and Reliability Conference – ESREL, Zurich (2015)
23. Drigo, E.S.; Avila, S.F. Organizational communication: discussion of pyramid model application in shift records. In: 8th International Conference on Applied Human Factors and Ergonomics (AHFE 2017)
24. Avila, S.F., Fonseca, M.N.E; Santos, A.L.A.; Santino, C.N.: Analysis of cognitive gaps: training program in the sulfuric acid plant. In: Proceedings of European Safety and Reliability Conference, Glasgow, Scotland (2016)
25. Reason, J.: Human Error. Cambridge University Press, Cambridge (2003)

Motion Analysis of Manufacturing of Large "Echizen Washi" Japanese Traditional Paper

Yuji Kawamori[1(✉)], Hiroyuki Nkagawa[1], Akihiko Goto[2],
Hiroyuki Hamada[3], Kazuaki Yamashiro[3], Naoki Sugiyama[3],
Mitsunori Suda[4], Kozo Igarashi[5], and Yoshiki Yamada[6]

[1] Taiyo Corporation, 34-6 Koizumi-cho, Hikone, Shiga 522-0043, Japan
imagineer138@gmail.com
[2] Osaka Sangyo University, 3-1-1 Nkagaito, Daito, Osaka 574-8530, Japan
[3] Kyoto Institute of Technology, Matsusaki, Sakyo-ku, Kyoto 606-8585, Japan
[4] Suda Shoten Corporation, 1-16-1 Kawakita, Fjiidera, Osaka 583-0001, Japan
[5] Igarashiseishi Corporation, 12-14 Iwamotocho, Echizen,
Fukui 915-0233, Japan
[6] Nishinosyoukai Corporation, 4-7-6 Sadatomocho, Echizen,
Fukui 915-0231, Japan
http://www.f-taiyo.jo.jp

Abstract. The Echizen Japanese paper has a long history, in the 4–5 century the paper was reported in Japan, it has been left in the ancient documents of "Shosoin" who had crowded the paper in Fukui Prefecture Echizen. Higher than 1500 years before the request is to have responded with high technology. High technology of craftsmen built by the inquisitive to the new technology and hard work that Persistent, leading to now. It is said to Echizen Japanese paper in Japan of handmade Japanese paper, to make the largest Japanese paper. Its work is carried out in a work called "Nagasisuki" and "Tamesuki". Work of this large-format sum papermaking is performed by the two people of the craftsman. Skilled artisans to indicate from the beginning of the work to determine the can of the product end. And the working process of a large-sized paper in this study, the behavior analysis of the skilled person performs, an object of the present invention is to analyze the differences in behavior of a person skilled two people.

Keywords: Echizen Washi · Japanese traditional paper ·
Washi Japanese paper · Handmade Japanese paper

1 Introduction

The history of Echizen Washi paper is old, and it is left in the ancient documents of Shosoin (The oldest warehouse in Japan) that the paper was papermaking in Echizen (now Echizen City, Fukui Prefecture) in the fourth-fifth century when the newspaper was told Japan. It has responded with a high level of technical power to a higher demand than 1500 years ago. In the Japan handmade washi paper, it is said to be

© Springer Nature Switzerland AG 2020
W. Karwowski et al. (Eds.): AHFE 2019, AISC 971, pp. 174–187, 2020.
https://doi.org/10.1007/978-3-030-20494-5_16

Echizen Washi paper that can make the largest Japanese paper, the work is done by using a large Nishino boat (plow) and drypoint-digit. This large-format Japanese paper-making work is done by two craftsmen. Skilled craftsmen judge the product from the beginning of the work and instruct the completion. The purpose of this study was to analyze the operation process of large-format Japanese paper and the operation of two skilled people.

2 Method

2.1 Outline of Experiment

This experiment was carried out in May 2018 at the paper making workshop in the Igarashi paper factory in Echizen City, Fukui Prefecture. The Echizen handmade Large-format Washi paper is subjected to the test, and the paper is made by making a sink. Originally handmade Washi paper uses Kozo (Koyo), Mitsumata (mitsualso), GANPI (cancer pi) as a raw material, but the purpose of this experiment was to evaluate the Washi paper which was finished by hand-made, 100% of the material with less variation in the pulp (used pulp, N-bsk Japan paper Chemical, N-bkp Loffton, abaca pulp, L pulp) and Trolo Aoi were used in the kneading material.

2.2 Experimental Participants

In this experiment, the subjects were asked to cooperate with Igarashi Paper Co., Ltd., which has been engaged in the paper industry for about 100 years in the founding of Iwamoto-cho, Echizen city. Although most of Echizen's handmade washi craftsmen are women, this paper is made by two female traditional craftsmen, especially in the case of Igarashi papermaking, and the products are used in a number of facilities and stores, particularly in the handmaking of large-format Japanese paper and the creation of washi paper. In this experiment, two traditional craftsmen (Table 1), who are ultra-skilled subjects, made replacement paper for each of the subjects' work positions. In the usual work, a super-skilled person directs work on the left side Fig. 1 the main position and makes judgment of the paper finish.

Table 1. Examinee

Participants	Gender	Career (years)	Dominant hand	Licence
B	Female	23	Right	Traditional craftsman
D	Female	46	Right	Traditional craftsman

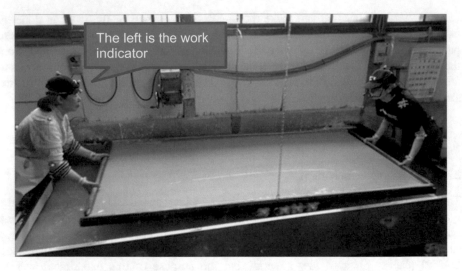

Fig. 1. C-A working position and papermaking motion

2.3   Working Method and Work Process

The paper making method of the most this experiment is a method of Nishino by dispersing the paper in water of the Fig. 7 boat (*Sukuifune*) and scooping it up with the drypoint digit Fig. 9. (*Sugeta*) Say "making" the central task (Suku). The paper, the Nishino, the small paper, and the large paper, respectively, are different, but both are put in the boat and stirred, and the start is made by adding the batter. The paper was made by the Fig. 2. The flow-making is a method of Kozo the paper, such as the GANPI, and the like, by adding the Nelli taken from Trolo Aoi and Noritsugi, and pumping the paper liquid several times with the drypoint girder, removing the drypoint from the digit when it reaches the desired thickness, and overlapping the wet paper made on the drypoint. In this method, even without cutting and careful beating of the fibers, the water leakage becomes slower than the reservoir of the paper liquid by adding a nelli, it is possible to make a paper layer slowly back and forth on the Drypoint draw the paper liquid many times. Nishino-digit (Fig. 8) is a tool that supports the Nishino drypoint (favorite) Fig. 9 and makes paper, combining Nishino drypoint with Nishino digits (the drypoint digit). Nishino Drypoint is like a net to form a paper layer on top of it by scooping up the paper liquid from the Nishino boat. The Nishino drypoint used in this experiment is the Taken which knitted the bamboo with a thin, round-sharpened lottery with silk or tail hairs. As shown in the figure of Fig. 2, the work process of the flow-forming, fig. 3 pumped, fig. 4 drypoint sinks, fig. 5, fig. 6 drypoint sink and work proceeds.

Fig. 2. Working process

Fig. 3. Kumiage. To scoop up the raw materials in the "Sukisu".

Fig. 4. Sunagashi. Flowing the raw materials on top of the "Sukisu". The first "Sunagashi" is flowing garbage on the "Sukisu", it will make the surface of the paper, When a thickness of hope to return to in the raw materials on the "Sukisu" "Sukibune". This flow of trash on the surface, creating the back of the paper.

Fig. 5. Suki

Fig. 6. Sunagashi

Fig. 7. Sukibune. Bath to keep dissolved stock to comb the paper.

Fig. 8. Sukisu. Equipment which scooped the paper stock solution to form a paper layer thereon from "Sukibune". The lottery shaved round thin bamboo is a Takesu braided tail hair of silk or horse.

Fig. 9. Sugeta. A tool that supports the "Skisu" and raises paper. The combination of "Sukisu" and "Skigeta" is called the "Sugeta".

2.4 Operation Measurement

The coordinates of each marker were measured using the optical motion capture system MAC3D Systems (motion Analysis Co., Ltd.). The sampling frequency was 120 Hz. The infrared reflective markers were pasted into the entire body of the experimental participants in 11 locations. The measurement landscape is shown in Figs. 1, 2, 3, 4, 5 and 6. The coordinate system was the y-axis, the longitudinal direction of the x-axis, the vertical direction and the z axis for the experimental participants. We measured the time between the working process, "Kumiage" → "Sunagashi" → "Suki" → to "Sunagashi". Time represents in seconds (Fig. 10).

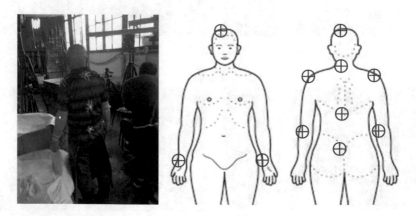

Fig. 10. The infrared reflective markers

2.5 Analysis Methods

In this study, we conducted three combinations of b-d and three d-b in the experiment. From the work process, the total work of subjects B and D were analyzed by focusing on the time of making, and the maximum angle and area of the drypoint-digit Nishino boat were determined (Fig. 11).

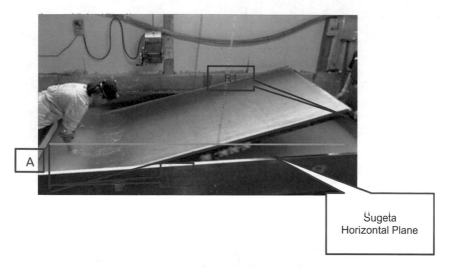

Fig. 11. Kumiage

2.6 Operating Time

In Fig. 12, the average of the time of the subject (suki) is shown in each. The "Suki" time of the subject B, B-D_1 (Table 2, Fig. 13) "Suki" number of times 4 times 18.5 s, B-D_2 (Table 3, Fig. 14) "Suki" 5 times 21.3 s, B-D_3 (Table 4, Fig. 15) "Suki" 5 times 20.6 s, D-B_1 (Table 5, Fig. 16) "Suki" number of times 7 times 25.4 s, D-B_2 (Table 6, Fig. 17) "Suki" number of times 7 times 22.0 s, To give the D-B_3 (Table 7, Fig. 18) "Suki" number of times 9 times 23.6 s of the results. In the case of B-D, B becomes the main operator. "Suki" time of the subject C, B-D_1 "Suki" number of times 3 times 18.6 s, B-D_2 "Suki" number of times 4 times 21.4 s, B-D_3 "Suki" number of times 4 times 18.4 s, D-B_1 "Suki" number of times 6 times 21.1 s, D-B_2 "Suki" number of times 7 times 21.4 s, To give the D-B_3 "Suki" number of times 8 times 21.1 s of the results. In the case of D-B, D is the main operator. B-D and D-B "Suki" number of times is the difference is, it is the difference due to the difference due to the density of the stock. From this result, the average of B "Suki" time is 21.9 s, D of "Suki" time is 20.3 s, Skilled person D is a difference of 1.5 s, also two skill was almost the same "Suki" time.

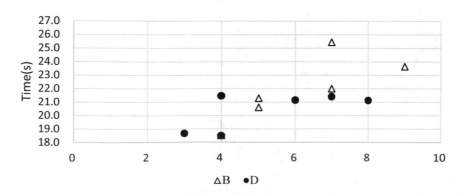

Fig. 12. Suki operation time

Table 2. B-D_1 process time

B-D_1	Worker	Time	Work name
1	C	00:00:04.95	Kumiage
2	C	00:00:02.97	Sunagashi
3	A	00:00:18.06	Suki
4	C	00:00:20.05	Suki
5	A	00:00:21.01	Suki
6	C	00:00:14.95	Suki
7	A	00:00:23.04	Suki
8	C	00:00:20.94	Suki
9	A	00:00:28.08	Suki
10	C	00:00:19.00	Suki
11	A	00:00:27.91	Suki
12	C	00:00:27.98	Suki
13	A	00:00:22.04	Suki
14	A	00:00:07.99	Suki
15	C	00:00:24.01	Suki
16	A	00:00:30.00	Suki
17	C	00:00:07.00	Sunagashi
		00:05:19.97	Time
		00:05:05.05	Suki Time

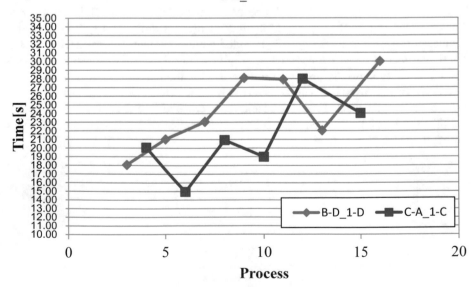

Fig. 13. B-D_1 process time

Table 3. B-D_2 process time

B-D_2	Worker	Time	Work name
1	C	00:00:04.01	Kumiage
2	C	00:00:03.90	Sunagashi
3	A	00:00:17.01	Suki
4	C	00:00:12.98	Suki
5	A	00:00:17.02	Suki
6	C	00:00:18.90	Suki
7	A	00:00:23.02	Suki
8	C	00:00:23.01	Suki
9	A	00:00:21.01	Suki
10	C	00:00:18.01	Suki
11	A	00:00:27.99	Suki
12	C	00:00:29.95	Suki
13	A	00:00:27.04	Suki
14	C	00:00:25.02	Suki
15	A	00:00:20.93	Suki
16	C	00:00:22.00	Suki
17	C	00:00:05.08	Sunagashi
		00:05:16.87	Time
		00:05:03.89	Suki Time

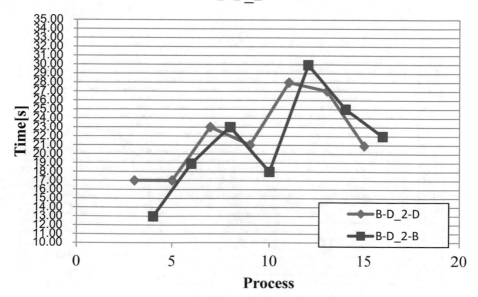

Fig. 14. B-D_2 process time

Table 4. B-D_3 process time

B-D_2	Worker	Time	Work name
1	C	00:00:04.00	Kumiage
2	C	00:00:03.94	Sunagashi
3	A	00:00:20.97	Suki
4	C	00:00:14.04	Suki
5	A	00:00:18.01	Suki
6	C	00:00:18.98	Suki
7	A	00:00:21.02	Suki
8	A	00:00:24.91	Suki
9	C	00:00:17.07	Suki
10	A	00:00:20.00	Suki
11	C	00:00:19.99	Suki
12	A	00:00:24.96	Suki
13	C	00:00:19.01	Suki
14	A	00:00:25.05	Suki
15	C	00:00:27.02	Suki
16	A	00:00:28.96	Suki
17	C	00:00:27.96	Suki
18	A	00:00:29.05	Suki
19	C	00:00:24.96	Suki
20	C	00:00:06.03	Sunagashi
		00:06:35.93	Time
		00:06:21.96	Suki Time

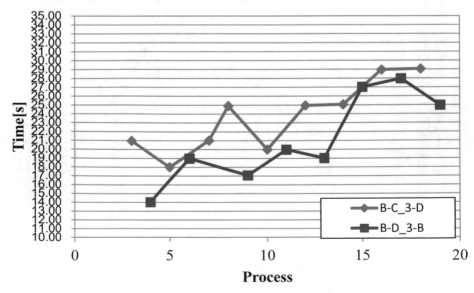

Fig. 15. B-D_3 process time

Table 5. D-B_1 process time

D-B_1	Worker	Time	Work name
1	D	00:00:05.08	Kumiage
2	D	00:00:02.96	Sunagashi
3	B	00:00:05.97	Sunagashi
4	D	00:00:17.08	Suki
5	B	00:00:16.93	Suki
6	D	00:00:21.05	Suki
7	B	00:00:17.95	Suki
8	D	00:00:22.99	Suki
9	B	00:00:21.11	Suki
10	D	00:00:12.94	Suki
11	D	00:00:07.96	Sunagashi
		00:02:32.02	Time
		00:02:10.04	Suki Time

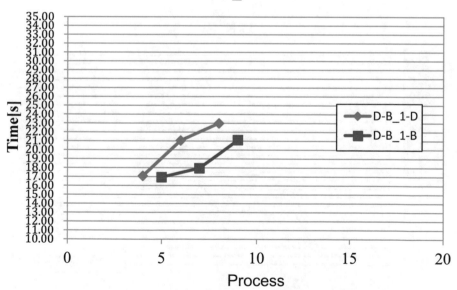

Fig. 16. D-B_1 process time

Table 6. D-B_2 process time

D-B_2	Worker	Time	Work name
1	D	00:00:03.06	Kumiage
2	A	00:00:02.99	Sunagashi
3	D	00:00:05.94	Sunagashi
4	D	00:00:12.03	Suki
5	D	00:00:04.96	Suki
6	B	00:00:20.04	Suki
7	D	00:00:16.96	Suki
8	B	00:00:21.10	Suki
9	D	00:00:15.96	Suki
10	B	00:00:25.94	Suki
11	D	00:00:22.99	Suki
12	B	00:00:26.04	Suki
13	D	00:00:30.00	Sunagashi
		00:03:28.02	Time
		00:02:46.03	Suki Time

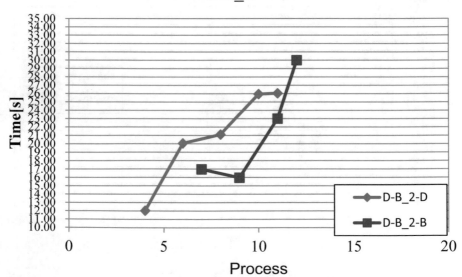

Fig. 17. D-B_2 process time

Table 7. D-B_3 process time

D-B_3	Worker	Time	Work name
1	D	00:00:03.95	Kumiage
2	D	00:00:02.99	Sunagashi
3	B	00:00:06.96	Sunagashi
4	D	00:00:13.06	Suki
5	B	00:00:13.96	Suki
6	D	00:00:21.97	Suki
7	B	00:00:13.10	Suki
8	D	00:00:22.93	Suki
9	B	00:00:14.96	Suki
10	D	00:00:21.09	Suki
11	B	00:00:31.94	Suki
12	D	00:00:24.06	Suki
13	B	00:00:02.98	Sunagashi
14	D	00:00:06.96	Sunagashi
		00:03:20.91	Time
		00:02:57.07	Suki Time

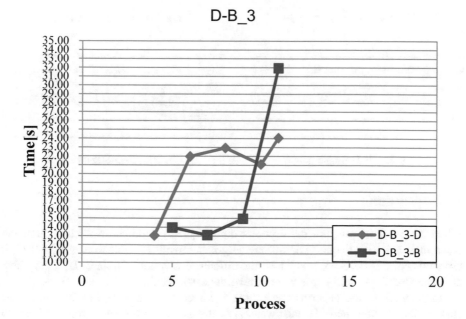

Fig. 18. D-B_3 process time

2.7   Sugeta Angle and Area

When performing work subject is pumped up, Among the "Sugeta" is "Sukifune", was determined "Sugeta" angle and the area at which most descended below (Fig. 11). Digit left corner and L2, shown in Table 8 to the digit right corner and R1. Figure 19 to the D-B "Sugeta" angle 74.21°, Of D-B in Fig. 20 "Sugeta" depth 295.69 mm, "Sugeta" to give the area 61853.21 mm². Of B-D in Fig. 21 "Sugeta" angle 73.85°, Of B-D in Fig. 22 "Sugeta" depth 314.85 mm, "Sugeta" to give the area 684474.99 mm².

Table 8. Sugeta angle and area

	Angle	A length (mm)	Area (mm²)
B	74.21°	295.69	618532.21
D	73.85°	314.85	684474.99

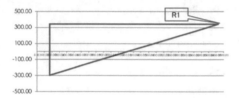

Fig. 19. D-B Sugeta angle

Fig. 20. D-B Sugeta length

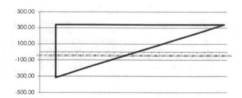

Fig. 21. B-D Sugeta angle

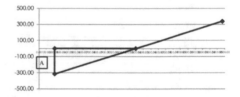

Fig. 22. B-D Sugeta length

3   Expedition

When the paper is pumped, the drypoint digit (Sugeta) depth is 295.7 mm from the horizontal, B is 314.9 mm. The difference got the result of 19.2 mm. drypoint digit (Sugeta) angle, D is 73.9°, B is 74.2°, the difference was 0.3°. Subjects D, b drypoint-digit (Sugeta) depth and angle were almost the same results.

In the work of the papermaking (Suki), the average time of D was 21.9 s, B was made at 20.3 s, and the difference was 1.5 s. It was almost the same time as the two skilled men. In the operation of the papermaking (Suki), it became almost the same operation as the skilled person D and B, I think that this is the result of the skill as a traditional craftsman, and succession of the skill of the Takumi is done.

B is how to carry the water only in the longitudinal direction on the long side of the drypoint digit (Sugeta), D is seen the movement of the horizontal and vertical sides to the long sides. In addition, when the drypoint is made, B will make a change to the other side by the end of the water only, but D will do the work of drawing the new paper again before passing the other digit. It is thought that there is a difference of several seconds in the time of making from this. D, the average time of the preparation time of B, was about the same time, when viewed in detail each of the time of the preparation, the operation time is increased for each increase in the number of times. I think that it goes to the finish, and it advances to a politer work.

4 Conclusion

This experiment was carried out focusing on the operation and working process skill B and D experiments. Skilled person B was almost the same behavior as the D is, I think and achievements of skilled as traditional craftsmen, and inheritance of craftsmanship has been carried out. In particular, B, D is a mother and daughter, from the fact that to study in the same workplace every day, think like tradition of the technique has been carried out. "Sunagashi" is an important work that determines the paper of facial expression (surface), and the purpose of passing a dust on the surface of the blind, is also a work filled the fiber in the eyes of the blind. It is required to flow evenly and quickly to the full width of the top of the order "Sugeta". This is a skilled person is also a factor that is born is a difference of formation by "Sunagashi" she says. Subject D from the results of experiments, B although both considered to flow an appropriate amount of the stock quickly. Especially skilled who have years of experience of 46 years, from the operation and the current result of papermaking and waste-free operation, considered as have done a politer.

References

1. Saito, I.: The story of Echizen Washi (1973)
2. The Mainichi Newspaper: Handmade Japanese Traditional Paper (1975)

Cognitive Analyses for the Improvement of Service Orders in an Information Technology Center: A Case of Study

Manuel Alejandro Barajas-Bustillos[1], Aide Maldonado-Macias[1(✉)],
Margarita Ortiz-Solis[1], Arturo Realyvazquez-Vargas[2],
and Juan Luis Hernández-Arellano[1]

[1] Department of Electrical Engineering and Computing,
Ciudad Juárez Autonomous University,
Ave. del Charro 450 Norte, Cd. Juárez, Chihuahua, Mexico
{al171528, al164612}@alumnos.uacj.mx,
{amaldona, luis.hernandez}@uacj.mx
[2] Technological Institute of Tijuana, Calzada Del Tecnológico S/N,
Fracc. Tomas Aquino, Tijuana, Mexico
arturo.realyvazquez@tectijuana.edu.mx

Abstract. This case of study takes place at a university's information technology center where it is possible to damage computers seriously during the initial checking due to procedural or human errors. Thirteen technicians were observed in this work. They perform tasks and subtasks suffering from mental and physical workload that can cause human errors since an average of 300 maintenance orders must be attended weekly. According to the participants, changing the power supply is the task that can cause the most severe damage on computers in case of error, so human error identification and mental workload evaluation techniques were applied using TAFEI and NASA TLX correspondingly. Results found that human errors are mainly due to the poor identification of the equipment waiting to be repaired, while Mental Demand is the highest source for mental workload. Recommendations are given to prevent human errors and reduce mental workload.

Keywords: Hierarchical Task Analysis (HTA) · Mental workload ·
NASA Task Load Index (NASA-TLX) ·
Task Analysis for Error Identification (TAFEI)

1 Introduction

At present, the use of computers or PCs in daily activities is increasing. It is becoming more common finding that in most workplaces, or even at home, the use of computers is one of the most important activities, so the development of different PCs repairing support facilities has become necessary. This work was carried out in an Information Technology Center (ITC) that attends an average of 300 service orders and works in two shifts weekly. In this center it is possible to damage computers seriously during the initial checking due to procedural or human errors causing economic losses.

© Springer Nature Switzerland AG 2020
W. Karwowski et al. (Eds.): AHFE 2019, AISC 971, pp. 188–198, 2020.
https://doi.org/10.1007/978-3-030-20494-5_17

For some authors, the prevention of human error and mental workload are main areas of study of cognitive ergonomics [1]. Through the analysis of human error, the possible types of failures that can originate in the workplaces can be considered, as well as their consequences, thus having a way of proposing recommendations to avoid them. The use of a mental workload assessment technique can help redesign the task so that technicians do not face stressful situations and can do their job correctly.

Based on the above, the objective of this work is carrying out cognitive analyses, through human error and mental workload, in a computer repair task which is carried in ICT to analyze the possible causes of damage on the equipment being repaired and suggest improvements in its operation to avoid future damage. The specific objectives are:

1. Develop a Hierarchical Task Analysis (HTA) to identify tasks and sub-tasks.
2. Identify human error by means of the method Task Analysis for Error Identification (TAFEI).
3. Evaluate the mental workload in the selected task through the NASA-TLX technique.
4. Recommend actions to reduce human error for the selected task.

The participants were observed performing this task over different days and hours.

1.1 Literature Review

Ergonomics has been present throughout human history. For example, it was already taken into account in tool design in the ancient Greece of circa the 5th century B.C. [2]. According to Leirós [3], the word ergonomics was first used in the middle of the 19th century in the philosophical treatise "Compendium of Ergonomics, or the Science of Work Based on Truths Taken from Nature" written by the naturalist philosopher Wojciech Bogumil Jastrzebowsk in 1857.

Cognitive ergonomics, the branch of ergonomics that deals with the cognitive part, as defined by the International Ergonomics Association [4], deals with mental processes such as perception, memory, reasoning and motor response. In addition to the aforementioned, Hollnagel [5] mentions that cognitive ergonomics is oriented to the psychological aspects of work, as well as how work affects the mind and in how the mind affects work. According to some authors, the main research topic of cognitive ergonomics is explaining, predicting and avoiding human error [6, 7].

The human error, according to Stanton, Salmon and Rafferty [8], has received constant attention among ergonomics experts and has been consistently identified as an important factor in a high proportion of incidents in dynamic and complex systems. Human error occurs on those occasions when a planned sequence of activities, mental or physical, does not achieve the desired result and only if these failures cannot be attributed to the intervention of causal agents [9].

The Task Analysis for Error Identification (TAFEI), developed by Baber and Stanton [10], is a method that can predict errors in the use of devices, modeling the interaction between the user and the device under analysis. This method assumes that the actions between the user and the device are limited by the state of the device at any point in the interaction and that the device provide information to the user about its functionality, therefore, the interaction between the user and the device progresses through several states and in each of these states, the user selects the most relevant action for his target, based on the total system image [8].

One of the most important steps in the application of TAFEI is the realization of the Hierarchical Task Analysis (HTA), which, according to Annett [11], is a process that consists of breaking down tasks into subtasks with the desired level of detail. The HTA is a powerful tool that gives the analyst an overview of how a process works since HTA is the analytical description of a process or activity that includes the realization of a hierarchy of objectives, sub-objectives, operations and task plans and was originally developed as a method for determining training requirements [8]. HTA is the oldest and best known task analysis technique, which is still valid, even if new methodologies have appeared [12].

On the other hand, according to Rizzo, Dondio, Delany, and Longo [13] the mental workload (MWL) can be described as the amount of cognitive work used in a given task over a given period of time. There are several methods of assessing mental load, according to Milan et al. [14] three are the most popular: NASA-Task Load Index (NASA-TLX), Subjective Workload Assessment Technique (SWAT) and Workload Profile (WP).

NASA TLX is a multidimensional subjective evaluation technique presented by Hart and Staveland [15], according to several authors it is one of the most used subjective techniques for the evaluation of CMT [8, 16–18]. This technique distinguishes six dimensions of subjective burden: mental demand, physical demand, temporal demand, effort, performance, and frustration level. This technique is usually performed immediately after the task has been completed.

2 Methodology

The methodology used in this work includes four stages, the selection of the task to analyze, the elaboration of an HTA, analysis of human error through TAFEI and the evaluation of mental load with NASA-TLX.

2.1 Selection of the Task to Analyze

At this stage of the methodology, a survey was conducted among the ITC staff to determine the task whose consequences would have the greatest impact and which, due to its poor execution, would cause irreparable damage to the equipment under repair.

2.2 HTA Development

For the development of a hierarchical task analysis, according to Stanton et al. [8], the following steps should be observed:

1. Define the task for analysis: The first step is to clearly establish the task to be analyzed and the purpose of the task analysis must also be defined.

2. Data collection process: Once the task to be analyzed is defined, task-specific data must be collected. Specific data should be collected on task steps, the technology used, man-machine interaction, team members, decision making, and task limitations.
3. Determine the overall goal of the task: The overall goal of the analyzed task must first be specified at the top of the hierarchy.
4. Determination of sub-goals of the task: Once the overall goal of the task has been specified, the next step is to divide this overall goal into significant sub-goals (usually four or five) that together form the tasks needed to achieve the overall goal.
5. Sub-goal decomposition: The analyst should then break down the sub-goals identified during the fourth step into other sub-goals and operations, according to the number of steps of the task in question. This process must continue until a proper operation is reached. The lower level of any branch in an HTA must always be an operation. While everything above an operation specifies goals, operations really say what needs to be done. Therefore, operations are actions that must be performed by an agent to achieve the associated objective.
6. Plan analysis: Once all sub-goals and operations have been fully described, plans need to be added. Plans dictate how goals are achieved. A simple plan would be: Do 1, then 2 and then 3. Plans do not have to be linear and exist in many forms, some examples are shown in Table 1.

Table 1. Types of plans for the HTA.

Plan	Example
Linear	Do 1, then 2, then 3
Nonlinear	Do 1, 2 and 3 in any order
Simultaneously	Do 1, then 2 and 3 at the same time
Bifurcation	Do 1, if X is present make 2, then 3, if X is not present Exit
Cyclic	Do 1, then 2 and then 3 and repeat until X
Select	Do 1, then 2 or 3

Source: Stanton et al. [8]

2.3   Analysis of Human Error Through TAFEI

To develop the TAFEI, three steps are necessary:

1. Develop a Hierarchical Task Analysis (prepared in the previous stage).
2. Create a space-time diagram.
3. Make a transition matrix.

Space-State diagrams
The Space-state Diagrams (SSDs) are constructions that represent the behavior of the device or product. Each of them represents one of the possible task states, listing the initial and final status, this is based on the HTA [19].

Transition matrix

The transition matrix is an important step in the TAFEI technique [19]. All possible states are inserted in this matrix. The transition states of the SSDs are placed in the cells of the table. Three approaches are adopted to complete the matrix [8]:

1. If the given transition is impossible, a dash is placed in the respective cell.
2. If a given transition is possible and desirable (I.e., the user is moving towards the target), it's a legal transition represented by L in the table.
3. If a given transition is possible but undesirable (deviation from the desired action), it's an illegal transition shown in the table, it's represented by an I.

When all possible intersections have been analyzed, the situations in which an illegal transition occurs (I) are analyzed.

Evaluation of the Mental Workload Through the NASA-TLX

For the evaluation of the MWL, before carrying out the task, participants were informed of the justification and reason for the method, as well as the basic principles of the mental workload, and the procedure for completing the form once the task was completed. In general, this technique involves two steps in sequential order, in the first step a comparison is made, in the form of pairs, between the 6 dimensions that distinguish this technique and then an assessment is made, on a scale from 1 to 100 with intervals every 5, of the role, played by each dimension in the assessment [15].

3   Results

Results found from the different stages presented in the methodology are provided in this section.

3.1   Selection of the Task to Analyze

The task selected for analysis, based on interviews among technicians is changing and test the computer power supply (PS), this is due to the PS is responsible for providing the proper electrical current to all elements of the computer (PC), so in case of an error and/or failure during this task the entire equipment is affected. According to technicians' comments, the PS failures represent 16% of the total failures of the computer equipment reviewed. However, this failure can cause a severe damage to computers. In this study, 13 participants were observed performing this task during different days and hours. All participants were men, of whom 46.2% were between the ages of 21 and 25, 23.1% were between the ages of 26 and 30, and the same percentage were between the ages of 31 and 35 and 7.7% of the participants were between the ages of 41 and 45.

3.2   HTA Development

For the elaboration of the HTA, which is shown in Fig. 1, and according to Stanton [8], the following points were observed:

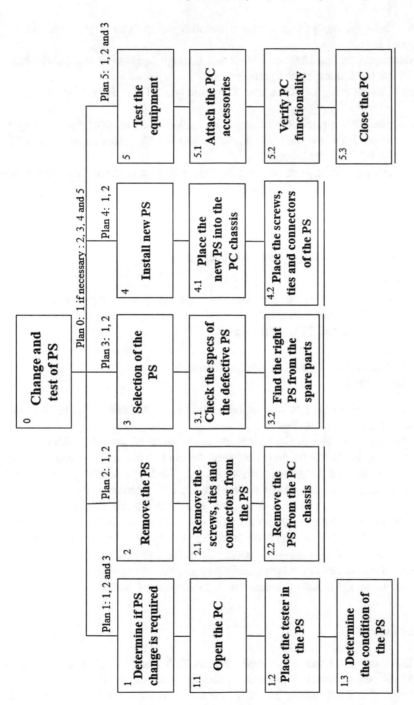

Fig. 1. HTA developed for the selected task.

1. Task definition for analysis: The task determined to analyze in this work is: PS change and test.
2. Data collection process: Data were collected through interviews and direct observation of the work done by the participants.
3. Determine the overall goal of the task: The main goal of the analysis is: Change of PS.
4. Determination of sub-goals of the task: The sub-goals proposed for the implementation of the HTA are: Test the PS, disconnect the PS, selection of the new PS, install a new PS, Test the equipment.
5. Sub-goal decomposition: each of the sub-goals was broken down into simple elements detailing the process of the task.
6. Plan analysis: a linear and selective plan was used to deploy 4 hierarchical levels for subtasks.

3.3   Analysis of Human Error Through TAFEI

Space-State Diagrams
From the HTA developed in the previous step, the SSDs were developed and are shown in Fig. 2. SSDs help to understand transitions during the human-artifact interaction. In this case, we had 11 different states.

3.4   Evaluation of the Mental Workload Through the NASA-TLX

NASA-TLX was applied once the execution of the task was finished, the results were summarized and are shown in Table 2.

As can be observed, Mental Demand is the main source of mental workload in the task, followed by Performance and Temporal demand. This implies that the task requires more mental and perceptual activity such as thinking and analyzing the harnesses disposition and deciding about the best manner of disconnect them to remove the defective PS and replace it with a new one. Additionally, technicians must remember the location of the set of connectors and deciding the correct specifications according the PS. Technicians commonly work into the try and error scheme that may determine the success of the goals of the task. About the Temporal demand, technician must complete the maintenance order according an established schedule; in this manner technician are subjected to time pressure and a rapid pace of work.

4   Discussion

Although literature about the use of both, NASA-TLX and TAFEI is limited, both methods can be used in a variety of areas separately; however, in this case study, they were used composed effectively to offer a more complete analysis of the task. According to Glendon, Clarke, and McKenna [20], the TAFEI technique can provide a useful picture of the interactions between human operators and machine components within a system, with respect to possible actions and errors. For example, TAFEI has

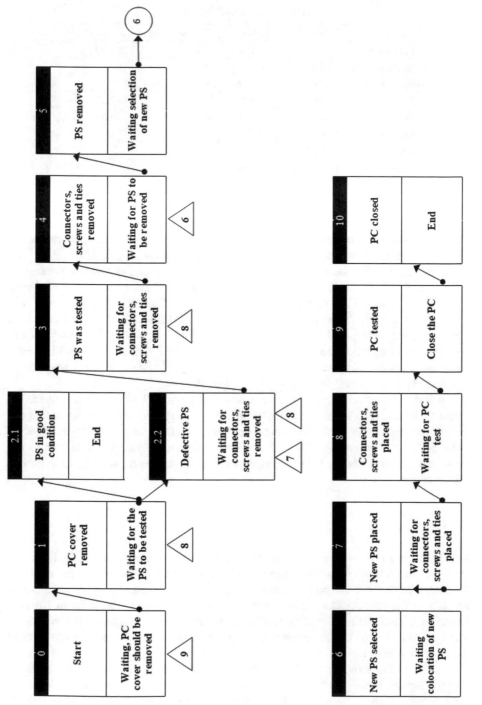

Fig. 2. SSDs developed for the selected task.

Table 2. NASA-TLX results.

Variable	Weight	Score	Converted score	Weighed score
Mental demand	3.00	16.00	80.00	240.00
Physical demand	1.77	6.31	31.54	55.80
Temporal demand	2.31	11.23	56.15	129.59
Effort	2.38	10.00	50.00	119.23
Performance	2.46	12.92	64.62	159.05
Frustration	3.08	7.54	37.69	115.98
Total	15.00			819.64
Average TLX	= 320/6	= 53.33		
Weighed TLX	= 819.64/15	= 54.64		

been used in the evaluation of many different applications, such as word processors, ATMs, automotive products, ticket vending machines, in the use of X-Ray machines, in meat grinding machines, orthopedic robots, among others [19, 21–23].

In the case of NASA-TLX, this technique achieved a similar level of concurrent validity and test-retest reliability as the other mental workload assessment techniques. In addition, NASA-TLX demonstrated a good level of sensitivity to different task demands. In a study by Rubio et al. [17] the WP, NASA-TLX, SWAT assessment techniques were compared in terms of intrusion, diagnosis, sensitivity, validity (convergent and concurrent) and acceptability, and was found that the NASA-TLX technique has a good level of sensitivity. On the other hand, NASA-TLX is fast and easy to use and requires minimal training of the analyst [8].

In this work, according to the results obtained through TAFEI, we can observe a series of situations in which human error is latent, and that in case of carelessness can result in damage to the equipment being repaired. In the case of the results obtained by NASA-TLX, it's observed that, in the experience of the participants, the task generates an important index of mental workload, reason why measures must be taken to correct this situation.

5   Conclusions

From these cognitive analyses, it is possible to recommend changes in the design of the task in such a way as to facilitate the correct identification of the state of the material. Based on the above, it can be considered that performing an analysis of potential errors, as well as determining the mental workload generated in the execution of the task provides valuable information that can be used at the time of designing and evaluating them, to ensure their correct implementation and then be able to reduce the level of errors of workers.

Based on the above, it can be concluded that the implementation of the methodology proposed in this work to reduce human error through cognitive analysis can be very useful for those in charge of designing and implementing tasks where human-device interaction is relevant to improve and enhance the service and satisfaction of clients and users of this ITC.

References

1. Carrillo-Gutierrez, T., Martínez, R.M.R., Rodríguez, J. de la R., Sanchez-Leal, J.: Relevant aspects of human error and its effect on the quality of the product. Study in the Maquiladora industry. In: Advances in Ergonomics of Manufacturing: Managing the Enterprise of the Future, pp. 475–485. Springer, Cham (2016)
2. Marmaras, N., Poulakakis, G., Papakostopoulos, V.: Ergonomic design in ancient Greece. Appl. Ergon. **30**, 361–368 (1999)
3. Leirós, L.I.: Historia de la Ergonomía, o de cómo la Ciencia del Trabajo se basa en verdades tomadas de la Psicología. Revista de Historia de la Psicología **30**, 33–53 (2009)
4. International Ergonomics Association: What Is Ergonomics: Definition and Domains of Ergonomics. http://www.iea.cc/whats/index.html
5. Hollnagel, E.: Cognitive ergonomics: it's all in the mind: Ergonomics. Ergonomics **40**, 1170–1182 (1997)
6. Cañas, J.J., Waerns, Y.: Ergonomía cognitiva: aspectos psicológicos de la interacción de las personas con la tecnología de la información. Ed. Médica Panamericana (2001)
7. Reyes-Martínez, R.M., Maldonado-Macías, A.A., O, R. de la, Riva-Rodríguez, J. de la: Theoretical approach for human factors identification and classification system in accidents causality in manufacturing environment. In: García-Alcaraz, J., Maldonado-Macías, A.A., and Cortes-Robles, G. (eds.) Lean Manufacturing in the Developing World, pp. 385–404. Springer, Cham (2014)
8. Stanton, N., Salmon, P.M., Rafferty, L.A.: Human Factors Methods: A Practical Guide for Engineering and Design. Ashgate Publishing, Ltd. (2013)
9. Reason, J.: Human Error. Cambridge University Press (1990)
10. Baber, C., Stanton, N.A.: Task analysis for error identification: a methodology for designing error-tolerant consumer products. Ergonomics **37**, 1923–1941 (1994)
11. Annett, J.: Hierarchical Task Analysis (HTA). In: Stanton, N.A., Hedge, A., Brookhuis, K., Sala, E., and Hendrick, H.W. (eds.) Handbook of Human Factors and Ergonomics Methods. pp. 33-1–33-7. CRC Press (2004)
12. Lorés, J., Granollers, T.: La Ingeniería de la Usabilidad aplicada al diseño y desarrollo de sitios web (2017)
13. Rizzo, L., Dondio, P., Delany, S.J., Longo, L.: Modeling mental workload via rule-based expert system: a comparison with NASA-TLX and workload profile. In: Artificial Intelligence Applications and Innovations, pp. 215–229. Springer, Cham (2016)
14. Milán, E.G., Salazar, E., Domínguez, E., Iborra, O., de la Fuente, J., de Córdoba, M.J.: Neurotermografía y Termografía Psicosomática. Fundación Internacional artecittà (2015)
15. Hart, S.G., Staveland, L.E.: Development of NASA-TLX (Task Load Index): results of empirical and theoretical research. In: Hancock, P.A. and Meshkati, N. (eds.) Advances in Psychology, pp. 139–183. North-Holland (1988)
16. Bommer, S.C., Fendley, M.: A theoretical framework for evaluating mental workload resources in human systems design for manufacturing operations. Int. J. Ind. Ergon. **63**, 7–17 (2018)
17. Rubio, S., Díaz, E., Martín, J., Puente, J.M.: Evaluation of subjective mental workload: a comparison of SWAT, NASA-TLX, and workload profile methods. Appl. Psychol. **53**, 61–86 (2004)
18. Young, M.S., Brookhuis, K.A., Wickens, C.D., Hancock, P.A.: State of science: mental workload in ergonomics. Ergonomics **58**, 1–17 (2015)

19. Mohammadian, M., Choobineh, A.R., Mostafavi Nave, A.R., Hashemi Nejad, N.: Human errors identification in operation of meat grinder using TAFEI technique. J. Occup. Health Epidem. **1**, 171–181 (2012)
20. Glendon, A.I., Clarke, S., McKenna, E.: Human Safety and Risk Management, Second Edition. CRC Press (2016)
21. Alferez-Padron, C., Maldonado-Macías, A.A., García-Alcaraz, J., Avelar-Sosa, L., Realyvasquez-Vargas, A.: Workload assessment and human error identification during the task of taking a plain abdominal radiograph: a case study. In: Advances in Neuroergonomics and Cognitive Engineering, pp. 108–119. Springer, Cham (2017)
22. Kuang, S.L., Hu, L., Zhang, S.T., Gao, D.H.: Applying TAFEI method to orthopaedic robot system's requirements analysis. In: 2009 16th International Conference on Industrial Engineering and Engineering Management, pp. 66–70 (2009)
23. Stanton, N.A., Baber, C.: Validating task analysis for error identification: reliability and validity of a human error prediction technique. Ergonomics **48**, 1097–1113 (2005)

Design of a Model of Assignment of Workers and Operations that Reduces the Biomechanical Danger in a Panela Productive Unit of Colombia

Y. A. Paredes-Astudillo$^{(\boxtimes)}$ and Juan P. Caballero-Villalobos

Department of Industrial Engineering, Pontificia Universidad,
Javeriana, Bogotá, Colombia
{yennyparedes, juan.caballero}@javeriana.edu.co

Abstract. Productive processes in agro-industries, in their majority, are still handmade and they keep hazardous tasks health. In Colombia the cane sugar exploitation has a relevant importance, it allows to produce sugar, ethanol, paper, and another popular product called "Panela". Panela's production is one of the most traditional and important agro-industries in Latin America, generating about 353.000 jobs in Colombia, it is classified as one of the riskiest agricultural industries, reporting 9.524 occupational disease cases per 100.000 workers in 2012.

This study propose a job assignment's strategy to minimize the maximum cumulative postural risk differences between each pair of workers, considering the study case of a small panela productive unit of Colombia. The problem has been represented by a Mixed Integer Linear Programing Model (MILP) and solutions were obtained by using exact methods and a proposed GRASP, from which it was got reductions of about 65% respect to the current situation.

Keywords: Panela · Job rotation · Ergonomics · Metaheuristic-Optimization

1 Introduction

Musculoskeletal disorders (MSD) refer to a variety of health conditions in which mainly tendons, ligaments, nerves and bones are affected [1], MSD related with work, are the consequence of the interaction of multiple factors such as, organizational factors, the social context and the individual characteristics [2]. During the last years the emphasis of epidemiology applied to ergonomics has focused on the mechanical demand of work, due in part because large part of the studies have demonstrated the importance of these factors as determinants in the appearance of MSD [3], and also because it is a controllable variable in the organization of work.

The work physical demand consists on manipulation of materials, inclinations, torsions, heavy physical work and total body vibration [4]. When works physical demand exceeds the tissues tolerance it generates changes in the structures that accumulate over time and finally trigger in a MSD [2].

© Springer Nature Switzerland AG 2020
W. Karwowski et al. (Eds.): AHFE 2019, AISC 971, pp. 199–208, 2020.
https://doi.org/10.1007/978-3-030-20494-5_18

MSD have a high impact on the economy and productivity, being indeed the main causes of absenteeism [5]. In Colombia, MSD correspond to the main occupational diseases diagnoses and the first cause of occupational morbidity in the occupational health system [6].

Productive processes in agro-industries, in their majority, are still handmade and they keep hazardous tasks for the health. According to INSHT [7], economic activities related with agriculture are one of the most exposed to ergonomic risks, as: high limb's repetitive movements, forced postures and load handling (58%). According to this situation, workers of this industry refer discomfort in lower back in 54% and neck in 31%. Between the agro-industries, the sugar cane exploitation has a relevant economic and cultural importance in several countries, because it allows to produce some well recognized products such as sugar, ethanol, paper, and another popular product called "Panela" in Colombia, "Jagerry" in India, "Piloncillo" in Mexico or "Raspadura" in Brazil. Panela's production is one of the most traditional rural agro-industries in Latin America and the Caribbean, and its market is growing and is gaining importance in the USA and Europe markets.

In Latin America, Panela's sector has high impact on the Colombian economy, which is the second largest producer in the world, contributing with 13.9% of world production, generating around 353.000 jobs, and being a source of income for more than 70.000 Colombian people [8].

Small productive units provide most of the Colombian production of panela, about the 83%. These productive units are characterized by having less than five hectares sown with cane, employ hand force that frequently are relatives or neighbors [8, 9].

The sugar cane exploitation activity is recognized as one of the highest risk sectors for workers in the agro-industry. According to the Colombian insurance federation [10], Panela industry is one of the riskiest agricultural industries, reporting 9.524 occupational disease cases per 100.000 workers in 2012. Although there are not specific records of MSD caused in panela's agro-industry, it is likely that, as others cane sugar related industries, it would have similar conditions and risks.

Several administrative techniques are used to reduce the risk of suffering MSD, among them, operation rotation is one of the most used for manual and repetitive work [11]. Work rotation in the context of risk control of MSD, refers to the strategy of alternating workers between tasks with different levels of exposure and occupational demand [12].

Some studies related with work rotation to control MSD have in consideration: Repetitive movements [13–15], which in their majority used OCRA, an observational method that allow to measure body posture and force for repetitive tasks [16]. Uncomfortable postures: [12, 15, 17], where used observational methods as REBA, which allows obtaining information related to body postures, force, type of movement, actions or repetitions [18], RULA one method focused on evaluation of risk for upper limbs [16]. By other hand studies related with load handling [17, 19–21], refers NOISH as tool of evaluation.

Most of the optimization approaches used to solve job rotation problems, are classified as Heuristics and Metaheuristics, because commonly are large models, difficult to solve by analytic methods due to the high computational consumption resources. Asensio-Cuesta [13], used genetic algorithms in a job rotation problem to minimize accumulative fatigue taking into consideration different type of tasks, each one with a physical demand, competences and abilities by the workers, Carnahan *et al.* [20] employed a genetic algorithm to solve a job rotation problem, considering biomechanical parameters as gender and the lifting worker capacity, Seçkiner and Kurt [22] used ant colony optimization to minimize difference between worker's workload. Bautista *et al.* [23] maximized the comfort in a mixed model assembly line solving the model with two different procedures: a Mixed Integer Linear Programming (MILP) and a Greddy Randomized Adaptative Search Procedure (GRASP).

This study pretends to develop a job assignment model to minimize the difference between the physical workload's workers in a panela productive unit for one week, using REBA as ergonomic tool in order to determine posture risk. REBA was used to this purpose due to it allows to be applied to different industrial sectors. As optimization tool it was used GRASP in order to provide solutions to the proposed mathematical model. GRASP is a multi-star metaheuristic that consists of an iterative process with two phases: constructive and local search. In the constructive phase a feasible solution is generated, whose neighbor is examined through a local search until a local minimum is reached. In the end the best solution found is left as the result to the problem [24].

2   Method

2.1   Study Case and Considerations

The Panela productive unit studied is located in the north of the department of Cauca – Colombia. There are six workers who work during production season for about eleven (11) hours, starting at 6:00 in the morning and ending at 5:00 in the afternoon, without defined resting periods. All workers in the productive unit accept to participate in this study, so there are six (6) subjects, all of them of masculine gender and age of 39 ± 12 years. Hand force is not polyvalent; each one of workers can participate in some operation, according with its experience and abilities.

Six (6) operations constitute the Panela productive process adopted by the unit, which are:

1. Extraction: the cane juice is obtained.
2. Bagasse collecting: bagasse obtained after the extraction of the juices and arranging it in the storage place.
3. Burner maintenance: Dry bagasse is introducing into the combustion chamber and keeping the fire lit.
4. Juices' cleaning and cooking: Binders are added to the juices and impurities are removed, then, clean juice is transferred to two boilers. Here the process of evaporation and concentration of the juice is carried out. The honeys reach an average temperature of 120 °C.
5. Mixed and molded: when honeys are ready, they are discharged in a wooden container where they are shacked until reaching consistency, after that the mixture is placed in wooden molds to cool.
6. Unmolded and packed: Once solidified, the panelas are removed from the molds and packaged in ecological paper according to the presentation.

2.2   Ergonomics Parameters

Postural workload was evaluated through the REBA method. Each person was evaluated in each of the operations in which he could participate. According to the level of risk, it was established that if a person was assigned to an operation with medium, high or very high level of risk, it would not be possible to go to a higher level of risk in the next period of rotation, it should go to operation with low or insignificant level of risk.

Worker's heart rate was obtained through a heart rate monitor during a period of rest for twenty minutes and for thirty minutes while working in each of the possible operations. Average worker's heart rate during work in each operation and at rest was calculated. Through Chamoux criterion, the cardiovascular load of each worker in each of the operations was assessed.

2.3   Model Mathematical

Sets:

I: *Workers*
J: *Operations*
T: *Job rotation periods*
V: *Body parts evaluated with REBA*
L: *Weekdays*

Parameters:

r_{jt} : *Personal requirment in operation j during period t*

$reba_{ij}$: *REBA score for each worker in operation j*

se_{ijv} : *Posture score for body part v for each work in operation j*

$cchm_{ij}$: *Chamoux classification for each work in operation j*

c_i : *Cost per work day*

presup : *Daily budget*

$bcchm_{ij}$: $\begin{cases} 1 : \textit{Operation j is heavy fow worker, according with Chamoux criterion} \\ \quad\quad\quad\quad 0 : \textit{Other} \end{cases}$

h_{ij} : $\begin{cases} 1 : \textit{worker has the ability to develop the operation j} \\ \quad\quad\quad 0 : \textit{Other} \end{cases}$

bse_{ijv} : $\begin{cases} 1 : \textit{There are harmful effects for workerˈs body v in operation j} \\ \quad\quad\quad\quad 0 : \textit{Other} \end{cases}$

Decision variables:

$$x_{ijtl} = \begin{cases} 1 : when\,worker\,is\,assigned\,to\,an\,operation\,j\,in\,period\,t\,on\,day\,l \\ \qquad\qquad 0 : Others \end{cases}$$

$$y_i = \begin{cases} 1 : when\,worker\,is\,hired \\ \quad 0 : Others \end{cases}$$

$w = Maximum\,difference\,in\,accumulative\,postural\,risk\,among\,workers$

Object Function:

$$Min\,Z = w \qquad\qquad (1)$$

Subject to:

$$\sum_{\forall j \in J} [(x_{ijtl} * bse_{ijv}) + (x_{ij(t+1)l} * bse_{ijv})] \leq 1 \qquad \begin{array}{l} \forall i \in I, \forall v \in V, \forall l \in L \\ \forall t \in T/t \neq Card(T) - 1 \end{array} \quad (2)$$

$$\sum_{\forall j \in J} x_{ijtl} \leq 1 \quad \forall i \in I, \forall t \in T, \forall l \in L \qquad\qquad (3)$$

$$\sum_{\forall j \in J} \sum_{\forall t \in T} x_{ijtl} * bcchm_{ij} \leq 1 \qquad \forall i \in I, \forall l \in L \qquad (4)$$

$$\sum_{\forall j \in J} x_{ijtl} \leq y_i \quad \forall i \in I, \forall t \in T, \forall l \in L \qquad\qquad (5)$$

$$x_{ijtl} \leq h_{ij} \qquad\qquad \begin{array}{l} \forall i \in I, \forall j \in J, \\ \forall t \in T, \forall l \in L \end{array} \quad (6)$$

$$\sum_{\forall i \in I} x_{ijtl} * h_{ij} \geq r_{jt} \qquad \forall j \in J, \forall t \in T,\, l \in L \quad (7)$$

$$\sum_{\forall i \in I} y_i * c_i \leq presup \qquad\qquad (8)$$

$$w \geq \sum_{\forall j \in J} \sum_{\forall t \in T} \sum_{\forall l \in L} reba_{ij} * x_{ijtl} - \sum_{\forall j \in J} \sum_{\forall t \in T} \sum_{\forall l \in L} reba_{sj} * x_{sjtl} \qquad \forall i \in I, \forall (s,p) \in K/i \neq s \quad (9)$$
$$-M * [1 - y_i] - M * [1 - y_s]$$

$$x_{ijtl} \in \{1,0\} \qquad\qquad \begin{array}{l} \forall i \in I, \forall j \in J, \\ \forall t \in T,\, l \in L \end{array} \quad (10)$$

$$y_i \in \{1,0\} \qquad\qquad \forall i \in I \qquad (11)$$

2.4   Restrictions' Description and Considerations

The objective function Eq. (1), minimizes difference on accumulative postural risk among workers per week. Constrains (3), (5), (6), (7), represent the conventional job rotation constraints. Equation (3) indicates that a worker should be assigned to only operation on each period. Equation (5) indicates that a worker could be assigned to an operation only if it was hired. Equation (6) indicates that one work can be assigned to an operation if it can work there. Equation (7) guarantees to satisfy the need of personnel in each operation. Equation (8) avoids overcoming daily budget. Equation (1) prevents workers from being assigned to consecutive operation that impose high workload on the same body part evaluated with REBA. Equation (4) indicates that one work only can be assigned to one operation with high cardiovascular load for one day. Equation (9) calculate the maximum difference on accumulative postural risk among workers.

2.5   Solution Methods and Results

Considering an instance of one week, this mathematical model was solved by using CPLEX in NEOS Server, the job assignment to minimizing the maximum difference between works load is showed in Fig. 1 and allows to get an objective function value of 112 points.

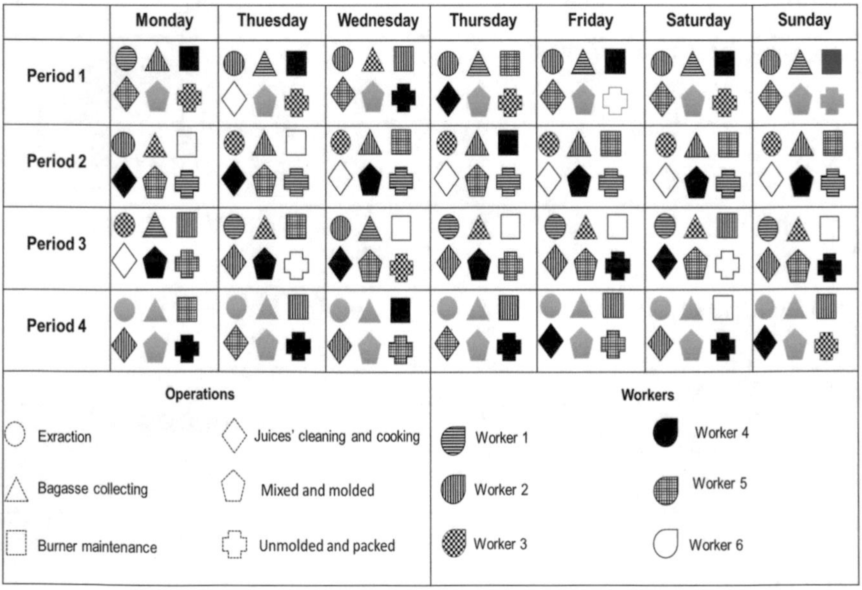

Fig. 1. Job assignment to minimizing difference between work's workload

But having in consideration that this type of problems have difficulties to be solved by analytics methods for large instances due to the high consumption of computer resources, was necessary to propose other tool to get the problem solved in that case. This study develops a Greedy Randomized Adaptive Search Procedures (GRASP). This algorithm is a mullti-start algorithm with two phases the initial phase where an initial feasible solution is built and the second phase where the solution is improvement [23]. For the initial phase, each worker started with an accumulated postural risk equal to zero; according to the assignment of the operation, the level of risk was modified. The value for the greedy parameter was 0.6, using similar studies as a reference. List of restricted candidates (RCL) allows to select the main candidates and finally the best candidate (11). Cost function allows the development of the construction phase (12), it is defined according to the accumulated risk for each worker.

$$f_{C(i,a,t)} = \left\{ \sum_{\forall j \in J} \sum_{\substack{\forall p \in T \\ /p \leq t}} \sum_{\forall l \in L} reba_{(i,a)j} * x_{(i,a)jpl} \right. \tag{12}$$

Once the cost function is calculated for all the elements that can form an initial feasible solution, the list of candidates that formed the LCL according to the expression was constructed.

$$RCL = \left\{ x/L \leq f_{C(x)} \leq L + \alpha(U - L) \right\} \tag{13}$$

Where:
$f_{C(x)}$: The cost function for the x element.
α: Greedy parameter.
L: The lowest value for the cost function.
U: The highest value for the cost function.

For second phase, in order to achieve the exploration of other regions in the initial solution, the change in the choice of workers who should rest in the first and last rotation period of each day was defined as a step. Solution obtained is updated with the change made in the previous phase and the solution with the best performance is selected, which in this case corresponds to the allocation program that minimizes the maximum ergonomic risk difference between each pair of workers.

Maximum accumulated risk difference between each pair of workers in one week, obtained with the proposed model, using the GRASP method, was 131 points.

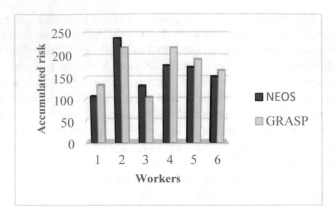

Fig. 2. Comparison NEOS solution vs GRASP solution

Figure 2 shows the results of accumulated weekly risk per worker, which shows a comparison between results obtained by NEOS and GRASP. The objective function value got by the second one is 14.5% below with respect to the first one.

3 Discussion

The model was developed using a MILP formulation looking for the minimization of the maximum cumulative postural risk differences between each pair of workers, considering an one-week planning horizon; the considered constraints by the model correspond to the classic rotation of personnel, allocation according to the worker's skills, restrictions that guarantee that the worker is only assigned to the operation with the highest cardiovascular load once a day and constraints that prevent a worker is assigned in consecutive periods of rotation to highly demanding operations for the same body segment.

To reduce workload levels among the workers, can be observed that some of them who has low levels reach higher levels with the proposed solution and tasks that get the lowest REBA score can be recovery tasks, but it was found that for this case, there are few workers who have the ability to develop it, situation that avoid to obtain even a lower deduction in workloads level.

4 Conclusions

Through the model, a distribution of the workload is achieved, given mainly by the postural risk that each of the workers presents in the possible operations to which it can be assigned. The model respects the constraints given by skills and risks per body segment that form the basis of the allocation criteria. The assignment is economically viable since it considers the budget of the productive unit and no additional expenses are incurred related to hiring personnel.

The scheme obtained from the model solution shows a reduction of 65.3% in the maximum cumulative postural risk difference among workers over a week. The results allow a distribution of the physical load among the hired workers, avoiding the overload that currently exists for certain workers.

References

1. Barrero, L.H., Caban-Martinez, A.J.: Muskuloskeletal disorders. En O. University, Oxford Textbook of Global Public Health, pp. 1–24. Oxford Medicine Online (2015)
2. Marras, W.: The future research in understanding and controlling work-related low back disorders. Ergonomics **48**, 464–477 (2005)
3. Winkel, J., Mathiassen, S.E.: Assessment of physical work load in epidemiologic studies: concepts, issues and operational considerations. Ergonomics **37**, 979–988 (1994)
4. Piedrahíta Lopera, H.: Evidencias epidemiológicas entre factores de riesgo en el trabajo y los desórdenes músculo-esqueléticos. Mapfre Medicina **15**, 212–221 (2004)
5. Osborne, R.H., N. M.: Prevalence and cost of musculoskeletal disorders: a population-based, public hospital system healthcare consumption approach. J. Rheumatol. **34**, 2466–2475 (2007)
6. Ministerio de la Protección Social.: Guía de Atención Integral Basada en la Evidencia para Desórdenes Musculoesqueléticos (DME) relacionados con Movimientos Repetitivos de Miembros Superiores. Bogotá (2006)
7. Instituto Nacional de Seguridad e Higiene en el Trabajo (INSHT). http://www.insht.es
8. Ministerio de Agricultura y Desarrollo Rural - República de Colombia.: El Sector Panelero Colombiano. Bogotá (2006)
9. Organización de las Naciones Unidas para la Agricultura y la Alimentación.: Producción de panela como estrategia de la diversificación en la generación de ingresos en áreas rurales de América Latina. Roma (2004)
10. FASECOLDA.: Riesgos laborales y el agro colombiano. Revista FASECOLDA, pp. 19–24. (2012)
11. Paul, P., Kuijer, F., Visser Bart, K., Han, C.: Job rotation as a factor in reducing physical workload at a refuse collecting department. Ergonomics **42**, 1167–1178 (1999)
12. Howarth, S., Beach, T., Pearson, A., Callaghan, J.: Using sitting as a component of job rotation strategies: are lifting/lowering kinetics and kinematics altered following prolonged sitting. Appl. Ergon. **40**, 433–439 (2009)
13. Asensio Cuesta, S., Diego Mas, J., Canós Darós, L., Andrés Romano, C.: A genetic algorithm for the design of job rotation schedules considering ergonomic and competence criteria. Int. J. Adv. Manuf. Technol. **60**, 1161–1174 (2012)
14. Asensio Cuesta, S., Diego Mas, J., Canós Darós, L., Andrés Romano, C.: A method to design job rotation schedules to prevent work-related musculoskeletal disorders in repetitive work. Int. J. Adv. Manuf. Technol., 1161–1174 (2010)
15. Jaturanonda, C., Nanthavanij, S., Chongphaisal, P.: A survey study on weights of decision criteria for job rotation in Thailand: comparison between public and private sectors. Int. J. Hum. Resour. Manage. **17**, 1834–1851 (2006)
16. David, C.: Ergonomic methods for assessing exposure to risk factor for work-related musculoskeletal disorders. Occup. Med. **50**, pp. 190–199 (2005)
17. Otto, A., Scholl, A.: Reducing ergonomic risks by job rotation scheduling. OR Spectrum **35**, 711–733 (2013)

18. Stanton, N., Hedge, A., Brookhuis, K., Salas, E., Hendrick, H.W.: Handbook of Human Factors and Ergonomics Methods. CRC Press, Washington, DC (2005)
19. Tharmmaphornphilas, W., Norman, B.A.: A methodology to create robust job rotation schedules. Ann. Oper. Res. **155**, 339–360 (2007)
20. Carnahan, B.J., Redfern, M.S.: Incorporating physical demand criteria to assembly line balancing. IIE Trans. **33**, 875–887 (2001)
21. Akyol, S.D., Baykasoğlu, A.: A multiple-rule based constructive randomized search algorithm for solving assembly line worker assignment and balancing problem under ergonomic risk factors. J. Intell. Manuf., 1–12 (2016)
22. Seçkiner, S., Kurt, M.: Ant colony optimization for the job rotation scheduling problem. Appl. Math. Comput. **201**, 49–160 (2008)
23. Bautista, J., Alfaro-Pozo, R., Batalla, C.: Maximizing comfort in Assembly Lines with temporal, spatial and ergonomic attributes. Int. J. Comput. Intell. Syst. **9**, 788–799 (2016)
24. Glover, F.W., Kochenberger, G. (eds.): Handbook of metaheuristics. Kluwer Academic Publishers, Boston (2003)

Job Strain Index by Gender Among Middle and High Managers of the Maquiladora Industry in Ciudad Juarez Mexico

Aidé Aracely Maldonado-Macías[1](✉), Margarita Ortiz Solís[1],
Oziely Daniela Armenta Hernández[2],
Karla Janeth Hernández Luna[1,2], and Jorge Luis García Alcaraz[2]

[1] Departamento de Ingeniería Industrial y de Manufactura,
Universidad Autónoma de Ciudad Juárez, Av. del Charro no. 450 Nte. Col.
Partido Romero, 32310 Ciudad Juárez, Mexico
amaldona@uacj.mx, {al164612, al131949}@alumnos.uacj.mx
[2] Departamento de Ingeniería Eléctrica y Computación,
Universidad Autónoma de Ciudad Juárez, Av. del Charro no. 450 Nte. Col.
Partido Romero, 32310 Ciudad Juárez, Mexico
al164439@alumnos.uacj.mx, jgarcia@uacj.mx

Abstract. Work stress has become a widespread problem in Mexico due to new forms of work organization, new work relationships, and new employment patterns especially in maquiladora industry. The objective of this research is to diagnose work stress by gender using the Job Strain Index in a sample of middle and high managers of the maquiladora industry in Ciudad Juarez, Mexico. As methods, the Job Content Questionnaire (JCQ) and the Job Strain Index were used and statistical tests for the verification of proportion differences by gender were applied. The sample is composed of 177 men and 55 women, of which the proportion that suffers stress (JSI > 1) is 19.2% for men and 38.2% for women. Results show that the proportion of stressed men is significant less than the proportion of stressed women. This means that there is a direct relationship between gender and work stress in middle and senior managers in this sample where women are more likely to suffer work stress than men are.

Keywords: Work stress · Job Content Questionnaire ·
Middle and high managers

1 Introduction

At present, Ciudad Juarez has 322 maquiladora industries according to statistical information from the IMMEX Program [1]. The maquiladora sector is recognized as the main source of employment in this city; accordingly, there is a growing trend of work stress and other related diseases affecting this sector.

Work-related stress has become an extensive problem due to the new forms of work organization, new work relationships and new employment patterns [2]. In recent years, there has been an increment in the number of industries in the world, which brings a greater impact on the work aspect of the population. Based on the concepts of the ILO

© Springer Nature Switzerland AG 2020
W. Karwowski et al. (Eds.): AHFE 2019, AISC 971, pp. 209–218, 2020.
https://doi.org/10.1007/978-3-030-20494-5_19

(International Labor Organization) [3] and the WHO (World Health Organization) [4], stress can be defined as the imbalance between the individual's abilities and knowledge and the demands or pressures that he/she must face.

1.1 Problem Statement

Nowadays, stress is considered the 21st century epidemic and it is necessary to study the existence of this disease, including its factors and its impact on workers' health as well as its impact on the productivity of the latter and companies. The WHO [5] points out that work-related stress is the response that people present when there is a non-correspondence between the demands that the work presents and the worker's abilities or skills to face such demands. When this occurs, psychosocial risks appear, causing stress [6].

Some authors have found some differences by gender with respect the response to work-stress [3]. However, there is lack of studies that can determine these differences among middle and high managers of maquiladora industry in México. In this manner, the main objective of this study is to determine the presence of work stress in a sample of middle and high managers of the maquiladora industry in Ciudad Juarez and determine differences by gender among them.

2 Literature Review

The work stress model of Karasek [7] establishes that psychological factors such as lack of control during work, a limited decision-making latitude, a poor use of skills, and imbalance of the work effort-reward scheme have been associated with the prevalence of work stress. Additionally, other effects of these factors are depression, metabolic malfunctioning, anxiety, fatigue, job dissatisfaction and Burnout prevalence [8–11]. Work stress and others risks of mental health can be present among workers who face high demands or psychological pressures of workload combined with low control over work or low decision latitude during work to satisfy those demands, accompanied with a low social support (supervisor and coworkers' support). The Job Content Questionnaire (JCQ) is one of the most popular instruments for evaluating psychosocial factors in work conditions. This is an instrument designed by Karasek [12], to measure the response to work-stress in a general manner that is applicable to all jobs. The most known dimensions that have been used to measure the high demand/low control model in the development of work stress are: Job demands, Skill discretion, and Job decision-making authority. These dimensions are also important for studies of work motivation, job satisfaction, absenteeism and job rotation. The instrument offers the Job Strain Index (JSI) arising from the interaction of high demands of work and low decision latitude.

3   Methodology

In this section we present the methodology used in this investigation in a detailed manner.

3.1   Stage 1: Design of the Questionnaire and Selection of the Sample

A review of instruments for the measurement of work stress was carried out and the Job Content Questionnaire was chosen as the instrument that would gather the information of participants. In addition, a section of questions was designed to gather sociodemographic data of the sample. With respect to the selection of the sample, it was directed to middle and high managers since it is vulnerable population subjected to high demanding jobs in terms of responsibilities and content of work in maquiladora industry. The adaptation of the questionnaire to Spanish language was made according to Cedillo and Karasek's version [13].

3.2   Stage 2: Application, Input and Validation of Data

The questionnaire is answered by group leaders, supervisors, technicians, and managers, administrative personnel of production, human resources, maintenance, quality, and engineering departments. Two modalities are used to obtain the information: direct interviews with respondents, for which a previous appointment was always made and the administration of questionnaire via email. The period of application of the survey is from August 2017 to May 2018.

A database is needed to input data and for analyzing them. For this, a database is created in SPSS 21® statistical software. In the same way, Excel book sheets are used to analyze and input data. The validation of the data is done through the SPSS 21® program, where a statistical analysis was carried out using the Cronbach's Alpha coefficient to determine the reliability of the instrument, that is, the internal consistency of the items. Validation has a range of 0 to 1, where a value greater than or equal to 0.70 is considered acceptable, but the most satisfactory result would be greater than or equal to 0.90.

3.3   Stage 3: Carrying Out the Descriptive Studies

First, a sociodemographic study is carried out with the information collected. For this stage a preliminary sample of 232 employees surveyed is obtained. For the purposes of this project, this database was divided by gender. Proper graphs and tables with sociodemographic information were obtained. Other sociodemographic variables are age, education, current job position, marital status among others. The second step of this stage is the determination of the proportion of individuals suffering of work stress according to the JSI. To obtain this index the complete database of 232 employees surveyed was used as well as the data bases of men and women separately. Equations 1, 2 and 3 were used from the JQC questionnaire to calculate the corresponding dimensions related to work stress. It is important to mention that the letter "q" means question. In order to apply these equations, variables have to be created for each of the scales.

$$Job\ skills\ discretion\ =\ [q1 + q3 + q5 + q7 + q9 + 5 - q2] * 2. \tag{1}$$

$$Job\ decision - making\ authority\ =\ [2 * (q4 + q6 + q8)] * 2. \tag{2}$$

$$Job\ demands\ =\ 3 * (q10 + q11) + 2 * (15 - q13 - q14 - q15). \tag{3}$$

$$Job\ decision\ Latitude\ =\ Use\ of\ skills\ +\ Authority\ to\ make\ decisions \tag{4}$$

According to Karasek [12], JSI can be determined based on Eq. 5:

$$Value\ of\ the\ JSI\ =\ (Demands * 2)\ /\ Decision\ Latitude. \tag{5}$$

A score greater than 1 could indicate work stress.

3.4   Stage 4: Conducting Statistical Analyses

This stage entails the procedure for the statistical tests that are carried out to determine significant differences by gender in work stress prevalence. The statistical test for the difference between two proportions was conducted; this test as carried out using commercial software. This statistical test is used to find out if the proportion of stressed men is greater than that of stressed women.

For this, it is necessary to calculate the proportions of individuals suffering work stress with the help of the following equations:

$$p1 = Men\ stressed\ (JSI > 1)\ /\ Total\ sample\ of\ men \tag{6}$$

$$p2 = Stressed\ women\ (JSI > 1)\ /\ Total\ sample\ of\ women \tag{7}$$

$$q1 = Men\ not\ stressed\ (JSI \leq 1)\ /\ Total\ sample\ of\ men \tag{8}$$

$$q2 = Women\ not\ stressed(JSI \leq 1)\ /\ Total\ sample\ of\ women \tag{9}$$

After carrying out the test, the null hypothesis is verified. Table 1 shows the hypotheses of an endpoint for the difference between the two proportions. The level of significance used is of 95% to be able to verify the hypothesis. Where P1 = proportion of stress men and P2 = proportion of stressed women.

Table 1. One-sided hypothesis for difference between portions

H_0:	$P_1 = P_2$
H_1:	$P_1 < P_2$

4 Results

With respect to the methodology outlined above, the results are presented for each of the stages.

4.1 Design of the Questionnaire and Selection of the Sample

As results of this stage, there is a questionnaire that includes the agreement confidentiality and participation consent of participants. Next, the instrument of the JCQ appears, where response instructions are provided and consists of 27 items; finally, a sociodemographic data sheet is the last section and will help obtain data such as the company, department, position, and seniority in the position, type of contract, gender, weight measurements, height and abdominal circumference, among others.

This research project was presented to the Association of Maquiladoras of Ciudad Juarez (IMMEX). Those responsible for the medical departments (medical professionals) who agreed to participate would take a preparation for the application of the questionnaire.

4.2 Results of Application, Input and Validation of Data

The questionnaire was applied in 6 different companies of Ciudad Juarez. After the questionnaires were collected, they were captured in the database. The sample size was of 232 participants. Validation was carried out by calculating Cronbach's alpha index to determine the internal consistency of the questionnaire items. This value should be the closest to 1, in this case a value greater than 0.7 was considered acceptable.

Table 2 shows the results of Cronbach's alpha value:

Table 2. Validation of the work content questionnaire

Cronbach's alpha	Elements
0.827	27

Table 3 shows that all elements of the JCQ questionnaire have a Cronbach's alpha value greater than 0.7, which means that the reliability of JCQ's dimensions are acceptable. Table 3 summarizes the reliability validation of all JCQ dimensions separately. As it can be seen, the first 3 dimensions show a Cronbach alpha value below the recommended reference value. On the other hand, the dimension of job insecurity has the lowest value of 0.094. The rest of the dimensions show acceptable reliability values.

Table 3. Validation of dimensions of the work content questionnaire

Dimensions	Items	Alpha Cronbach
Skill discretion	1, 2, 3, 5, 7, 9	0.579
Decision latitude	4, 6, 8	0.553
Job demands	10, 11, 13, 14, 15	0.475
Physical demands	12	N/A
Co-worker support	17, 18, 19, 20	0.786
Supervisor support	21, 22, 23, 24	0.905
Job insecurity	25, 27, 16, 26	0.094

4.3   Results of the Descriptive Studies

In this section, the sociodemographic information obtained from the sample is presented first. Then, the job strain index is presented for the general sample. Subsequently, the general sample was divided by gender to show the frequency and then the proportion of individuals suffering from work-stress.

The sociodemographic study of this sample presents the characteristics obtained from the database. Regarding the characteristics of the sample, this was made up of a total of 232 employees surveyed from 6 different companies of which the percentage of the sample per turn is presented in Table 4.

The range of age with the highest frequency in the sample is of 21 to 30 years old with 45.3%. The marital status with the highest percentage of the population is divorced. The majority of the participants have a degree of 57.8%. Regarding the job position of the participants, the majority occupy middle management production positions (36.2%) followed by technician with (32.3%). On the other hand, in Table 5, the frequencies of male and female participants are shown by percentage. The sociodemographic data can be seen in Table 6. In addition, almost the half of participants have weekly hour's journeys of 45 h per week (45.3%).

Table 4. Variety of participating companies

Industries	Frequency	Percentage	Valid percentage	Accumulated percentage
1. Automotive	82	35.3	35.3	35.3
2. Automotive	41	17.7	17.7	53.0
3. Printers	1	0.4	0.4	53.4
4. Automotive	35	15.1	15.1	68.5
5. Medical	50	21.6	21.6	90.1
6. Automotive	23	9.9	9.9	100.0
Total	232	100.0	100.0	

To determine the proportion of individuals suffering work stress, the (JSI) is calculated for each participant. Table 8 shows, in terms of frequencies, that 76.3% does not present work stress. While 55 employees (23.7%) do have work stress according to the JSI.

Table 5. Gender of participants.

Gender	No. participants	Percentage
Male	177	76.3
Female	55	23.7

Table 6. Sociodemographic data

Category of the variable	Frequency	Percentage
Variable: Age (Years)		
Under 21	4	1.7
21 a 30	105	45.3
31 a 40	57	24.6
41 a 50	54	23.3
51 a 60	12	5.2
Variable: Marital status		
Single	97	41.8
Married	117	50.4
Free union	12	5.2
Divorced	6	2.6
Variable: Education		
High school	13	5.6
High school	56	24.1
Bachelor's degree	134	57.8
Master's degree	27	11.6
PhD.	2	0.9
Variable: Job Position		
Manager	16	6.9
Supervisor	42	18.1
Technician	82	32.3
Group leader	8	3.4
Production planner/coordinator	84	36.2

Table 8. Presence of work stress in full sample according to the JCQ.

Job Stress Index (JSI > 1)	Frequency	Percentage	Valid percentage	Accumulative percentage
Yes	55	23.7%	23.7%	23.7%
No	177	76.3%	76.3%	100.0%
Total	232	100.0%	100.0%	

4.4 Results of the Statistical Tests

In this stage, the statistical tests necessary for the verification of the hypotheses defined in the methodology of this document is presented.

The two proportions test was carried out with the help of the Minitab® software. The results of JSI of Table 9 of the sample of men and women were used.

Table 9. Percentage of JSI in men and women

			Gender		Total
			Woman	Man	
Job Stress Index (JSI)	No	Frequency	34	143	177
		Percentage	61.8%	80.8%	76.3%
	Yes	Frequency	21	34	55
		Percentage	38.2%	19.2%	23.7%
Total		Frequency	55	177	232
		Percentage	100.0%	100.0%	100.0%

The hypotheses used for the test are:

$$H0 : P1 = P2 \tag{10}$$

$$H1 : P1 < P2 \tag{11}$$

Where: P1 = Proportion of men suffering work stress
P2 = Proportion of women suffering work stress
The decision criterion was:

If $p \leq 0.05$, Ho is rejected.

After the analysis using Minitab software, the results of Table 10 were obtained.

Table 10. Statistical Test using JSI

Sex	X Frequency of stressed individuals	N Frequency	Proportion
Men	34	177	0.1920
Women	21	55	0.3818
Difference = p (men) − p (women)			
Estimation of the difference: −0.1897			
Upper limit 95% of the difference: −0.07147			
Test for the difference = 0 vs. < 0:	Z = −2.64		Value p = 0.004

Table 10 shows the proportions of men and women with the studied condition in the sample, this mean individual who present a job strain index superior of one (JSI > 1). Also, the difference between these two proportions was −0.3818 and the upper limit of −0.07147.

Based on the result of the p-value equal to 0.004 (p = 0.004), which is lower than the significance value of 0.05 (p < 0.05), there is statistical evidence to reject the null hypothesis. In this way it can be concluded that there is enough statistical evidence to declare that the proportion of stressed men is smaller than the proportion of stressed women, in other words, the proportion of stressed women is greater than the proportion of stressed men and there is significant difference by gender on the prevalence of work stress in the sample.

5 Conclusions and Recommendations

According to the results obtained after carrying out the JCQ studies in the sample, it was concluded that from the 232 employees surveyed, 23.7% (55 people) presented work stress based on the obtained Job Strain Index, JSI > 1.

Results show that women suffering from work stress presented a greater proportion than the one of men with the same condition in the sample. It can be concluded that, there is a direct relationship between gender and work stress in middle and senior managers. Where women are more likely to suffer work stress than men are. The statistical analysis for two proportions test with a level of significance of 95% showed a value of p < 0.05; therefore, the null hypothesis is rejected and significant difference is found between the proportions, where the proportion of men suffering work stress is less than the proportion of women suffering it.

Finally, it is recommended that maquiladora industries in Mexico take preventive measures that contribute to reduce work stress of their employees and the costs associated with it. These measures may include effective strategies through individual or group psychological interventions; also, providing internal or external consultation of professional services of mental health physicians and specialists on occupational health with no cost for employees. Additionally, companies must invest in work stress prevention and management programs that may include physical and recreational activities among employees.

References

1. Instituto Nacional de Estadística y Geografía: Estadística mensual del programa de la industria Manufacturera, maquiladora y de servicios de exportación (IMMEX). Recuperado de. http://www.inegi.org.mx/ (2011)
2. Organización Internacional del Trabajo: Estrés en el trabajo: un riesgo colectivo. Ginebra, Suiza (2016)
3. Organización Internacional del Trabajo: Guía del formador SOLVE: Integrando la promoción de la salud a las políticas de SST en el lugar de trabajo (Forastieri, Ed.) (2nda ed.). OIT, Ginebra, Suiza (2012)

4. Organización Mundial de la Salud: La organización del trabajo y el estrés. Instituto de trabajo, salud y organizaciones. *Serie protección de la salud de los trabajadores* no. 3, 37 (2004)
5. Organización Mundial de la Salud: Recuperado de http://www.who.int/topics/obesity/es/ (2017)
6. Gil-Monte, P.R.: Riesgos psicosociales en el trabajo y salud ocupacional. Rev. Peru. Med. Exp. Salud Publica **29**(2), 237–241 (2012). Recuperado de http://www.scielo.org.pe/pdf/rins/v29n2/a12v29n2.pdf
7. Karasek, R.A.J.: Job demands, job decision latitude, and mental strain: implications for job redesign. Adm. Sci. Q. **24**(2), 285–308 (1979). https://doi.org/10.2307/2392498
8. D'Souza, R.M., Strazdins, L., Lim, L., Broom, D., y Rogers, B.: The effects of hospital restructuring that included layoffs on individual nurses who remained employed: a systematic review of impact. J. Epidemiol. Community Health **57**, 849–854 (2003)
9. Kuper, H., Singh-Manoux, A., Siegrist, J., y Marmot, M.: When reciprocity fails: effort-reward imbalance in relation to coronary hear t disease and health functioning with the Whitehall II study. Occup. Environ. Med. **59**(11), 777–784 (2002)
10. Mausner-Dorsch, H., Eaton, W.E.: Psychosocial work environment and depression: epidemiologic assessment of the demand-control model. Am. J. Public Health **90**(11), 1765 (2000)
11. Peter, R., Siegrist, J.: Psychosocial work environment and the risk of coronary heart disease. Intern. Arch. Occup. Environ. Health **73**, 41–45 (2000)
12. Karasek, R.A.J.: Job content questionnaire and user's guide. University of Massachusetts, Lowell, 1–66 (1985). Recuperado de http://scholar.google.com/scholar?hl=en&btnG=Search&q=intitle:Job+Content+Questionnaire+and+User+'+s+Guide#0
13. Cedillo, L., y Karasek, R.: Reliability and validity of the Spanish version of the job content questionnaire among maquiladora women workers. JCQ-CENTER www.Jcqcenter.org (2003)

The Difference Between Expert and Non-expert Skill in Beauty Services Based on Eye Tracking Technology

Kohei Okado[1(✉)], Kazuyuki Tanida[2], Hiroyuki Hamada[3],
Akihiko Goto[4], and Yuka Takai[4]

[1] Vories Gakuen Omi Brotherhood Junior High School,
Omihachiman, Shiga 523-0851, Japan
cowhey@gmail.com
[2] Graha-Inc, Chuo-ku, Kobe-shi, Hyogo 651-0097, Japan
kau.tanida02@gmail.com
[3] Kyoto Institute of Technology, Matsugasaki,
Sakyo-ku, Kyoto 606-6585, Japan
hhamada@kit.ac.jp
[4] Osaka Sangyo University, Nakagaito, Daito-shi, Osaka 574-8530, Japan
{gotoh,takai}@ise.osaka-sandai.ac.jp

Abstract. In this study, we analyze the techniques of experts and non-experts as a hairdresser and compared their skills based on eye tracking technology. The first purpose of this research is to clarify the differences between the beauty skills of experts and non-experts. The second purpose is to provide the findings to assist in the development of building educational programs for beauty skills of non-experts. In this study, the characteristics of experts were indicated based on video recording, eye tracking technology, and interview survey from experts.

Keywords: Based on eye tracking technology · Skill in beauty services

1 Introduction

To be a hairdresser in Japan, it is necessary to take the national exam of beauty license and pass it. The qualification to take this national exam is having completed the course for a hairdresser at a training facility. Also, those who want to be a hairdresser need to attend a beauty school. The national exam for the mastery of basic skills in beauty technology is examined by technical tests and written tests (Japan Center for Hairdressing and Beauty Education, 2009).

Therefore, regarding the basic skills, passing the national exam will prove the acquisition of beauty skills. However, in order to work as a hairdresser, it is not enough to only pass the national exams, it is also necessary to undergo training in an actual beauty salon. Generally, after about 2 to 3 years of training, the trainee will be a qualified to work as hairdresser.

In Japan, the rate of turnover is 70 to 80% during 3 years after entering a beauty salon. This training is done for each beauty salon, and an expert hairdresser gives guidance. In order to solve the problem of the high rate of turnover in beauty salons, it

© Springer Nature Switzerland AG 2020
W. Karwowski et al. (Eds.): AHFE 2019, AISC 971, pp. 219–224, 2020.
https://doi.org/10.1007/978-3-030-20494-5_20

is necessary to analyze the beauty technology and develop a program to assist in the training of beauty techniques, which is useful for each beauty salon.

In this study, we analyze the techniques of experts and non-experts as a hairdresser and compared their skills. The work process was divided into two processes, the process of arranging hair (Preparatory work for cutting), and the process of cutting the hair with scissors (cutting operation). The first purpose of this research is to clarify the differences between the beauty techniques of experts and non-experts. The second purpose is to provide the findings to assist in the development of building educational programs for beauty technology of non-experts.

2 Method

2.1 Equipment Used

Tobii Pro Glasses 2 were used to record the movement and work process of the subjects point of focus during work.

2.2 Experimental Method

Subjects were A: non-expert (0 year after beauty school), B: non-expert (5 years of work experience in a beauty salon), C: expert (15 years of work experience in a beauty salon). Tobii Pro Glasses 2 was attached to each subject, and cutting of a doll's hair (Fig. 1) was recorded. In advance, three subjects confirmed the common style (Fig. 2). And three subjects used same instrumentals (Fig. 3). In Fig. 3 "B1", "B2" were scissors, "S1", "S2", "S3" were thinning scissors, "H" were hair clips, "C" were combs.

Fig. 1. The doll for cutting hair.

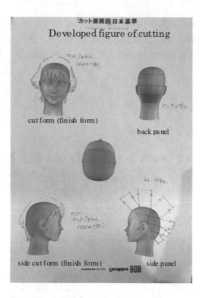

Fig. 2. Developed figure of cutting.

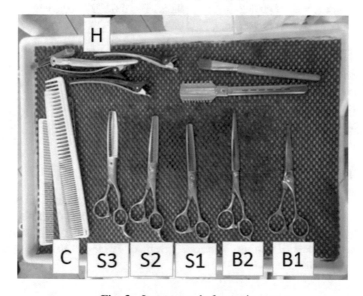

Fig. 3. Instrumentals for cutting.

2.3 Interview Survey

The interview survey was conducted from experts (15 years of work experience in a beauty salon).

In this interview survey, the characteristics of experts were investigated based on video recording, eye tracking technology. Moreover, in this interview survey the results of analyze of skill was confirmed based on video recording, eye tracking technology.

3 Results

3.1 Comparison of Work Process

Table 1 shows comparison of work process between expert and non-experts.

In the working time, the total working time and the working time of preparations of C is shorter than A and B. However in the working time of cutting, A and C are shorter than B. In the number, the number of cutting and preparations of B and C is more than A. In the average time of cutting, there was no difference between the three subjects. However in the average time of preparations, B and C are shorter than A.

Table 1. Comparison of work process between expert and non-experts

	A: 0 year	B: 5 years	C: 15 years
Total working time	0:34:37	0:34:25	0:32:47
Working time of cutting	0:05:14	0:07:02	0:05:49
Working time of preparations	0:29:23	0:27:23	0:26:57
Number of cutting	133	251	225
Number of preparations	132	250	224
Average time of cutting	0:00:02	0:00:02	0:00:02
Standard deviation of cutting	0:00:02	0:00:01	0:00:01
Average time of preparations	0:00:13	0:00:07	0:00:07
Standard deviation of preparations	0:00:14	0:00:11	0:00:12

3.2 Eye Tracking Technology

From the results of eye tracking technology, Subject C who is an expert tended to look closely at the hair at the starting part of cutting and ending part of cutting. However, Subject A and B who are non-experts tended to focus on the hair from starting part of cutting to ending part of cutting. Moreover, Subject A tended to gaze upon various parts of the hair during cutting.

3.3 Interview Survey

In the interview survey, it is indicated that the work of cutting for beauty service can be classified to three working sentences; Working to create a design outline, Working to thin out the hair, and Working to complete the design.

From the results of cutting based on eye tracking technology, in the working of creating a design outline, Subject C tended to look the starting part of cutting and ending part of cutting. However Subject A and B tended to maintain focus from starting of cutting to ending of cutting.

About this, an expert said following. A Nun-expert does not have sufficient skills on how to use scissors, so they need to look carefully during the entire procedure. Experts have sufficient skills on how to use scissors, so it is enough to check the starting part of cutting and ending part of cutting. Instead of maintain focus during cutting, the expert could prepare for the next cutting work by gazing at the surroundings. As results, in the outline creation work, experts can proceed quickly, so they could use time to prepare for activities.

4 Discussion

4.1 The Difference Between Expert and Non-expert Skill

This study indicated the difference between expert and non-expert skills. In the working time, the time spent cutting and preparing of expert is shorter than non-experts. It can be achieved through the skills of cutting and of predicted for cutting. This was confirmed by analysis based on eye tracking technology and interview survey.

Based on the above results, it is useful for non-expert to educate their skills of cutting and of planning skills for cutting.

Moreover, to assess the status of skills of non-experts, it could be said that analysis based on eye tracking technology is useful.

4.2 Eye Tracking Technology for Beauty Service

This study indicated that expert has skills of pre-meditating the cutting. An expert can predicted the process from start to end of beauty service. Moreover, an expert can understand the necessary work required from the prediction. This skill is acquired through experience.

Based on the above results, in the education for non-experts, the video of experts' working is beneficial. And the video of eye tracking technology is furthermore beneficial for educating non-experts.

In order to solve the problem of the high rate of turnover in beauty salons, we have to collect video of beauty service using eye-tracking technology. In the future study, we want to collect a variety of video and we want to provide the video as programs of education for non-experts of beauty service. Moreover, through the accumulation of data and analysis, we take aim to provide educational support system using augmented reality that fuses virtual space and real space.

5 Conclusion

In this research, we analyze the techniques of experts and non-experts as a hairdresser and compared their skills based on eye tracking technology.

This study indicated the difference between expert and non-expert skills. This results could indicate the difference between expert and non-expert skill in beauty service. Furthermore, It is expected that these new and important findings on the difference between expert and non-expert based on video recording, eye tracking technology, and inter-view survey from expert, could to contribute for education in beauty service.

However, this research could identify only three subjects persons. Therefore, the applicability of these findings to understanding expert could be limited. It is suggested that future studies should include more subject persons of experts in beauty service.

In future study, we have to research more subject person of experts in beauty service. Moreover, we would like to investigate the beauty techniques adapted to the type of hairstyle and the nature of the hair.

Reference

1. Japan Center for Hairdressing and Beauty Education, "Related regulations/systems" (2009)

Evaluation of "Jiai" of Large "Echizen Washi" Japanese Traditional Paper

Hiroyuki Nakagawa[1]([✉]), Yuji Kawamori[1], Akihiko Goto[2],
Hiroyuki Hamada[3], Kazuaki Yamashiro[3], Naoki Sugiyama[3],
Mitsunori Suda[4], Kozo Igarashi[5], and Yoshiki Yamada[6]

[1] Taiyo Corporation, 34-6 Koizumi-Cho, Hikone City, Shiga 522-0043, Japan
imagineerl38@gmail.com
[2] Osaka Sangyo University, 3-1-1 Nkagaito, Daito City, Osaka 574-8530, Japan
[3] Kyoto Institute of Technology, Matsusaki, Sakyo-ku, Kyoto 606-8585, Japan
[4] Suda Shoten Corporation, 1-16-1 Kawakita, Fjiidera City,
Osaka 583-0001, Japan
[5] Igarashiseishi Corporation, 12-14 Iwamotocho, Echizen City,
Fukui 915-0233, Japan
[6] Nishinosyoukai Corporation, 4-7-6 Sadatomocho, Echizen City,
Fukui 915-0231, Japan
https://www.f-taiyo.jo.jp

Abstract. In the paper of the world, there is a word "Jiai" the evaluation criteria of the paper. It is expressed as "Jiai is good" a good paper. Fiber dispersion state of the "Jiai", Unevenness is felt visually when the light is transmitted, or it refers to the characteristic expressions such as uniformity of the paper. In general, the Japan Federation of Printing Industries" Symmetry degree of fibrous tissue that make up the paper. Look through the paper on the entire surface is a good thing symmetry "Jiai". That there is a partial thickness thin non-uniformity is bad "Jiai". It is defined as. Among the evaluation of "Echizen Washi" Japanese traditionalpaper, the total thickness of the paper refers to the uniformity of such density. In this experiment, the method of evaluation of handmade "Washi" Japanese traditional paper, Comparative study the evaluation criteria, Whether to the evaluation of what kind of Japanese paper by evaluators for the purpose.

Keywords: Echizen Washi · Japanese traditional paper ·
Washi Japanese paper · Handmade Japanese paper · Jiai

1   Introduction

In the paper-making world, there is a word "Jiai" on the evaluation standards of papers. A good paper as "Jiai is good". "Jiai" is felt visually when the dispersion state of the fiber and the light are transmitted. Unevenness is felt visually when the light is transmitted, or it refers to the characteristic expressions such as uniformity of the paper. In the Association of Japan Printing Industry, Symmetry degree of fibrous tissue that make up the paper. Look through the paper on the entire surface is a good thing symmetry "Jiai". That there is a partial thickness thin non-uniformity is bad "Jiai". It is defined as. Among the evaluation of "Echizen Washi" Japanese traditional paper.

© Springer Nature Switzerland AG 2020
W. Karwowski et al. (Eds.): AHFE 2019, AISC 971, pp. 225–232, 2020.
https://doi.org/10.1007/978-3-030-20494-5_21

The total thickness of the paper refers to the uniformity of such density. In this experiment, the method of evaluation of handmade "Washi" Japanese traditional paper, Comparative study the evaluation criteria, Whether to the evaluation of what kind of Japanese paper by evaluators for the purpose.

2   Method

2.1   Outline of Experiment

This experiment was carried out papermaking in the operation analysis of large-format paper. August 29, 2018, in Fukui Prefecture Echizen City Corporation Igarashi papermaking workshop, was evaluated of Echizen.

2.2   Test Target

In this experiment, the subject test Echizen handmade large-format paper. Subject to ultra-skilled persons, unskilled and inexperienced persons. Ultra-skilled person is a qualified person of the traditional craftsmen. Handmade work position, shown in Table 2. In each combination, it was handmade permutation large format paper in each of the left and right working position. Ten out of the 22 sheets were extracted at random. The large-format paper of 2,200 mm × 1,500 mm, cut to 350 mm × 300 mm, It was

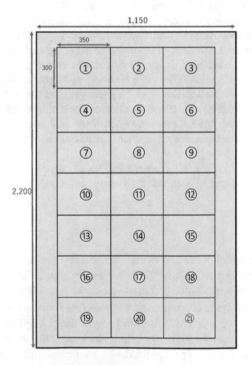

Fig. 1. Large Echizen washi test pieces

the 21 test pieces. In addition, three pieces from among the Using the ⑤ and ⑪ and ⑰ shown in Fig. 1. 3 pieces 10 sheets ×, it was carried out a total of 30 sheets of evaluation (Table 1).

Table 1. Examinee

	Sex	Year of experience	Dominant arm	Licence
A	Men	52 years	Right	
B	Women	46 years	Right	Traditional craftsmen
C	Men	30 years	Right	
D	Women	23 years	Right	Traditional craftsmen
E	Men	0 years	Right	

Table 2. Examinee combination

	No. 2	No. 5	No. 8	No. 11	No. 14	No. 17	No. 19	No. 20	No. 21	No. 22
組合せ	D–C	C–D	D–B	B–D	B–C	C–B	D–A	A–D	E–D	D–E

The alphabet on the left is the work indicator

2.3 Evaluation Participants

In this experiment, we were evaluated by three people from Igarashi Paper Corporation, which is engaged in the paper industry in Echizen City as the standpoint of the Japanese paper maker, and the Japanese paper wholesale business is managed as a standpoint that sells Washi paper 1 person from Nishino Shokai Co., Ltd., 1 person from Taiyo Corporation, one from the contractor as a standpoint of using washi paper. I had the cooperation of one university student of the Japanese painting major who had visited Igarashi paper by an intern. The evaluation person is shown in Table 3. Since three of the members of the evaluation have made experimental specimens, it was evaluated so that they did not know who made the work paper.

Table 3. Rater

Participants	Gender	Occupation	Licence
A	Men	Paper industry	
B	Women	Paper industry	Traditional craftsmen
D	Women	Paper industry	Traditional craftsmen
F	Men	Paper wholesale	
G	Men	Building contractors	
H	Men	Paper wholesale	
I	Women	Japanese painting major college students	

2.4   Evaluation Method and Interview

The evaluation experiment of washi paper in this experiment was evaluated in three phases of the test sample from the upper part of the same day in the same condition and the superior, good, and acceptable at the same date. In the top evaluation of Yu's paper, we evaluated the ranking further.

From the evaluated subjects, we conduct a hearing survey of the selected Standard, and compare the evaluation method of Japanese paper of experienced person and inexperienced person. The evaluation method is shown in Fig. 2.

Fig. 2.   Evaluation

2.5   Paper Evaluation

Table 4 and Fig. 3 graph show the evaluation results of producers of Japanese paper. Table 5 and Fig. 4 graph show those of sales people, and Table 6 and Fig. 5 graph show those of users. Figure 6 graph shows the total number of respondents of the above three groups, who evaluate as the top rank (so-called "Yu" in Japan).

Table 4. Paper evaluation (manufacturer)

Selecter	A			Selecter	B			Selecter	D		
	Excellent	Good	Fair		Excellent	Good	Fair		Excellent	Good	Fair
①	8-⑪	14-⑰	14-⑤	①	8-⑪	17-⑰	17-⑤	①	20-⑤	17-⑤	5-⑪
②	20-⑰	14-⑪	11-⑤	②	20-⑰	2-⑤	11-⑪	②	8-⑪	17-⑰	22-⑤
③	8-⑰	11-⑪	8-⑤	③	5-⑤	2-⑰	17-⑪	③	5-⑤	8-⑤	11-⑪
④	21-⑤	5-⑪		④	22-⑰	8-⑤	2-⑪	④	22-⑰	19-⑤	2-⑪
⑤	20-⑪	2-⑪		⑤	19-⑤	21-⑤	14-⑰	⑤	22-⑪	11-⑤	17-⑪
⑥	19-⑰	21-⑰		⑥	11-⑰	20-⑪	5-⑪	⑥	19-⑰	19-⑪	14-⑰
⑦	20-⑤	19-⑪		⑦	22-⑤	8-⑰	11-⑤	⑦	21-⑰	20-⑰	14-⑤
⑧	22-⑰	17-⑪		⑧	20-⑤	22-⑪	5-⑰	⑧	21-⑤	5-⑰	14-⑪
⑨	22-⑤	2-⑰		⑨		19-⑰	14-⑪	⑨	11-⑰	20-⑪	
⑩	2-⑤	5-⑰		⑩		19-⑪	14-⑤	⑩	2-⑤		
⑪	22-⑪	17-⑰		⑪		21-⑰		⑪	2-⑰		
⑫		11-⑰		⑫		21-⑪		⑫	19-⑪		
⑬		21-⑪		⑬				⑬	8-⑰		
⑭		19-⑤		⑭				⑭			
⑮		17-⑤		⑮				⑮			
⑯		5-⑤		⑯				⑯			

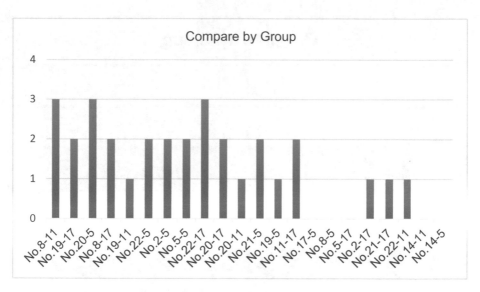

Fig. 3. Paper evaluation (manufacturer)

Table 5. Paper evaluation of the (seller)

Selecter	F			Selecter	H		
	Excellent	Good	Fair		Excellent	Good	Fair
①	20-⑤	5-⑤	21-⑪	①	8-⑪	14-⑰	11-⑤
②	20-⑰	21-⑤	11-⑪	②	8-⑰	14-⑪	2-⑪
③	8-⑰	19-⑤	14-⑪	③	19-⑪	20-⑪	22-⑰
④	19-⑪	8-⑪	14-⑤	④	11-⑰	20-⑤	21-⑰
⑤	20-⑪	22-⑤	11-⑰	⑤	19-⑤	20-⑰	5-⑪
⑥	19-⑰	22-⑪	17-⑰	⑥	17-⑤		22-⑪
⑦	8-⑤	14-⑰	17-⑪	⑦	14-⑤		5-⑤
⑧	2-⑤	22-⑰	2-⑪	⑧	19-⑰		2-⑤
⑨	17-⑤	11-⑤	5-⑪	⑨	8-⑤		22-⑤
⑩		22-⑰		⑩			21-⑪
⑪		5-⑰		⑪			22-⑰
⑫		21-⑰		⑫			17-⑰
⑬				⑬			17-⑪
⑭				⑭			21-⑤
⑮				⑮			5-⑰
⑯				⑯			22-⑪

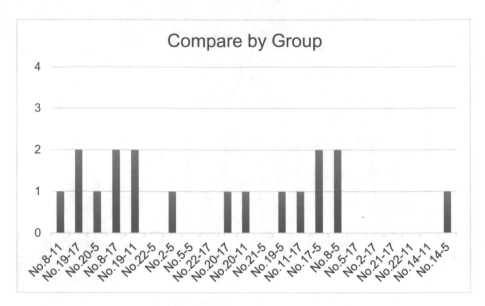

Fig. 4. Paper evaluation (seller)

Table 6. Paper evaluation (usage)

Selecter	G			Selecter	I		
	Excellent	Good	Fair		Excellent	Good	Fair
①	19-⑤	8-⑤	11-⑪	①	5-⑰	8-⑤	17-⑰
②	19-⑰	21-⑪	5-⑪	②	17-⑤	11-⑤	19-⑪
③	19-⑪	2-⑤	2-⑪	③	8-⑪	22-⑪	5-⑪
④	5-⑤	22-⑪	17-⑪	④	20-⑤	20-⑰	14-⑤
⑤	21-⑤	21-⑰	17-⑰	⑤	22-⑤	21-⑤	14-⑰
⑥	22-⑤	22-⑰	14-⑤	⑥	20-⑪	17-⑪	8-⑰
⑦	5-⑰	20-㊲	11-⑰	⑦	2-⑤	19-⑤	11-⑰
⑧	2-⑰	14-⑰	11-⑤	⑧	5-⑤	2-⑰	21-⑪
⑨	8-⑪	8-⑰	17-⑤	⑨	19-⑰		11-⑪
⑩	14-⑪	20-⑤	20-⑪	⑩			21-⑰
⑪				⑪			22-⑰
⑫				⑫			2-⑪
⑬				⑬			14-⑪
⑭				⑭			
⑮				⑮			
⑯				⑯			

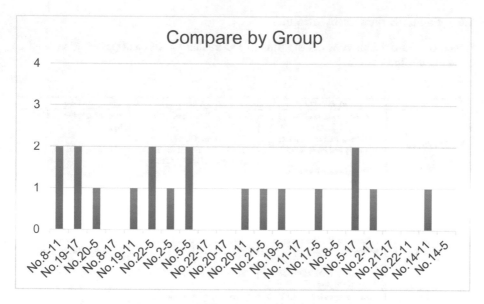

Fig. 5. Paper evaluation (usage)

The highest rated paper in Yu was the result of no, 8 and no, 19. All of these two washi papers were evaluated by the students. Next, the evaluation of the A, B, D, and F which is the maker and the seller showed the result of a roughly same evaluation.

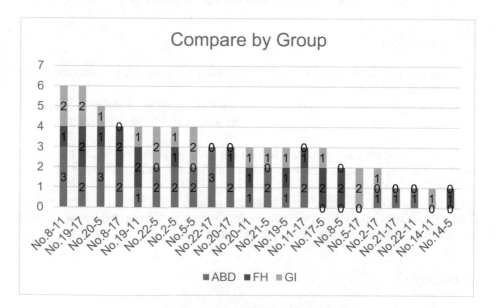

Fig. 6. Paper evaluation by group

2.6   Paper Evaluation by Group

It was carried out interviews at the time of the evaluation of each paper. It showed its contents to Fig. 7.

	Manufacturer	Seller	User
①How do you see the paper?	·Looking watermark to light. ·We have seen in the paper vertically and horizontally. ·When viewed through the paper on, looking at the average degree of fiber density.	·Looking watermark to light. ·When viewed through the paper on, looking at the average degree of fiber density.	·Looking watermark to light. ·Looking in hand.
②Please tell me the evaluation criteria	·And make sure that they are uniformly throughout the. ·Check whether the plow up without plaques. ·Evaluated in the whole of the fibers of the flow.	·Make sure that they are uniformly throughout the. ·It has been evaluated by the flow of fiber. ·Evaluated in thickness.	·Have confirmed the thickness of the paper. ·It has been evaluated by the uniformity of appearance.
③Otherwise	·Come also changed sentiment in the difference of the materials used. ·Look watermark, have confirmed the thickness, and the like of suspicious I thought points.	·Handmade, even whether or not the hungry and the gloves, the difference appears in the paper. ·You can see the difference just with the Japanese paper is.	

Fig. 7. Hearing about the paper evaluation

3   Conclusion

In this experiment, when the Japanese paper is evaluated, everyone confirms the Washi paper by watermarking to light.

However, the manufacturer, the seller has confirmed the inner surface of the paper (fiber dispersibility) from the face of Washi, the person is determined to emphasize the external viewpoint.

When interviewing the place where the manufacturer is looking at the watermark, the fiber density, fiber is not become plaques, it is confirmed whether the fibers are uniform.

In the future, we will study the relationship between the behavior analysis and the formation of the production method and clarify the criteria of the manufacturer (expert) and strive to elucidate good washi paper and good soil.

References

1. Iwao Saito: The story of Echizen Washi 1973
2. The Mainichi Nespaper: Handmade Japanese Traditional Paper 1975

Three-Dimensional Motion Analysis of Mochi Pounding

Akihiko Goto[1], Naoki Sugiyama[2], Daigo Goto[3], Tomoko Ota[4(⊠)], and Hiroyuki Hamada[2]

[1] Osaka Sangyo University, Nakagaito, Daito, Osaka 574-8530, Japan
gotoh@ise.osaa-sandai.ac.jp
[2] Kyoto Institute of Technology,
Matsugasaki, Sakyo-ku, Kyoto 606-8585, Japan
nsugiyama.u@gmail.com, hhamada@kit.ac.jp
[3] Yuzen, 2-14-22purezidentoko-To3F, Yamamotodori,
Chuo-ku Kobe-shi, Hyogo 650-0003, Japan
goto@cold-storage.jp
[4] Chuo Business Group, Funakoshi-cho, Chuo-ku, Osaka 540-0036, Japan
promotl@gold.ocm.ne.jp

Abstract. Mochi is Japanese traditional rice cake which is very close relation to Japanese persons daily life and culture. Steamed rice is subjected impact energy by wooden hammer. Properties finished Mochi were different between that made by expert and non-expert. Motion analysis of Mochi pounding was conducted. The hammer motion was analyzed. Stick pictures were obtained as basic data.

Keywords: Rice cake · Motion analysis · Hammer

1 Introduction

Rice cake, Mochi in Japanese, is a food material which is close connection with Japanese rice culture and faith. Kojiki, Japan's oldest history book, mentioned that is a traditional Japanese food. Rice cake also has been accepted as a gift to the gods. Rice cake making, Mochi Tsuki in Japanese, has been historically carried out as a religious event. Particularly, Mochi Tsuki is an event in New Year and celebrations, so that a joint event in a large family or local community gathering.

It is said that taste of Mochi is different by the Mochi Tsuki technique expert and non-expert. Therefore, in this research, measurement was carried out by three-dimensional motion analysis using a motion capture system with the aim of recording and quantifying the Mochi Tsuki motion for preserving Japanese culture and preserving technology.

© Springer Nature Switzerland AG 2020
W. Karwowski et al. (Eds.): AHFE 2019, AISC 971, pp. 233–243, 2020.
https://doi.org/10.1007/978-3-030-20494-5_22

2 Method

2.1 Test Subjects

Mochi Tsuki is carried out in pairs of "hands Tsuki Te" and "pair of hands AinoTe". The former pours rice cakes with hammer and the latter turns rice cakes according to movement of hands. In this research, hammer motion was analyzed such as raise up / swing down. The subject was a total of 4 "hands", one expert and three non-expert people with experience years 0 years. The expert was 62 years old of male, 43 years' experience of Mochi Tsuki, 180 cm height, 82 kg weight and Right of dominant hand. Three non-expert people were male and right hand dominant, and 180 cm, 181 cm and 173 cm height, and 74 kg, 55 kg and 62 kg weight respectively. For "pair of hands", experts with experience years equivalent to those of the expert were in charge of turning rice cakes in all measurements.

2.2 Measurement Conditions

In the experiments the same tools, the hammer /dies, were used. The hammer was made of wood and the die was made of stone. Also, glutinous rice was steamed for each subject's measurement. The subjects were instructed to continue the work from the beginning to the end. The end means the completion of rice cakes. The timing of the end of Mochi Tsuki was determined by "pair of hands", when he felt that the rice cake was completed during the series of work.

2.3 Measurement Method

Three-dimensional motion analyzer (MAC3D SYSTEM, manufactured by Motion Analysis Co., Ltd.) and six infrared cameras (Raptor-H, manufactured by Motion Analysis) were used for the measurement of Mochi Tsuki motion. Infrared reflective markers were attached to the subjects on the head, both shoulders, both elbows, both wrists, both hips, both knees, both ankles, and the back as shown in Fig. 1 [1]. As shown in Fig. 2, an infrared marker was affixed to the hammer and die used for this measurement. The positional information was measured. In addition, since infrared markers could not be affixed to the striking surface, infrared markers were affixed to the hammer face which was in a fixed state beforehand. The experimental views are shown in Fig. 3.

Fig. 1. Position of infrared reflection markers [1]

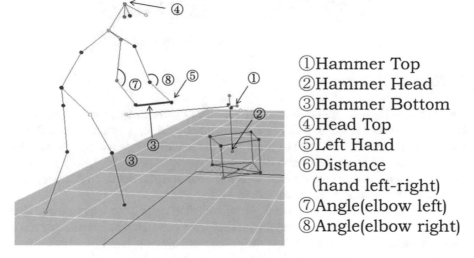

①Hammer Top
②Hammer Head
③Hammer Bottom
④Head Top
⑤Left Hand
⑥Distance
  (hand left-right)
⑦Angle(elbow left)
⑧Angle(elbow right)

Fig.2. Maker position of hammer and die

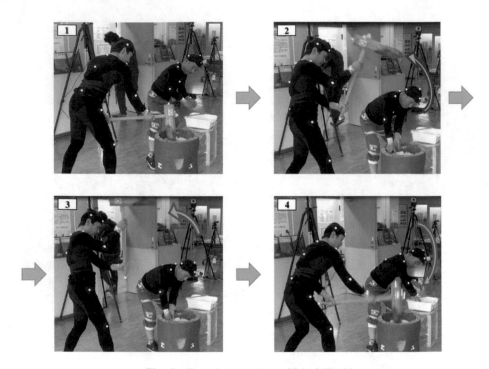

Fig. 3. Experimental view of MochiTsuki

3 Results

First, the working time from start to finish, number of strokes of punch were measured. For the expert working time was 138 s and number of strokes was 55 times. For the non-expert person1 297 s and 114 times, for the non-expert person2 376 s and 146 times, and for the non-expert person3 321 s 147 times respectively. Compared to the three non-expert people, the expert's working time and the number of strokes were short and less frequent, which were almost half.

Figures 4, 5, 6 and 7 show stick picture obtained from the experiments. Each makers position was measured, and the makers were connected to indicate posture of the participants. Figures 4 and 5 show the posture of swing up, and Figs. 6 and 7 show that of swing down. Figures 8 and 9 show the trajectory of hammer. Comparing these figures gives us an interesting awareness. For example, when comparing Figs. 4 and 5, it can be seen that the position of the hammer when swung up is different and the angle to the die was also different. In the comparison between Figs. 6 and 7, this is a stick picture at the time of swing down. For example, the position of the waist was different. Non-expert people may say that their waist was falling. To the contrary, expert has waist near the die, waist was coming forward. Examine the trajectory of the hammer in

Figs. 8 and 9. The non-expert seems to have a large trajectory drawing an arc. Expert the hammer goes down towards Die straightly and it seems to be headed for a short distance. Furthermore, as described above, it is well understood that the position of the waist is different between the expert and the non-expert. Using such a stick picture, qualitative evaluation was possible. In the following sections, quantitative evaluation was performed.

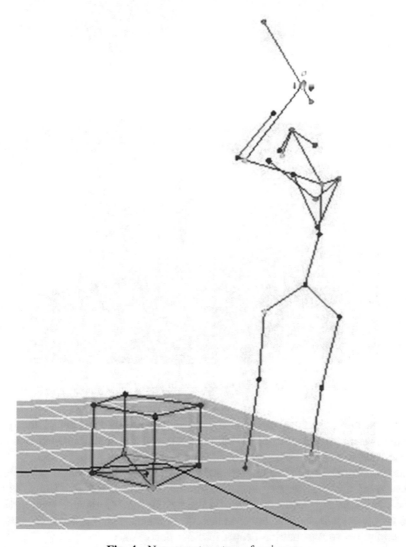

Fig. 4. Non-expert posture of swing up

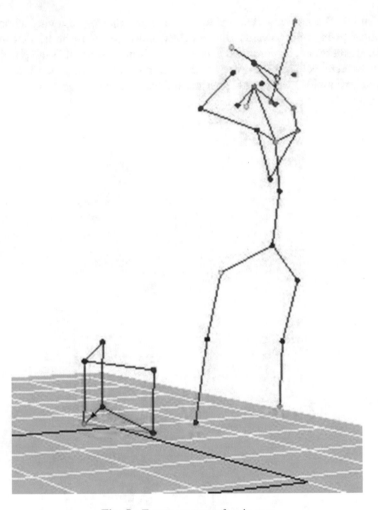

Fig. 5. Expert posture of swing up

In this section, three values were discussed; the lowest position of the hammer head, the maximum position of the hammer top and the maximum position of the hammer when it was swinging down.

The lowest position of the hammer head was described. This position was measured by the expert and non-expert swing down the hammer process. The expert was 322 mm, and non-expert was 327 mm, 345 mm, 338 mm. Compared with three non-expert people, expert showed lower values. This means that the hammer which the expert swung down was putting deeper into the rice cake.

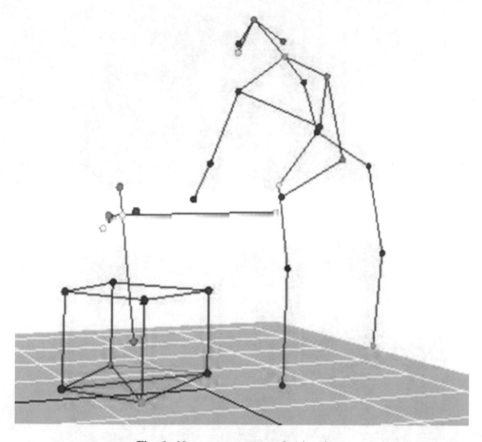

Fig. 6. Non-expert posture of swing down

It was confirmed that the position of the hammer was lowered by completely sprinkling and shifting to the swing down motion after reaching the highest position once swinging up the hammer. Therefore, in discussing applied energy to the die we need both the maximum position of the hammer top and the maximum position when swinging down. The maximum position of the hammer top was measured. The expert was 2087 mm, and non-expert was 2141 mm, 2065 mm, 1868 mm. The expert showed higher values than non-expert 2 and 3, but compared with non-expert 1, it was confirmed that the hammer was positioned higher than the expert.

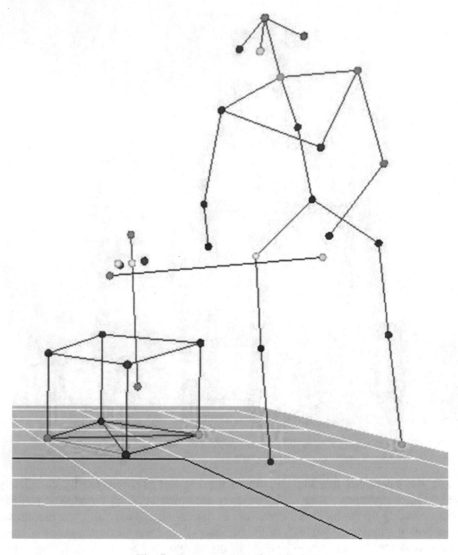

Fig. 7. Expert posture of swing down

The maximum position when swinging down. The expert was 2096 mm, and non-expert was 1976 mm, 2065 mm, 1868 mm. As compared with the non-expert at the time of swinging down, the expert knows that the pencil is being swung down from a higher position. The small waveform occurring before the high waveform shows the acceleration at the time of swinging up but the small waveform generated after the high waveform shows the action of pencil breaking immediately after hitting the punch against the rice cake.

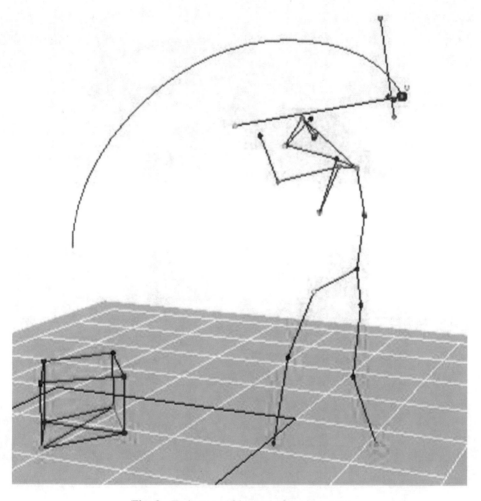

Fig. 8. Trajectory of hammer for non-expert

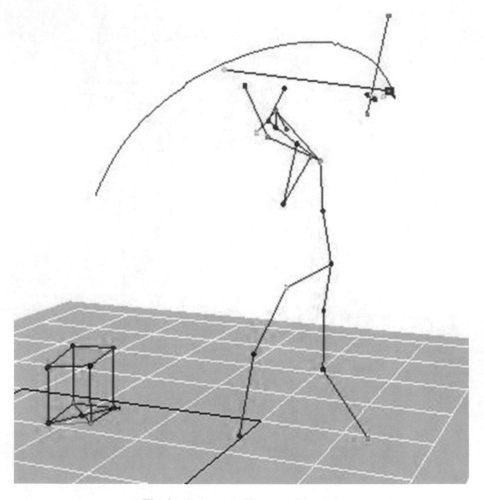

Fig. 9. Trajectory of hammer for expert

4 Discussion and Conclusion

In this study, the work time until the skilled worker finished the rice cake was short, and the stroke number of the punch was small. As shown by the lowest position of the striking face, it is thought that the expert made digging in the rice cake more deeply, and it was thought that the stronger rice cake was contributing to this. Regarding how to firmly add rice cake, it is thought that influence of high punch tip acceleration of expert and higher punch tip height when swinging than in non-expert person. More quantitative analysis such as velocity, acceleration is needed.

Reference

Matsumuro, K., Kobayashi, T., Takai, Y., Goto, A.: Hiroyuki Hamada: three-dimensional motion analysis of mochi pounding. Jpn. Soc. Mech. Eng. (2015)

Organizational Culture Requirements for the Achievement of World Class Manufacturing

Zbigniew Wisniewski[✉], Michal Paszkowski,
and Malgorzata Wisniewska

Faculty of Management and Production Engineering,
Lodz University of Technology, ul. Wolczanska 215, 90-924 Lodz, Poland
{zbigniew.wisniewski,
malgorzata.wisniewska}@p.lodz.pl,
michpasz@interia.pl

Abstract. In the situation where organizations have become participants in the global competition war and are looking for different ideas to allow them to prevail, one of such ideas is World Class Manufacturing (WCM) which requires improvement of quality, cost, and order/service lead time [1]. On the other hand, however, what complicates the matter is the concept of culture since it is difficult to define but also crucial for success.

That article is an attempt to present the World Class Manufacturing approach and what it requires of the organization. Additionally, the concept of culture is discussed as are some of its important elements to consider. In conclusion, those elements of culture that are the most relevant for WCM are identified.

The article is a part of the co-author's dissertation research which in its empirical section provides an analysis of companies with a variety of national backgrounds but in the same industry with respect to the influence of certain factors on progressing towards the implementation of World Class Manufacturing in the organization. The analysis was performed in the SYDYN research group.

Keywords: Organization culture · National culture · Values · Heroes · Rituals · Values and assumptions · Cultural levels · Artifacts · World Class Manufacturing

1 Background Information

The advancement of humanity was possible due to the formation of groups and undertaking collective efforts. As early as in the ancient history of China, Egypt, Greece, India, and Mesopotamia, organizational processes are evident in the constitution of hierarchies, establishment of rights and obligations [2]. The Industrial Revolution [3] was a transitional event which marked a shift from the economy based on agricultural production, handicraft, small manufactures to mechanized factory production on a large industrial scale [4]. Organizational units emerged as a result of the following three factors: education, specialization, and technology [2].

© Springer Nature Switzerland AG 2020
W. Karwowski et al. (Eds.): AHFE 2019, AISC 971, pp. 244–253, 2020.
https://doi.org/10.1007/978-3-030-20494-5_23

From this history transpires a developmental trend that could practically confirm Stefan Szuman's claim that a human is a being who builds itself, creates its own design, and, most importantly, strives to achieve goals that are beyond its capabilities [5]. However, humans do not act in these processes on their own but in a group, for as S. Jobs, the founder of Apple Inc., once said: *"Great things in business are never done by one person. They're done by a team of people"* [6]. Therefore, in order to understand fully how this reality functions today and to be able to find one's place in it, it is necessary to understand how people function in a group - something that is best described by reference to culture, where culture is what personality is to an individual [7].

It needs to be pointed out that nowadays, cooperation between people is strongly determined by the two phenomena: technological advancement and globalization.

As for technological revolution, which began with the Industrial Revolution, its effect on the cooperation is such that the time required for communication has been significantly reduced and the physical distance has become a less relevant factor in seeking cooperation, cooperating, and expanding operation of the organization.

Globalization on the other hand comes into play when individuals of different nationalities, speaking different languages, practicing different religions, upholding different traditions are brought together in an organization. Additionally, organizations established in a certain country but carrying out global projects are compelled to operate in different from the native cultural realities. The co-author's collaboration during the MBA program may serve as an interesting illustration of the complexity of the problem: an international team was expected to prepare an implementation of Management By Objectives for an Italian company operating in Poland where the team members came from Yemen, Nigeria, Kenya, Poland, and Tanzania. The collaboration brought to light differences in the process of arriving at collective solutions and in time management during the project completion. Furthermore, were these globalization processes to be coupled with ambitions for growth, one could come to believe that concepts such as World Class Manufacturing are nothing out of the ordinary and are simply a crystallization of those ambitious goals that Szuman writes about [5]. Nonetheless, whether the WCM concept can be effectively applied in any cultural reality is a cause for a deliberation. Some controversy arises from the fact that a few authors including G. Hofstede, G.J. Hofstede, M. Minkov [8] cast doubt on the possibility of transplanting Japanese management methods that like the Toyota Production System tally with the concept of World Class Manufacturing (WCM) [9].

This article provides a review of the literature on the subject matter of organizational culture requirements in the implementation of World Class Manufacturing, and a starting point for empirical research in this respect.

2 World Class Manufacturing

The term 'world-class' was first coined in 1984 by R.H. Hayes and S.C. Wheelwright [9, 10], who presented a comparison of Japanese and German organizations to American organizations in order to develop guidelines for the achievement of a world-class level of production equal to theirs and to prevail over Japanese and German competitors [11]. The term was popularized in practice by Hajime Yamashina, professor at Kyoto University, member of the Royal Swedish Academy of Engineering Sciences and of the Royal Society of London [9], who, among other things, adapted the WCM concept to the requirements of the Fiat Group [12], the eighth largest car manufacturer in the world [13].

Organization which develop their own models based on the generic concept aim at outstripping others in terms of quality, costs and lead times [9]. WCM seeks the highest level of excellence of international production, accomplishment of the four zeroes, that is zero waste (unnecessary motions, wait, shortages, inventory, and transportation, overproduction, irrelevant processes and improvements), zero injuries, zero defects/faults/deficiencies, zero breakdowns [9].

WCM applies methods presented in Fig. 1. World Class Manufacturing in Fiat Group Automobiles Source: [14] and aligned with the objectives discussed by Piasecka-Głuszak [9].

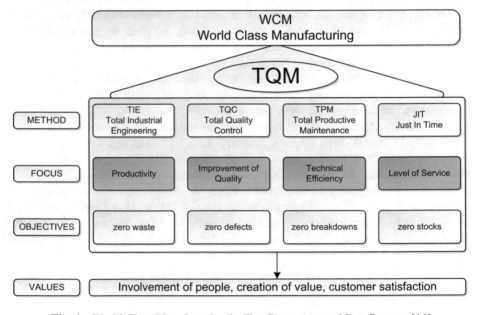

Fig. 1. World Class Manufacturing in Fiat Group Automobiles. **Source:** [14]

TQM, TPM, JIT, together with Employee Involvement and Simplicity are the core component parts of the WCM model developed by R.J. Schonberger [14].

A definition of TQM is provided in ISO 8402 and it draws attention that this method of organizational management, focused on quality, requires the participation of all members of the organization for sustainable success [15, 16]. However, success comes when the customer is happy and when members of the organization and the society also reap benefits of organizational management [17 as cited in 18]. A TQM approach demands that the focus be on the system where 94% of problems have their source [J.M. Deming as cited in 19].

Total Productive Maintenance (TPM) refers to the management of plant equipment. Its objectives include zero breakdowns, defects, injuries, and shortages that could be attributed to the operation of a machine [20]. All of them have a direct bearing on the quality and efficiency of work [21, 22].

Total Industrial Engineering (TIE) means industrial management to reduce work overload: in Japanese Muri (unnatural operation), Mura (irregular operation), and Muda (non-value-added operation), all of which negatively impact on the effectiveness of processes [9].

Calvasina et al. [23] define Just in Time (JiT) as a production control system directed at the minimization of raw materials and work-in-progress. This system clearly refers to Lean Manufacturing (LM), a concept concentrated on the elimination of defects, stabilizing production, streamlining processes. Also important in JiT are flexible workers with extensive knowledge and a wealth of experience.

WCM is based on technical and managerial pillars [9, 12, 14]. Piasecka-Głuszak claims [9] that the managerial pillars provide guidelines for the implementation of the technical pillars, whereas each of the technical pillars is divided into what is called steps spelled out in terms of objectives. The following technical pillars and their corresponding managerial pillars [9, 14] - evidently deep-rooted in the Toyota Production System - can be distinguished [9, 24]:

- safety-hygiene & workplace organization (technical pillar)-
  management commitment (managerial pillar)
- cost deployment
  clarify objectives
- focused improvement-
  route map to WCM
- professional maintenance-
  allocation of highly qualified people to model areas
- quality control
  organization commitment
- logistics and customer service
  competence of organization towards improvement
- early equipment and product management
  time and budget
- people development
  level of detail
- environment management
  motivation of operators

The implementation has three major phases. The first phase - reactive, is a response to emerging problems known also as 'narrow escapes'; the second - preventive phase concentrates on the analysis of problems and preventing their occurrence; and the third - proactive phase relies on risk analysis and taking preventive actions. Activities and projects are assigned to each of the phases [9].

At the outset of a WCM implementation in an organization, activities are only undertaken in a designated area of the organization with an especially appointed team of people to develop optimal standards and practices to be finally implemented across the entire organization. To monitor the progress towards WCM implementation internal and external audits of individual pillars are performed during which they are rated by the award of points, for example on a scale from 0 to 5 where 0 means lack of activity, whereas 5 means world-class. To conjure up a better visual representation, points are calculated into index numbers e.g. the IIM (Index Implementation Methodology), and these into medals: bronze, silver, and gold [9]. A bronze medal for a plant means that the implementation has started, and its preliminary effects have been observed. A silver medal signifies operation in conformity with the WCM principles where numerous projects bring tangible results. A gold medal recognizes and spotlights the company as a champion and paragon of improvement and innovation [12].

The benefits following from a WCM implementation are multiple: financial, social, related to quality, organizational, and related to processes. An organization which achieves the WCM class enjoys lower costs and better financial results, provides better workplace conditions, which impacts on worker productivity and motivation, reduces the number of defects, boosts the safety of work, and facilitates communication. As relates to customers, it ensures that they are serviced according to their requirements in terms of specific expectations, lead times, and deadlines [9].

3 Organizational Culture

The Latin word *cultura* meant agriculture, cultivation [25], and it was E. Jacques [26 as cited in 27] who first applied the term metaphorically to enterprises when he stated that culture of the factory was *'its customary and traditional way of thinking and doing of things, which is shared to a greater or lesser degree by all its members, and which new members must learn, and at least partially accept, in order to be accepted into service in the firm'*. Sułkowski concedes [28] that the available definitions of culture are vague and heterogeneous. He specifies [29 as cited in 25] seven ways to define the term 'organizational culture': as organization itself, as one of culture circles, social rules of the game, corporate identity, philosophy of the organization, a system of shared meanings and values, and as models or standards of behavior.

The body of knowledge in the field of organization and management includes two ways of approaching organizational culture: the first is to look inward - at the organization's interpersonal processes and group dynamics; the second is to look outward - at the organization's environment and therefore, at its national culture or the community in which the organization operates [27]. The authors' professional practice in an international environment allows them to claim that both these approaches are very meaningful and need to be taken into account in the pursuit of ambitious goals of WCM.

The inward frame of reference is useful for understanding the concept of culture in individual cases and for the perception of its manifestations. It provides an insight into what culture means for members of the organization, which is relevant for effective functioning in the organization or managing it. In her work, Furmańczyk [25] employs the concept of levels to present different elements of culture: those most overt and clearly manifested (visible), those partially manifested and perceived, and those that are covert (invisible) and unperceived. To better explain the concept, various models and typologies of organizational culture have been developed and if only two of them were mentioned: E. Schein's and G. Hofstede's, which are regarded as the most cited [28] and most influential [30, 31], the following elements of organizational culture could be identified: see Table 1. Models of culture by E. Schein and G. Hofstede.

Table 1. Models of culture by E. Schein and G. Hofstede

Culture level	Schein's model	Hofstede's model
Outermost	Artifacts (objects made by humans)	Symbols, heroes, rituals
Intermediate	Espoused values, beliefs, norms, rules,	Practices
Innermost	Basic, underlying assumptions	Values

Source: [authors' analysis based on 8, 32]

According to Schein [32], the outermost layer of culture are artifacts, objects made by human beings that can be seen, heard, and felt, and which, to be understood require presence in and observation of a group. Beliefs, values, norms, and rules that shape everyday work and behavior patterns and are a point of reference for what is and what should be can also be analyzed. They are important for group formation or when a group comes across a new task or a problem, and if it turns out that they have not failed the group during task completion and problem solving, they become core beliefs. Underlying assumptions (premises) enable full deciphering of organizational standards and prediction of future behaviors of the organization; they are taken for granted by the entire organization. They are also uncontestable and extremely difficult to modify.

In Hofstede's model [8], symbols include words, gestures, images, and objects, which in practice are language, jargon, dress code, hairstyles, trademarks, flags, status markers. They are meaningful because they are recognizable for the members of the culture and the most overt. Heroes are contemporary, historical, real or fictional figures characterized by traits that are appreciated by the culture; they are paragons of conduct in this culture. Rituals, that is collective activities, are perceived as social imperatives: for example, greetings, ways of expressing respect and esteem, ceremonies, solemnities, business meetings, ways of exchanging views and expressing opinions. Practices in this construct mean human activity and routines taking into account the symbols people respond to, heroes they admire, and rites and rituals in which they participate. At the core of this concept are values understood as leanings that determine the choices people make.

Moving on, peculiar to the second, outward frame of reference [27] is that of its parts which refers to national cultures. Barwiński [33], following in the footsteps of other authors, while attempting to formulate a definition of 'nation', points out that a nation is a result of historical development and that it has some of the qualities characteristic of a certain community. These idiosyncratic factors influence each nation's perceptions of reality and their relation to them. With models of national organizational cultures in mind, Furmańczyk [25] draws attention to single-variable models such as E. T. Hall's and F. Fukuyama's as well as multi-dimensional ones such as G. Hofstede's and R.R. Gesteland's. Each of these authors, based on their studies and their thoughts, put forward certain criteria by which to distinguish cultures of individual national groups. Some of them are presented below.

According to Hall, culture is strictly associated with communication, whereas the factors that differentiate cultures are context, time, and space. On this account, the following can be distinguished:

- high context cultures (e.g. use fewer words and depend more on context to understand the meaning of a message) and low context cultures (e.g. where all the necessary information needs to be verbally explicitly stated),
- monochronic (where a great deal of attention is given to the planning of procedures, sequencing of tasks, and precision) and polychronic (where multitasking is common and where people are more important than deadlines),
- collectivist (where group identification and adherence to its rules are important) and individualistic (where self-reliance and personal responsibility are important).

Fukuyama links national culture with trust, which allows him to classify cultures into:

- low trust (where people outside the group are not given much trust; rules, procedures and hierarchies are important) and high trust cultures (where elaborate social bonds and relationships exist as do mutual trust, teamwork, and the delegation of authorities).

Hofstede's model conceptualizes the following five dimensions of national cultures:

- power distance (may be low, in which case any inequalities in the distribution of power are opposed; it may also be high, in which case inequalities in the society are rationalized and expected),
- collectivism/individualism (in individualistic cultures, an individual is the fundamental, independent component part of the society, whereas in collectivist cultures, individuals belong to groups and groups are entrusted with decision-making; groups bear responsibility for their members),
- uncertainty avoidance (nations characterized by high uncertainty avoidance perceive the future to be threatening, whereas those characterized by low uncertainty avoidance see it as a challenge, a possibility, and a welcome opportunity),
- masculinity/femininity (refers to the extent to which the culture requires the perpetuation of frequently stereotypical, socially constructed gender roles),
- short-term/long-term orientation (refers to orientation towards the past, the present, and shorter-term goals or towards the future and longer-term goals).

The last of the models, the R.R. Gesteland model, uses four criteria to distinguish:

- relationship-focused cultures (characterized by a reluctance to business transactions with foreigners/strangers) and deal-focused (direct and open to cooperation regardless of the degree of familiarity or a lack thereof),
- formal (the most prominent feature of which is hierarchy) and informal cultures,
- monochronic and polychronic (essentially referring to Hall's construct), and

- expressive cultures (characterized by highly animated communication and disregard for personal space) and conservative/reserved cultures (typical of which are restraint and high regard for personal space).

4 Organizational Culture and World Class Manufacturing

Considering the previous discussion concerning the WCM concept and the concept of organizational culture, and taking into account the context outlined in the background information section, the following appear to be vital for the achievement of the goal referred to in the title:

- becoming cognizant of the existence of an organizational culture and defining it, what it is like, what part of it contributes to the growth of the organization and what part of it hinders that growth,
- taking account of the role and the presuppositions rooted in the national background of the organization and in the environment in which its branches operate, and next, acting on its strengths while counteracting its weakness,
- considering how tomorrow's and today's goals have been defined, whether by reference to a longer time frame because WCM implementation is a multi-stage undertaking, whereas preferences for specific time frames are also inherently cultural,
- analysis of cultural collectivism and individualism in action because WCM implementation calls for collective commitment to the accomplishment of real goals,
- reflecting upon leaders' attitudes because they have a direct bearing on effectiveness and efficiency, but they also tend to be culturally conditioned by power distance,
- determining the level of organizational identification and employee loyalty to the organization because WCM, drawing upon the Japanese culture of Toyota, shows that workers who have a strong sense of belonging in the organization have stronger intrinsic motivation to contribute to the company's success.

May Hofstede's opinion concerning cultural awareness serve as a summary and a word of caution: top management should know what culture is lest they should fail [34, 35]. Therefore, raising awareness of the existence of a culture and its effects on the organization's performance is imperative both for regular members of the organizations as well as for those who have been invested with the responsibility of being their leaders.

References

1. Wisniewski, Z.: Wdrazanie zarządzania jakoscia w przedsiebiorstwie produkcyjnym-uwarunkowania i bariery. Politechnika Lodzka, Lodz (2002)
2. Starbuck, W.H.: Oxford Handbook of Organization Theory. Oxford University Press (2003)
3. Hoppit, J.: Understanding the industrial revolution. Hist. J. **30**, 211–214 (1987)

4. The Editors of Encyclopaedia Britannica. Industrial Revolution - Britannica Online Encyclopaedia. https://www.britannica.com/event/Industrial-Revolution. Accessed 15 Jan 2019
5. Szuman, S.: Natura, osobowosc i charakter człowieka. Wydawnictwo WAM, Warszawa (1995)
6. Jobs, S.: Steve Jobs: His Own Words and Wisdom. Silicon Valley Press, Cupertino (2011)
7. Hofstede, G.: Culture and organizations. Int. Stud. Manage. Organ. **10**, 15–41 (1981)
8. Hofstede, G., Hofstede, G.J., Minkov, M.: Kultury i organizacje. Zaprogramowanie umyslu. PWE, Warszawa (2011)
9. Piasecka-Gluszak, A.: Implementacja World Class Manufacturing w przedsiebiorstwie produkcyjnym na rynku polskim. Ekonomia XXI Wieku **16**, 52–65 (2017)
10. Fekete, M.: World Class Manufacturing – the concept for performance increasement and knowledge acquisition. Trendy v Podnikani, Zapadoceska Univerzita v Plzni (2013)
11. Salaheldin, S.I., Eid, R.: The implementation of world class manufacturing techniques in Egyptian manufacturing firms: an empirical study. Ind. Manage. Data Syst. **107**, 551–566 (2007)
12. Walczak, M.: Dyfuzja produkcji w klasie swiatowej (ang. World Class Manufacturing) wewnatrz lancucha tworzenia wartosci (na przykładzie Fiat Auto Poland SA). Przedsiebiorstwo i Region **7**, 113–122 (2015)
13. Minnock, O.: Top 10 biggest car manufacturers in the world 2017. https://www.manufacturingglobal.com/top10/top-10-biggest-car-manufacturers-world. Accessed 26 Jan 2019
14. De Felice, F., Petrillo, A., Monfreda, S.: Improving operations performance with world class manufacturing technique: a case in automotive industry. Operations management, InTech (2013)
15. Walaszczyk, A.: Bariery we wdrazaniu systemu zarzadzania jakoscia w przedsiebiorstwach sektora MSP w Polsce. Przedsiebiorczosc i Zarzadzanie **19**(9), 361–374 (2018)
16. Walaszczyk, A.: Risk management of processes in the quality management system. Annales Universitatis Mariae Curie-Skłodowska, Sectio H Oeconomia **52**(1), 201–209 (2018)
17. Penc, J.: Strategiczny System Zarzadzania. Placet, Warszawa (2001)
18. Rajkiewicz, M., Mikulski, R.: Tendencje zmian w systemach zarzadzania problemy integracji oraz wdrozenia. Wydawnictwo Politechniki Lodzkiej, Lodz (2016)
19. Karaszewski, R.: Systemy zarzadzania jakoscia największych korporacji swiata i ich dyfuzja:(zjawisko, rozwój, znaczenie). Wydaw. Uniwersytetu Mikolaja Kopernika, Torun (2003)
20. Furman, J.: Wdrazanie wybranych narzedzi koncepcji lean manufacturing w przedsiebiorstwie produkcyjnym. Oficyna Wydawnicza PTZP, Opole (2014)
21. Blaszczyk, A., Wisniewski, Z.: Requirements for IT systems of maintenance management. Adv. Intell. Sys. Comput. **793**, 531–539 (2019)
22. Wisniewski, Z., Blaszczyk, A.: Changes in maintenance management practices - standards and human factor. Adv. Intell. Sys. Comput. **606**, 348–354 (2018)
23. Calvasina, R.V., Calvasina, E.J., Calvasina, G.E.: Beware the new accounting myths. Manage. Account. **12**, 41–45 (1989)
24. Hawrysz, L., Foltys, J.: Environmental aspects of social responsibility of public sector organizations. Sustainability **8**, 1–10 (2016)
25. Furmanczyk, J.: Kulturowe uwarunkowania przywodztwa w miedzynarodowych przedsiebiorstwach branzy motoryzacyjnej w Polsce. Wydzial Gospodarki Międzynarodowej Uniwersytetu Ekonomicznego, Poznan (2011)
26. Jacques, E.: The Changing Culture of a Factory. Tavistock Publications, London (1951)

27. Aniszewska, G.: Geneza pojęcia "kultura organizacyjna". Przeglad Organizacji **10**, 17–20 (2003)
28. Sulkowski, L.: Czy warto zajmować sie kultura organizacyjna. Zarzadzanie Zasobami Ludzkimi **6**, 9–25 (2008)
29. Sulkowski, L.: Kulturowa zmiennosc organizacji. Polskie Wydawnictwo Ekonomiczne, Warszawa (2002)
30. Kundu, K.: Influence of organizational culture on the institution building process of an organization. CURIE J. **2**, 48–57 (2010)
31. Stankiewicz-Mroz, A.: Approach to the issues of leadership in the processes of companies' acquisitions. Procedia Manuf. **3**, 793–798 (2015)
32. Schein, E.H.: Organizational Culture and Leadership. Jossey-Bass (2004)
33. Barwinski, M.: Pojecie narodu oraz mniejszosci narodowej i etnicznej w kontekscie geograficznym, politycznym i socjologicznym. In: Acta Universitatis Lodziensis, Folia Geographica Socio-Oeconomica, vol. 5, pp. 59–74 (2004)
34. Mnich, J., Wisniewski, Z.: Strategy and structure in public organization. Adv. Intell. Sys. Comput. **783**, 351–358 (2019)
35. Kozlowski, R., Matejun, M.: Characteristic features of project management in small and medium-sized enterprises. E M Ekon. Manage. **19**, 33–48 (2016)

Research on Thermal Comfort Equation of Comfort Temperature Range Based on Chinese Thermal Sensation Characteristics

Rui Wang[1(✉)], Chaoyi Zhao[1], Wei Li[1], and Yun Qi[2]

[1] Ergonomics Laboratory, China National Institute of Standardization,
No. 4 Zhi Chun Road, Haidian District, Beijing, China
{wangrui, zhaochy, liwei}@cnis.gov.cn
[2] China Standard Certification Co., Ltd., Beijing, China
qiyun@csc.gov.cn

Abstract. Thermal comfort is a subjective experience in which the human body feels satisfied with the thermal environment. The human thermal sensation is mainly related to the heat balance of the whole body. This balance is not only affected by environmental parameters such as air temperature, average radiant temperature, wind speed and air humidity, but also by human activities and dress. Local thermal discomfort includes evaluation indexes such as draft, vertical temperature difference, ground heating and cooling, and radiation asymmetry. When these parameters are estimated or measured, the overall thermal sensation of the person can be predicted by calculating the predicted mean thermal sensation index (PMV). Internationally, the currently accepted standard for predicting and evaluating indoor thermal environment comfort is ISO 7730-2005 and American Standard ASHRAE 55-2013. In the ISO 7730-2005 standard, the comfort is evaluated for the steady-state and non-steady-state thermal environments. The evaluation of the steady-state environment mainly evaluates the PMV, PPD indicators and local thermal discomfort. The unsteady environment is mainly evaluated from indicators such as temperature cycle and drift. In terms of thermal comfort evaluation, after extensive experiments, the researchers tried to link the four elements that make up the thermal environment (air temperature, humidity, wind speed and ambient average radiant temperature) to the thermal sensation of the human body. However, heat balance is not a sufficient condition for human thermal comfort. The Fanger thermal comfort equation and its evaluation indicators have their limitations. Studies have shown that the actual thermal sensation of the test subjects is significantly higher than the PMV through the thermal comfort survey of winter and summer residential buildings, and pointed out that there is a lack of equations. To this end, this paper carried out a test for the thermal sensation characteristics of Chinese people, conducted thermal comfort tests on nine environmental conditions of refrigeration and heating through 40 subjects of different ages, and determined the differences in thermal sensory characteristics of human body at different ages, analyze the test results and correct the PMV equation according to the results to make it more in line with the physiological characteristics of Chinese people. The test results show that the thermal sensation index of children and the elderly is significantly different, the children's thermal sensation is obviously cold, while the elderly are obviously warm, and the thermal sensation

© Springer Nature Switzerland AG 2020
W. Karwowski et al. (Eds.): AHFE 2019, AISC 971, pp. 254–265, 2020.
https://doi.org/10.1007/978-3-030-20494-5_24

characteristics of adults with different genders are also different. Men are more like cold than women. When testing for the elderly and children, gender has little effect on the characteristics of thermal sensation, and there is no obvious difference between hot and cold in different genders.

Keywords: Thermal comfort equation ·
Chinese thermal sensation characteristics · PMV

1 Introduction

Thermal comfort is a subjective experience in which the human body feels satisfied with the thermal environment. The human thermal sensation is mainly related to the heat balance of the whole body. This balance is not only affected by environmental parameters such as air temperature, average radiant temperature, wind speed and air humidity, but also by human activities and dress. When these parameters are estimated or measured, the overall thermal sensation of the person can be predicted by calculating the predicted mean thermal sensation index (PMV). Internationally, the currently accepted standards for predicting and evaluating the comfort of indoor thermal environments are ISO 7730-2005 and American Standard ASHRAE 55-2013. In the ISO 7730-2005 standard, the comfort is evaluated for the steady-state and non-steady-state thermal environments. The evaluation of the steady-state environment mainly evaluates the PMV, PPD indicators and local thermal discomfort. Local thermal discomfort includes evaluation indexes such as draft, vertical temperature difference, ground heating and cooling, and radiation asymmetry. The unsteady environment is mainly evaluated from indicators such as temperature cycle and drift. American Standard ASHRAE55-2013 also gives a method to determine the thermal environment of comfort. It mainly evaluates thermal comfort from indicators such as temperature (also referred to as the calculated temperature, the operating temperature, that is, the uniform temperature in the imaginary black enclosure, the amount of radiant heat transfer and convective heat exchange in the black enclosure is the same as the amount of heat exchange in the actual non-uniform environment), humidity, elevated air speed, local thermal discomfort and temperature change with time. The local thermal discomfort assessment content is the same as ISO 7730-2005. The change of temperature with time is evaluated from two aspects: temperature change cycle and drift.

The domestic research standard for comfort GB/T 18049-2017 "Medium thermal environment PMV and PPD index determination and thermal comfort conditions" is equivalent to the standard content of ISO 7730-2005; GB/T 33658-2017 "Indoor indoor thermal comfort environment requirements and evaluation methods" mainly refer to ASHRAE55 and ISO7730-2015 test items and test methods, but the specific test point position and other considerations of Chinese human body size data. The PMV equations in GB/T 18049 and GB/T 33658 are the same as the international standards.

In terms of thermal comfort evaluation, after a lot of experiments, the researchers tried to link the four elements (air temperature, humidity, wind speed and ambient average radiant temperature) that make up the thermal environment with the thermal sensation of the human body, so that there is an equivalent temperature and effective Some indicators for measuring the thermal environment, such as temperature and standard effective temperature [1], are used as the basis for dividing the comfort zone. Fanger also proposed the thermal comfort equation for the energy balance of the human body under steady state conditions. The proposed Predicted Mean Vote (PMV) index is the most accepted thermal comfort evaluation index. Due to individual differences in people, a 100% thermal environment that satisfies everyone's comfort requirements is impossible. Therefore, he also proposed the Predicted Percent Dissatisfied (PPD) indicator, any indoor climate must meet the comfort requirements of most people as much as possible. Fanger's research results have become the basis for the American Society of Heating, Refrigerating and Air-Conditioning Engineers (ASH-RAE) and the International Organization for Standardization (ISO) to develop and evaluate indoor thermal comfort standards: ASHRAE 55-2004 [2] and ISO 7730-2005 [3]. The thermal environment standards used by countries also generally refer to these standards, including China's GB/T 18049-2000 "Medium thermal environment PMV and PPD index determination and thermal comfort conditions". The PMV predictions are basically consistent with the ASHARE comfort zone for office work with the new effective temperature ET*. Despite this, the two thermal comfort standards ASHRAE 55-2004 and ISO 7730-2005 do not fully consider regional and climate change factors, so their applicability has been questioned by scholars from various countries.

With the development of technology and the deepening of research, many scholars have conducted extensive discussions on its applicability. In 1989, Cao et al. [4] proposed that heat balance is not a sufficient condition for human thermal comfort. The Fanger thermal comfort equation and its evaluation indicators have their limitations. In 1996, Cabanac systematically described the difference between thermal sensation and thermal comfort. Thermal comfort should not be defined as a subjective neutral state of the thermal environment [5]. In 2002, Humphreys et al. [6] pointed out that the actual range of application of PMV variables is much narrower than that indicated in the ISO7730 standard, and there is a bias in predicting the average thermal comfort of personnel under normal conditions in buildings. Especially in warm environments, it is also pointed out that PMV can be modified to improve the effectiveness of its predictions. In 2003, Charles [7] proposed Fanger's thermal comfort model for the testees wearing normal indoor clothing, in a sedentary state, without the thermal neutrality of the excessive air flow rate, leaving the state results in deviations. In 2007, Li Shigang et al. [8] proposed that Professor Fanger's theory is based on experimental results under specific state parameters, and that the conditions of use should be noted. In 2009, Becker et al. [9] found that the actual thermal sensation of the subjects was significantly higher than that of PMV through a survey of thermal comfort in winter and summer residential buildings, and pointed out that there is a lack of equations. In 2011, Singh et al. [10] found that PMV always overestimated or underestimated the perception of the thermal environment through a thermal comfort survey in three climate zones in northeastern India, and suggested that adaptive models could be used for correction.

To this end, this paper carried out tests for Chinese people's thermal sensation characteristics. Through 40 subjects of different ages, the thermal comfort test was carried out on nine environmental conditions of refrigeration and heating to determine the difference of thermal sensory characteristics of human body at different ages. The test results were analyzed and the PMV equation was corrected according to the results. It is more in line with the physiological characteristics of the Chinese.

2 Test Methods

The test was carried out in the Thermal Comfort Laboratory of the China National Institute of Standardization, which simulates the home environment. The internal compartment has a structural size of 8 (m) × 3.5 (m) × 2.6 (m), and the temperature, humidity and wind speed are adjustable.

Temperature parameter test range: −25–55 °C, test accuracy: ±0.5 °C, response time: ≤1 s; humidity parameter test range: 20–80% RH, test accuracy: ±3% or less, response time: 1 s; wind speed parameters Test range: 0.05–5 m/s, test accuracy: ±(0.05 + 0.05 v) m/s, response time: 0.5 s.

There are two symmetrical inner compartments, each of which has 9 (length) × 9 (width) × 7 (height), a total of 567 temperature test points in the horizontal plane, and a distance of 0.4 m between the water level temperature measurement points. The vertical surface temperature measurement points are 0.1 m, 0.5 m, 0.9 m, 1.3 m, 1.7 m, 2.1 m from the ground and 0.05 m from the ceiling; 9 humidity test points, which are in the plane of 1.2 m from the ground in the room; 9 wind speed test points, but when testing the wind speed, according to the requirements of ASHREA55-2013, the distance from the ground is 0.1 m, 0.6 m, 1.1 m, and the distance from the air outlet of the air conditioner is about 2 m, and the average of the three numbers is taken. The indoor temperature is taken as the average of 567 temperature measurement points, and the indoor humidity is taken as the average of 9 humidity test points.

3 Thermal Comfort Equation Correction Method

In order to correct the existing PMV equation according to the Chinese human thermal sensory characteristics, the modified PMV equation is named PMV', and the thermal sensing characteristic coefficient and the thermal sensitive characteristic coefficient are introduced in the fitting process.

The PMV' thermal comfort equation is fitted as follows:

$$PMV' = a \times HeatLoad + b \qquad (1)$$

In the formula:

a——the thermal sensitivity characteristic coefficient indicates the sensitivity of the human body to the thermal environment. The larger the coefficient, the more sensitive the thermal environment changes under this working condition, and the comfort interval will be narrower;

b——the thermal sensory characteristic coefficient indicates the degree of deviation of the human body's thermal sensation from the theoretical comfort temperature. The larger the absolute value of the coefficient, the greater the deviation of the Linlin comfort temperature under this condition. The positive coefficient is expressed as cold and heat, and the negative coefficient means heat and fear of cold.

Combining formula (1) with the PMV equation, the fit formula is:

$$PMV' = a \times \{(M-W) - 3.05 \times 10^{-3} \times [5733 - 6.99(M-W) - p_a] - \}$$
$$\{0.42 \times [(M-W) - 58.15] - 1.7 \times 10^{-5} M(5867 - P_a) - \}\{0.0014M(34 - t_a) \quad (2)$$
$$- 3.96 \times 10^{-8} f_{cl} \times \}\{\} + b$$

4 Test Results and PMV Equation Correction

Considering the difference in thermal resistance of clothing under different working conditions, the correction coefficient is given according to the winter working condition and the summer working condition when the PMV equation is corrected. The test is also tested according to the winter working condition and the summer working condition.

4.1 Summer Condition

The comfort interval determined according to the pre-experimental test results was changed, and the final summer test conditions were determined to be 18, as shown in Table 1.

Table 1. Summer test conditions

Season	Clothing thermal resistance (clo)	Test sequence number	Wind speed (m/s)	Temperature (°C)	Humidity
Refrigeration season	0.55	1	0.15	27.00	40%
		2	0.20	27.00	40%
		3	0.25	27.00	40%
		4	0.20	27.00	70%
		5	0.25	27.00	70%
		6	0.30	27.00	70%
		7	0.10	26.50	40%
		8	0.15	26.50	40%
		9	0.20	26.50	40%
		10	0.15	26.50	70%
		11	0.20	26.50	70%
		12	0.25	26.50	70%
		13	0.05	26.00	40%
		14	0.10	26.00	40%
		15	0.15	26.00	40%
		16	0.10	26.00	70%
		17	0.15	26.00	70%
		18	0.20	26.00	70%

The thermal resistance of the garment during the test was tested by the warm-body dummy. The male jacket has a regular long-sleeved shirt + vest, trousers + boxer briefs, a single leather shoe + ordinary cotton socks; the female jacket has a common long-sleeved shirt + camisole + bra, trousers + underwear, single shoes + ordinary cotton socks, tested clothing thermal resistance is about 0.55clo.

During the test, the subjects used the method of sitting quietly, and did not work externally. The metabolic rate was calculated according to 1.2M. The indoor environment has no heat source, and the average radiant temperature is very close to the indoor average air temperature. The same value is used when calculating the PMV. Substituting environmental conditions into the PMV calculation formula yields an expected thermal sensation index for all operating conditions. Through the subjective test, the subjective test results of the working condition were fitted to the calculated PMV, and the PMV correction equation under summer conditions was obtained, and the correction coefficient was obtained. The PMV correction factor and equation are shown in Figs. 1, 2, 3, 4 and 5.

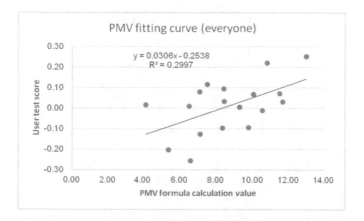

Fig. 1. "Everyone" PMV fitting curve for refrigeration conditions

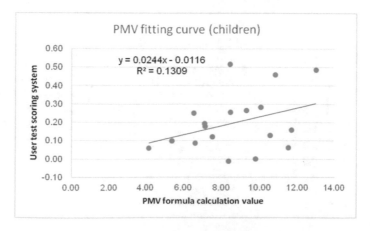

Fig. 2. Refrigeration condition "child" PMV fitting curve

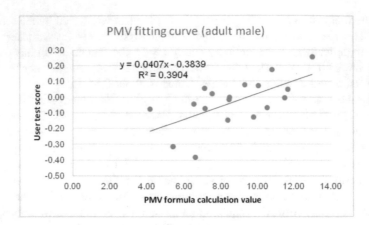

Fig. 3. PMV fitting curve of youth (adult) male in refrigeration conditions

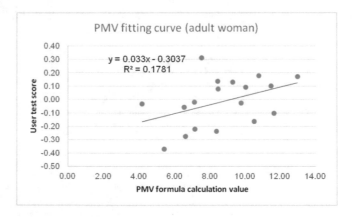

Fig. 4. PMV fitting curve of the youth (adult) female in the refrigeration condition

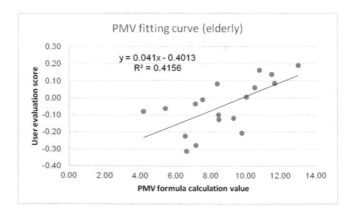

Fig. 5. PMV fitting curve of the "old man" in the refrigeration condition

The PMV equation fitting coefficients and correlation results of different groups and all people are counted, as shown in Table 2:

Table 2. Fitting results of PMV equation for refrigeration conditions

Working condition	Coefficient	Everyone	Child	Young (adult) male	Young (adult) female	Old man
Refrigeration condition	a	0.0306	0.0240	0.0407	0.0330	0.0410
	b	−0.2538	−0.0116	−0.3839	−0.3037	−0.4013
	R^2	0.2997	0.1309	0.3904	0.1781	0.4156

The results show that the elderly have the highest thermal sensitivity coefficient and the highest thermal sensitivity coefficient, indicating that the elderly are most enthusiastic when cooling, and are most sensitive to cold. Children have the lowest thermal sensitivity coefficient and the lowest thermal sensitivity coefficient, indicating that children prefer a cooler environment and are less sensitive to ambient temperature and have a wider range of acceptable comfort environments. The test results of adult males and females are small. The thermal sensory characteristic coefficient of all the people's test results was negative, indicating that the Chinese human thermal sensation index is significantly lower than the calculated value of the PMV equation proposed by Professor Fanger, which means that the Chinese are warmer.

4.2 Heating Condition

According to the pre-experimental test results, the comfort interval was changed, and the final winter test conditions were 9 in total, as shown in Table 3.

Table 3. Winter thermal comfort test environment conditions

Season	Clothing thermal resistance (clo)	Test sequence number	Wind speed (m/s)	Temperature (°C)	Humidity
Heating season	0.83	1	0.05	23.50	50%
		2	0.10	23.50	50%
		3	0.15	23.50	50%
		4	0.10	24.00	50%
		5	0.15	24.00	50%
		6	0.20	24.00	50%
		7	0.15	24.50	50%
		8	0.20	24.50	50%
		9	0.25	24.50	50%

The thermal resistance of the garment during the test was tested by the warm-body dummy. The male jacket has a common long-sleeved shirt + vest + sweater, trousers + warm pants + boxer briefs, single shoes + ordinary cotton socks; Cardigan + ordinary long-sleeved shirt + camisole + bra, trousers + warm pants + underwear, single shoes + ordinary cotton socks, tested clothing thermal resistance is about 0.83clo.

During the test, the subjects also used the meditation method, and did not work externally. The metabolic rate was calculated according to 1.2M. The indoor environment has no heat source, and the average radiant temperature is very close to the indoor average air temperature. The same value is used when calculating the PMV. Substituting environmental conditions into the PMV calculation formula yields an expected thermal sensation index for all operating conditions. Through the subjective test, the subjective test results of the working condition were fitted to the calculated PMV, and the PMV correction equation under summer conditions was obtained, and the correction coefficient was obtained. The PMV correction factor and equation are shown in Figs. 6, 7, 8, 9 and 10.

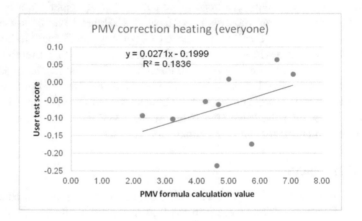

Fig. 6. "Everyone" PMV fitting curve for heating conditions

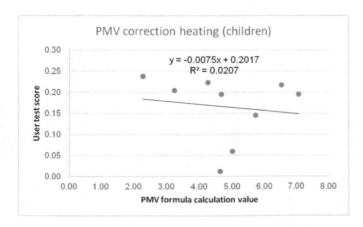

Fig. 7. Childhood PMV fitting curve for heating conditions

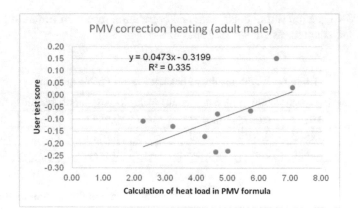

Fig. 8. PMV fitting curve of "young (adult) male" in heating condition

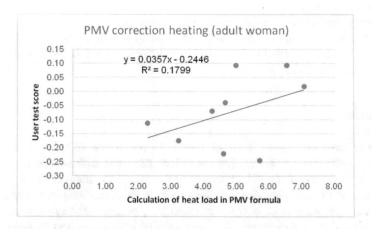

Fig. 9. PMV fitting curve of "young (adult) female" in heating condition

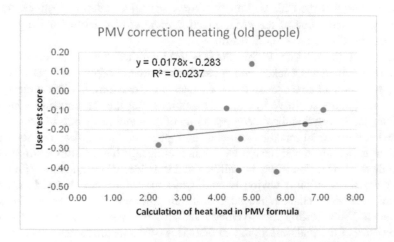

Fig. 10. PMV fitting curve of "old man" in heating condition

The PMV equation fitting coefficient and correlation results of different groups and all people are counted, as shown in Table 4:

Table 4. Fitting results of heating system PMV equation

Coefficient	Everyone	Child	Young (adult) male	Young (adult) female	Old man
a	0.0271	−0.0075	0.0473	0.0357	0.0178
b	−0.1999	0.2017	−0.3199	−0.2446	−0.2830
R^2	0.1836	0.0271	0.3350	0.1799	0.0237

The results show that the correlation in winter test results is worse than that in summer, mainly because the clothes in winter are thicker, which causes the human body to be insensitive to the thermal sensation changes caused by changes in the external environment. Older people are the most sensitive in cooling conditions, but the least sensitive in heating conditions, indicating that the elderly are afraid of cold and not afraid of heat. Children have the lowest thermal sensitivity coefficient, indicating that children are not sensitive to ambient temperature and have a wide range of acceptable comfort environments.

5   Conclusion

The elderly have the highest thermal sensitivity coefficient and the highest thermal sensitivity coefficient, indicating that the elderly are most enthusiastic when cooling, and are most sensitive to cold. Children have the lowest thermal sensitivity coefficient and the lowest thermal sensitivity coefficient, indicating that children prefer a cooler environment and are less sensitive to ambient temperature and have a wider range of acceptable comfort environments. The test results of adult males and females are small. The thermal sensory characteristic coefficient of all the people's test results was negative, indicating that the Chinese human thermal sensation index is significantly lower than the calculated value of the PMV equation proposed by Professor Fanger, which means that the Chinese are more warm.

The correlation in winter test results is worse than that in summer, mainly because the clothes in winter are thicker, which causes the human body to be insensitive to the thermal sensation changes caused by changes in the external environment. Older people are the most sensitive in cooling conditions, but the least sensitive in heating conditions, indicating that the elderly are afraid of cold and not afraid of heat. Children have the lowest thermal sensitivity coefficient, indicating that children are not sensitive to ambient temperature and have a wide range of acceptable comfort environments.

The thermal sensation index of children and the elderly is obviously different. The child's thermal sensation is obviously cold, while the elderly are obviously warm, and the thermal sensation characteristics of adults are different. Men are more like women than women. When testing for the elderly and children, gender has little effect on the characteristics of thermal sensation, and there is no obvious difference between hot and cold in different genders.

Acknowledgments. This research is supported by "Special funds for the basic R&D under-takings by welfare research institutions" (522018Y-5984 and 522017Y-5276) and "2017 National Quality Infrastructure (2017NQI) project" (2017YFF0206605).

References

1. Wang, R.: Research and Development Report on the Progress of Refrigeration. Science Press (2007)
2. ASHRAE. ANSI/ASHRAE 55-2004 Thermal environmental conditions for human ccupancy. American Society of Heating, Refrigerating and Air Conditioning Engineers Inc., Atlanta (2004)
3. ISO.BS EN ISO 7730-2005 Ergonomics of the thermal environment - Analytical determination and interpretation of thermal comfort using calculation of the PMV and PPD indices and local thermal comfort criteria. International Standards Organization, Geneva (2005)
4. Cao, J., Zhou, Y.: Thermal comfort equation of Fanger and its application. J. Beijing Inst. Cloth. Technol. **2**, 96–101 (1989)
5. Cabanac, M.: Pleasure and joy, and their role in human life. In: Proceedings of Indoor Air' 96, Nagoya (1996)
6. Humphreys, M.A., Fergus Nicol, J.: The validity of ISO-PMV for predicting comfort votes in every-day thermal environments. Energy Build. **34**(6), 667–684 (2002)
7. Charles, K.E.: Fanger's Thermal Comfort and Draught Models. irc.nrc-cnrc.gc. ca/ircpubs, 2003-10-10
8. Li, S., Lian, Z.: Discussions on the application of Fanger's thermal comfortable theory. Shanghai Refrigeration Institute Academic Annual Conference (2007)
9. Becker, R., Paciuk, M.: Thermal comfort in residential buildings Failure to predict by Standard model. Build. Environ. **44**(5), 948–960 (2009)
10. Singh, M.K., Mahapatra, S., Atreya, S.K.: Adaptive thermal comfort model for different climatic zones of North-East India. Appl. Energy **88**(7), 2420–2428 (2011)
11. Wang, Z., Zhao, J., Liu, J.: Indoor Air Environment. Chemical Industry Press, Beijing (2006)
12. Liu, J.: Study on indoor thermal environment and human thermal comfort of natural ventilation buildings in hot summer and cold winter areas. ChongQing University, ChongQing (2007)

Study on the Perception Characteristics of Different Populations to Thermal Comfort

Rui Wang[1], Wei Li[1(✉)], Chaoyi Zhao[1], and Yun Qi[2]

[1] Ergonomics Laboratory, China National Institute of Standardization,
No. 4 Zhi Chun Road, Haidian District, Beijing, China
{wangrui, liwei, zhaochy}@cnis.gov.cn
[2] China Standard Certification Co., Ltd., Beijing, China
qiyun@csc.gov.cn

Abstract. Physiological studies have shown that when people are in thermal comfort, their thinking, observation and operation skills are in the best state. At present, the relatively perfect models of human thermal comfort are based on the research of physiology and psychology of Westerners. However, there are great differences in thermal regulation parameters such as body size and composition between Chinese and Westerners. For example, there is a big difference between the standard human body model in China and the standard human body model in the West. Moreover, the differences in social and natural environment between China and the West lead to some differences in thermal psychology and preferences between Chinese and Westerners. These differences will affect the thermal sensation of the human body in the environment. Therefore, the establishment of a human body model based on the thermal physiological and psychological characteristics of the Chinese people is of great value to improve the prediction accuracy of the thermal comfort state of the Chinese people. Through collecting a large number of basic data of human thermal sensation in China, this paper studies the difference of demand for environmental thermal comfort among different populations. The results show that children feel the hottest in the same thermal environment, followed by "young (adult) men", "young (adult) women" and "old people" feel the coldest. Generally speaking, the test results are correct. The lower the temperature, the greater the air velocity, the colder the feeling, and the less the influence of humidity on comfort.

Keywords: Thermal comfort · Different populations ·
Perception characteristics

1 Introduction

According to statistics, more than 80% of people's life is spent indoors. Indoor environmental quality such as sound, light, thermal environment and indoor air quality will have a significant impact on people's physical and mental health, comfort and work efficiency [1]. Physiological studies show that when people are in thermal comfort, their thinking, observation ability and operation skills are in the best state [2].

© Springer Nature Switzerland AG 2020
W. Karwowski et al. (Eds.): AHFE 2019, AISC 971, pp. 266–275, 2020.
https://doi.org/10.1007/978-3-030-20494-5_25

AHSRAE defines thermal comfort in its industry standards -ASHRAE Standard 55-2013 as: The human body is satisfied with the thermal environment [3]. The definition of thermal comfort has been widely accepted, but it does not clearly define what is "mental state" or "satisfaction", but emphasizes that the judgment of thermal comfort level belongs to a human cognitive process, which is affected by many factors, including physiological, psychological and other aspects. The influence of thermo-physiological factors belongs to the research field of human thermal regulation, while thermo-psychological factors involve the influence of social and natural environment. Therefore, the establishment of human thermal comfort evaluation needs systematic knowledge of bio-heat transfer, human physiology, environmental psychology and other discipline. Establishing thermal comfort model is an important means to evaluate and predict human thermal comfort state, and accurate thermal comfort model can also provide valuable reference for the design of air conditioning environment. However, the relatively perfect human thermal comfort models are based on the research of physiology and psychology of Westerners. There are great differences between Chinese and Westerners in thermal adjustment parameters such as body size and composition. For example, there are great differences between Chinese and Western standard human models. Further differences in social and natural environment between China and the West lead to differences in thermal psychology and preferences between Chinese and Westerners. These differences will affect the thermal sensation of the human body in the environment [4]. Therefore, the establishment of a human body model based on the thermal physiological and psychological characteristics of the Chinese people is of great value to improve the prediction accuracy of the thermal comfort state of the Chinese people.

In the study of thermal comfort, it is often necessary to evaluate the thermal environment. The main methods involved are subjective evaluation and objective evaluation [5]. By processing and synthesizing the subjective evaluation scale, the subjective evaluation of the thermal environment can be obtained accordingly. The results are usually discrete [6]. Subjective evaluation method can intuitively and clearly evaluate the comfort of the current environment. The objective evaluation results are based on the test results, which have good stability and consistency, making the test results based on different thermal environment laboratories comparable and repro-ducible [7]. Therefore, the combination of subjective evaluation and objective obser-vation is the main method to study environmental thermal comfort.

With the development of economy and the improvement of people's living stan-dard, people's requirement for comfort of environment has been gradually raised. The design criterion of air conditioning system is based on the thermal comfort model of foreign countries. Whether such a model can really accurately predict the thermal comfort state of Chinese people has been questioned [8]. Acquiring the actual thermal sensation data of Chinese population is the basis of accurately predicting the thermal comfort state of Chinese people. In this paper, the basic thermal sensation data of Chinese people of different ages and sexes are collected, and the differences between the predicted results and the experimental results of the widely used PMV model are compared and analyzed, and the PMV equation is modified.

2   Testing Conditions

The test recruited 44 subjects, including 11 children under 12 years old, 11 young men (adults) aged 13–49 years old, 11 young women (adults) aged 13–49 years old and 11 elderly people over 50 years old. When recruiting subjects, try to cover all age groups, and take into account the proportion of people in the southern and Northern regions.

The test was carried out in the Thermal Comfort Laboratory of China Institute of Standardization. The laboratory simulates the home environment. The size of the interior compartment is 8(m) * 3.5(m) * 2.6(m). Temperature, humidity and air velocity can be adjusted.

Relative humidity is set at 40% and 70% in refrigeration condition, 50% in heating condition, and 0.05 m/s as starting point and 0.05 m/s as step length to adjust air velocity. All test conditions are shown in Table 1.

Table 1. Testing conditions of thermal comfort in refrigeration season and heating season

Season	Clothing insulation (clo)	Testing number	Air velocity (m/s)	Temperature (°C)	Humidity
Refrigeration season	0.55	1	0.15	27.00	40%
		2	0.20	27.00	40%
		3	0.25	27.00	40%
		4	0.20	27.00	70%
		5	0.25	27.00	70%
		6	0.30	27.00	70%
		7	0.10	26.50	40%
		8	0.15	26.50	40%
		9	0.20	26.50	40%
		10	0.15	26.50	70%
		11	0.20	26.50	70%
		12	0.25	26.50	70%
		13	0.05	26.00	40%
		14	0.10	26.00	40%
		15	0.15	26.00	40%
		16	0.10	26.00	70%
		17	0.15	26.00	70%
		18	0.20	26.00	70%
Heating season	0.83	1	0.05	23.50	50%
		2	0.10	23.50	50%
		3	0.15	23.50	50%
		4	0.10	24.00	50%
		5	0.15	24.00	50%
		6	0.20	24.00	50%
		7	0.15	24.50	50%
		8	0.20	24.50	50%
		9	0.25	24.50	50%

In the process of testing, the subjects are required to uniformly dress according to the requirements of Table 2, which can make the clothing thermal resistance consistent, and the test results are convenient for comparison and statistics. The subjects were tested for thermal comfort according to the requirements of Fig. 1. Pictures of the field test are shown in Fig. 2.

Table 2. Men's and women's test dress information table

Number	Categories	Cloth	Clothing insulation (clo)
1	Men wear summer clothes	Tops: Plain long sleeve shirt + vest Pants: Trousers + Boxer underwear Feet: Single-layer shoes + Plain cotton socks	0.55
2	Women wear summer clothes	Tops: Plain long sleeve shirt + Bra + Camisole Pants: Trousers + underwear Feet: Single-layer shoes + Plain cotton socks	0.55
3	Men wear winter clothes	Tops: Summer wear + Woolen sweater (Medium thickness) Pants: Summer wear + Warm pants (Medium thickness)	0.83
4	Women wear winter clothes	Tops: Summer wear + Woolen sweater (Medium thickness) Pants: Summer wear + Warm pants (Medium thickness)	0.83

3 Test Results and Analysis

According to the ASHRAE 55-2013 Appendix F for cold and heat perception scoring method for different parts of the body (Whole body, head, forebreast, back, arm, front of thigh, back of thigh, front of shin, back of shin, feet) thermal sensation score. That is: +3 hot; +2 warm; +1 slightly hot but comfortable; 0 comfortable; −1 slightly cool but comfortable; −2 cool; −3 cold.

44 questionnaires were conducted by 44 subjects under each test condition shown in Table 1. Under each working condition, 44 data were weighted average according to the five groups of "all people", "children", "young (adult) men", "young (adult) women" and "old people". The scores of thermal comfort, blowing sensation and humidity of different groups under each working condition were obtained.

During the data processing, it was found that some of the subjects were insensitive to heat. More than two-thirds of the thermal sensation data in 18 refrigeration conditions filled in the same value. These data were regarded as invalid data. In order to ensure the accuracy of the data, the data of the invalid subjects were deleted in the subsequent data processing.

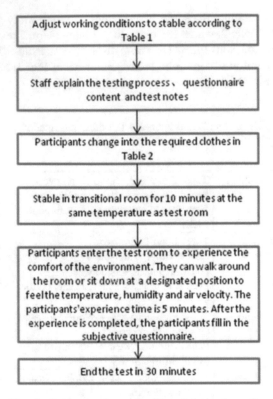

Adjust working conditions to stable according to Table 1

↓

Staff explain the testing process、 questionnaire content and test notes

↓

Participants change into the required clothes in Table 2

↓

Stable in transitional room for 10 minutes at the same temperature as test room

↓

Participants enter the test room to experience the comfort of the environment. They can walk around the room or sit down at a designated position to feel the temperature, humidity and air velocity. The participants' experience time is 5 minutes. After the experience is completed, the participants fill in the subjective questionnaire.

↓

End the test in 30 minutes

Fig. 1. Subjective test procedure of thermal comfort

Fig. 2. Field test photo

3.1 Rationality Analysis of Evaluation Data

Two typical examples are selected to analyze the rationality of the evaluation data.

3.1.1 Fixed Relative Humidity, Air Velocity, Changed Temperature

The rationality of the evaluation data is analyzed by taking the thermal sensation score of different people as an example when the relative humidity is 40%, the air velocity is 0.15 m/s, the temperature is 26, 26.5 and 27 °C. See Fig. 3 for details.

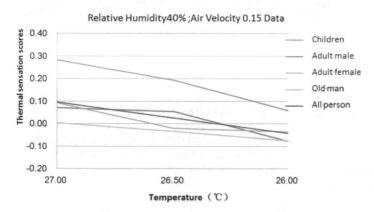

Fig. 3. Thermal comfort score curve of different crowds with fixed relative humidity, air velocity and variable temperature

Theoretically speaking, When the relative humidity is fixed and the air velocity is fixed, the lower the temperature is, the cooler people will feel, that is, the lower the score of thermal sensation. As shown in Fig. 3, when the temperature changes from 27 to 26 °C, the thermal sensation scores of the five groups of subjects decrease gradually, which is consistent with the theory. Under the conditions of relative humidity 40% and air velocity 0.15 m/s, the temperature of 26.5 °C is more comfortable.

3.2 Fixed Relative Humidity, Temperature and Variable Air Velocity

Under refrigeration conditions, when the relative humidity is 40%, the temperature is 27 °C, the air velocity is 0.15 m/s, 0.20 m/s and 0.25 m/s, the rationality of the evaluation data is analyzed by taking the thermal sensation score of different people as an example, as shown in Fig. 4.

In theory, when the relative humidity is fixed and the temperature is fixed, the higher the air velocity, the cooler people will feel, that is, the lower the score of thermal sensation. As shown in Fig. 4, when the air velocity changes from 0.15 m/s to 0.25 m/s, the thermal sensation scores of the five groups of people decrease gradually, which is consistent with the theory. The air velocity of 0.2 m/s is more comfortable under the conditions of relative humidity 40% and temperature 27 °C.

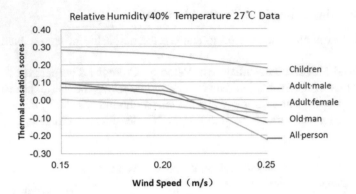

Fig. 4. Thermal comfort scoring curve for different populations with fixed relative humidity, temperature and variable air velocity

3.3   Different People Have Different Thermal Sensations

Although the sample size of the subjects in this test is small, the subjective evaluation of thermal comfort can explain some problems. The thermal sensation scores of five groups of people and 18 groups under refrigeration conditions are given. The results are shown in Fig. 5.

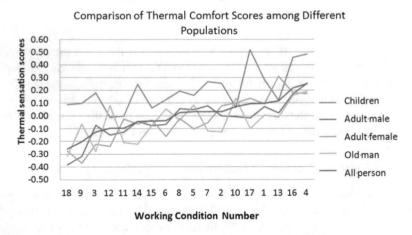

Fig. 5. Comparison of thermal comfort scores of different populations under refrigeration conditions

In Fig. 5, the abscissa is the test serial number, which is drawn in the order from low to high of everyone's thermal sensation score. As can be seen from Fig. 5, the blue curve representing "children" is far higher than other curves, which indicates that the score of thermal sensation is higher under the same working condition, that is, "children" will feel hotter under the same working condition, while the Yellow curve representing "old people" is below the "all people" pink curve under most working conditions, that is, under the "all people" pink curve. Under the same working condition, the "old man" will feel cooler; the curve of "young (adult) male" and "all" is basically the same; the "situation (adult) female" feels cooler under most working conditions, but higher than the "old man".

Therefore, in the same thermal environment, children will feel the hottest, followed by "young (adult) men", "young (adult) women", "old people" feel the coldest. If the product is subdivided, the most suitable thermal comfort parameters are set for different groups of people.

The results of the "children" test show that although the overall trend is correct, the subjective differences of children are quite large. Seen from Fig. 5, the blue curve representing "children" fluctuates greatly from top to bottom. On the one hand, there are great differences among different children. On the other hand, some children are not able to completely subjective evaluate the thermal comfort performance of their environment.

From Fig. 6, it can be seen that the higher the air velocity, the lower the thermal comfort score, when the relative humidity and temperature are constant in theory. However, for the thermal sensation scores of the "elderly" and "children", the air velocity at the maximum air velocity of 0.2 m/s is higher, which is inconsistent with the theory. This phenomenon also occurs under other working conditions. The reason for this problem is that winter clothes are thicker. The difference of each working condition is relatively small. So when the subjects wear winter clothes, the smaller difference of environmental thermal comfort can not be reflected in the thermal sensation score, and the difference of thermal sensation is not obvious.

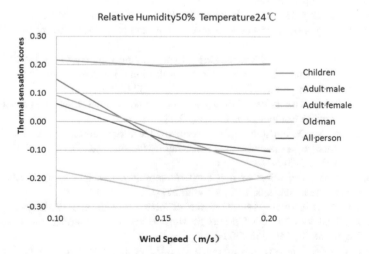

Fig. 6. Fixed relative humidity and temperature, changing air velocity and thermal comfort score curve of different crowd

4 Conclusion

Through the analysis of the test results, we can see that:

In the same indoor environment, children feel significantly warmer, while older people feel significantly colder. The difference between adult males and females is small, and males feel slightly warmer than females. For the whole population, the comfortable combination of temperature, humidity and air velocity should be adjusted according to 26.5 C, 0.05 m/s air velocity and 40% humidity under the condition of sitting in summer clothes and slight mental activity. The combination of comfortable temperature, humidity and air velocity is regulated by 23.5 C, 0.05 m/s air velocity and 40% humidity under mild mental activity in winter.

Acknowledgments. This research is supported by "Special funds for the basic R&D undertakings by welfare research institutions" (522017Y-5276, 522018Y-5984) and General Administration of Quality Supervision, Inspection and Quarantine of the People's Republic of China (AQSIQ) science and technology planning project for 2017 (2017QK157).

References

1. Xiaolin, Xu, Baizhan, Li: Influence of indoor thermal environment on thermal comfort of human body. J. Chongqing Univ. **4**(28), 102–105 (2005)
2. Li Shigang, Lian Zhiwei: Discussions on the application of Fanger's thermal comfortable theory. In: Shanghai Refrigeration Institute Academic Annual Conference. (2007)
3. Holmér, I., Nilsson, H., Bohm, M., et al.: Thermal aspects of vehicle comfort. Appl. Hum. Sci. J. Physiol. Anthropol. **14**(4), 159–165 (1995)
4. Hai, Ye, Runbai, Wei: Evaluation indices of thermal environment based on thermal manikin. Chin. J. Ergon. **11**(2), 26–28 (2005)
5. Tanabe, S., Zhang, H., Arens, E.A., et al.: Evaluating thermal environments by using a thermal manikin with controlled skin surface temperature. Ashrae Transactions **100**, 39–48 (1994)
6. Zhu Yingxin: Building Environment[M]. China Building Industry Press. (2010)
7. Nilsson, H.O., Holmér, I.: Comfort climate evaluation with thermal manikin methods and computer simulation models. Indoor Air **13**(1), 28–37 (2003)
8. Zhaohua, Zhang: Thermal manikin application in the thermal comfort evaluation. China Personal Protective Equipment **1**, 23–25 (2008)
9. Foda, E., Kai, S.: Design strategy for maximizing the energy-efficiency of a localized floor-heating system using a thermal manikin with human thermoregulatory control. Energy Build. **51**(8), 111–121 (2012)
10. Barna, E.: Combined effect of two local discomfort parameters studied with a thermal manikin and human subjects. Energy Build. **51**(4), 234–241 (2012)
11. Bogerd, C.P., Brühwiler, P.A.: The role of head tilt, hair and wind speed on forced convective heat loss through full-face motorcycle helmets: a thermal manikin study. Int. J. Ind. Ergon. **38**(3), 346–353 (2008)
12. Oliveira, A.V.M., Gaspar, A.R., Francisco, S.C., et al.: Analysis of natural and forced convection heat losses from a thermal manikin: comparative assessment of the static and dynamic postures. J. Wind. Eng. Ind. Aerodyn. **132**, 66–76 (2014)

13. Cheong, K.W.D., Yu, W.J., Kosonen, R., et al.: Assessment of thermal environment using a thermal manikin in a field environment chamber served by displacement ventilation system. Build. Environ. **41**(12), 1661–1670 (2006)
14. Elabbassi, E.B., Delanaud, S., Chardon, K., et al.: Electrically heated blanket in neonatal care: assessment of the reduction of dry heat loss from a thermal manikin. Elsevier Ergon. Book **05**, 431–435 (2005)
15. Matsunaga, K., Sudo, F., Yoshizumi, S., et al.: Evaluating thermal comfort in vehicles by subjective experiment, thermal manikin, and numerical manikin. Jsae Rev. **17**(4), 455 (1996)
16. ISO 14505, Ergonomics of the thermal environment—Evaluation of thermal environments in vehicles (2007)

Strategic Decision Making Models in Manufacturing and Service Systems

A Component Based Model Developed for Machine Tool Selection Decisions

Yusuf Tansel İç[1]([✉]) and Mustafa Yurdakul[2]

[1] Faculty of Engineering, Department of Industrial Engineering,
Baskent University, 06810 Etimesgut, Ankara, Turkey
yustanic@baskent.edu.tr
[2] Faculty of Engineering, Department of Mechanical Engineering,
Gazi University, 06570 Maltepe, Ankara, Turkey
yurdakul@gazi.edu.tr

Abstract. Machine tools are widely used in manufacturing sectors; such as automotive industry, metal cutting industry, aerospace industry etc. Purchase of a machine tool is a long-term capital investment decision and requires a high initial investment cost. Machine tool producers offer a wide-ranging types and models of machine tools. On the other hand, expectations and requirements of the manufacturing companies differ depending on the parts produced and their strategic objectives. High stiffness, rigidity, metal cutting performance, surface finish and low tolerance range are common expectations from machine tools. This paper aims to develop a technical evaluation model to help manufacturing companies in their machine tool purchasing decisions. In the proposed model, first components used in machine tools are analyzed and based on this analysis a technical evaluation model is developed. The application of the developed model is illustrated by making a selection among nine different machine tool alternatives.

Keywords: Technical evaluation · Machine tools · Machine tool selection · Model · Machine tool performance criteria · Machine tool components

1 Introduction

Investments in machine tools are capital investment projects that improve the manufacturing performance and contribute to the sales revenue of a company in its market. Without proper investment in machine tools achieving high competitiveness and getting maximum profit is extremely difficult for manufacturing companies. In addition, manufacturing companies become more flexible and productive within their manufacturing activities with the correct machine tool selections. The selection of appropriate machine tools is critical for survival of manufacturing companies.

Today, modern machine tools, such as machining centers, turning centers and mill-turn centers, do not require human workers even for loading and unloading work parts and perform milling, turning, drilling, and other machining operations along with quality measurement activities in the same machining cycle automatically. They provide the benefits of specialized and flexible (non-dedicated) machine tools together so

© Springer Nature Switzerland AG 2020
W. Karwowski et al. (Eds.): AHFE 2019, AISC 971, pp. 279–288, 2020.
https://doi.org/10.1007/978-3-030-20494-5_26

that various dissimilar operations can be performed on a work-part resulting low machining cost and manufacturing lead time.

The diversity of machine tools in terms of technical capabilities makes the selection decision a difficult problem. Furthermore, the budget restrictions of companies effect the selection decision. In a selection decision, technical factors such as; axis size, power, spindle speed and tolerances and other factors such as machine price, service availability after sales have to be considered [1–3]. In this paper, the impact of alternative machine tools on the technical criteria (stiffness, damping capacity, thermal stability, accuracy and cutting speed) are evaluated and measured qualitatively. The qualitative scores are then combined to obtain a final score for each alternative. The alternative with the highest score is recommended for selection.

In the machine tool selection literature, there are various studies. For example, Tabucanon et al. [4], and Wang et al. [5] studied the machine selection problem for flexible manufacturing systems (FMS). Hence, Arslan et al. [6], Lin and Yang [7] and Oeltjenbruns et al. [8] proposed analytic hierarchy process (AHP) for machine tool selection problem. On the other hand, Ic and Yurdakul [9] studied a decision support system (DSS) form machining center selection based on fuzzy AHP and fuzzy Technique for Order Preference by Similarity to Ideal Solution (TOPSIS) approaches. Yurdakul [10] presented a selection model which links machine alternatives to manufacturing strategies in its decision hierarchy. Beard [11] and Vasilash [12] developed a computer program namely "machine tool selector" which makes its selection in its own database. In another study, Sun [13] presented a machine tool selection methodology which uses Data Envelopment Analysis (DEA) method. Davedzic and Pap [14] presented a model which uses rigidity criterion in ranking machine tools. Georgakellos [15] developed a scoring model using technical and commercial criteria in its study.

Majority of the selection models available in the literature utilize the data presented in the alternative machine tools' catalogues. However, the presented data in the catalogues may not reflect the actual usage of the machine tools. The machine tools are generally required to operate under heavy loads for long periods of time without compromising their machining performance under various environmental conditions so that a robust performance is preferred from machine tools. For this reason, in this study, critical machine elements that constitute the main structures of the machining centers are classified and their impact on the machining performance of machining tools. The machining performance is described in terms of stiffness, damping capacity, thermal stability, accuracy and cutting speed criteria. The min-max approach described in Naik and Chackravarty [16] is applied to link machine tool components to machining performance for machine tool alternatives.

2 Component Based Description of Machine Tools

A machine tool (such as a machining center or a turning center) has three main sections namely, structure (frame), drive system and CNC unit. Machine tool's structure consists of fixed components (columns portals and beds) and moving elements (guides, saddles, tables). The drive systems provide torque, cutting force, speed and acceleration with its motor and transmission elements and its spindle and bearings. The CNC controls start and stop motions in rotational and axial directions, adjust speed and acceleration in a coordinated way to obtain the desired shape on the work-piece [3].

In a machine tool, damping capacity provides chatter stability and depends on the structure material types and joining types such as welding or bolt-nut etc. If damping capacity in a machine tool is low, surface quality of machined work part will also be low [2].

The whole structure of a machine tool must ensure high accuracy in the motions and high rigidity, both static and dynamic, against the cutting and other forces acting on the structure. Cutting forces generate elastic deformations in the machine-tool structure, and these deformations cause dimensional and shape deviations in the finished work-piece [2].

Machine tool producers continuously improve machine tools to obtain higher stability and accuracy in the cutting process. For example, a new design approach namely, driven at the center of gravity is being applied in developing new machine tools. The principle is based on the idea of 'something has to be pushed at its center of gravity; otherwise it will spin and become unstable'. In the application of this principle, the center of gravity is enclosed with two drive points (ball screws) on either side. The parallel line passing through the middle of the two ball screws passes through the center of gravity of the work part being moved.

The most common structure materials in a machine tool are gray or ductile cast iron, polymer concrete, ceramics and granite-epoxy composites. Gray iron has the advantages of low cost and good damping capacity, but it is heavy. Polymer concretes are a mixture of crushed concrete and plastic. Although they have good damping capacity, polymer concretes have low stiffness and poor thermal conductivity. Ceramic structures are being preferred in modern advanced machine tools for their high strength, stiffness, corrosion resistance, and good thermal stability. On the other hand, composites, made of polymer-metal or ceramic-matrix with various reinforcing materials, are expensive and presently limited to high speed machining (HSM) applications. A typical example for composite material is granite-epoxy combination which consists of 93% crushed granite and 7% epoxy binder. This composite combination provides high stiffness, damping capacity, thermal stability and resistance to environmental degradation [3].

Guides form three separate main groups namely, box type, linear (roller) type and hydrostatic type. Box ways consist of precision-ground surfaces and operate by sliding along the reference edges. The advantages of box ways are high stiffness and rigidity, simplicity in their design and operation. The disadvantages are high friction, low speed capability and difficulty in their maintenance and repair activities. There has been a transition in recent years from conventional sliding guide ways (box ways) to rolling guide ways in machine tools [17]. Rolling guide ways are available with balls or cylindrical rollers, and each has different characteristics in terms of load capacity, rigidity, vibration damping, accuracy, friction, life, and noise. Although roller guide ways offer higher stiffness, load ratings, rigidity and lower deflection during travel, most machine builders prefer ball guide ways because of their lower purchase prices [17]. Roller guide ways also generate low noise levels in the high-frequency range to which people are sensitive, resulting in quiet machine operation [17].

Hydrostatic bearings are preferred over other bearing types for their capability in reaching higher speed because of their lower friction coefficients by employing a fluid film between the guide surfaces. The hydrostatic bearing configuration requires a very elaborate, potentially expensive fluid handling system to maintain appropriate pressure and volume levels [18, 19].

The discussion provided above illustrates that guides, spindle/bearing, feed drive, and structure are the four components that are critical in determining a machine tool's robust performance. Five criteria, namely stiffness, damping capacity, thermal stability, speed capacity, and accuracy, are selected to represent the robust performance of a machine tool this study. The selection (performance) criteria and the machine tool components are linked together as shown in Fig. 1. The effects of components' types on the robust performance criteria are summarized in Table 1 [20–28].

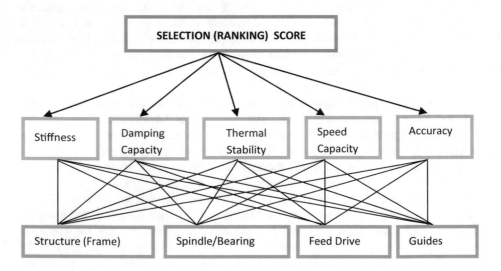

Fig. 1. The developed decision hierarchy in machine tool selection problem

3   Application of the Min-Max Approach to Determine the Selection Score

In the literature, Naik and Chakravarty [16] illustrated application of the min-max approach in selection of advanced manufacturing technologies (AMT). In the approach, importance is measured with five alpha-numeric symbols namely, N (None Important), L (Low Important), M (Medium), H (High Important), V (Very Important). The min-max approach is applied below for machine tool selection problem and results are provided in the following section (Tables 2, 3, 4, 5, 6).

Table 1. Component types and their performance levels [28]

Component name	Interaction/ Functions	Component types	Performance level at selection criteria
Guides	Feed Motion	Box-Type	High stiffness, good damping capacity, high friction, high wear, low speed, difficult maintenance and repair
		Linear-Ball Bearing	Very low friction, good stiffness, good damping capacity, high accuracy, very high speed, very easy maintenance and repair
		Linear-Cylindrical Bearing	High speed, high stiffness, low friction, very good damping capacity, very high accuracy, easy maintenance and repair
		Hydrostatic	No wear, very low friction, very high loading capacity, difficult and expensive maintenance and repair
Spindle/Bearing	Tool-Work-part Motion	Ball Bearing	Good accuracy, high speed, low friction
		Angular Contact Ball Bearing	High accuracy, high friction, high loading capacity
		Cylindrical Bearing	High accuracy, high damping capacity, good speed capacity, high stiffness, very high loading capacity, high friction
		Hybrid Bearing	High accuracy, high damping capacity, very fast speed
Feed drive	Table and Head Stock Motion	Ball-Screw-Single Nut	Very good accuracy, high speed, low heating, low friction
		Ball-Screw-Double Nut	Very high stiffness, quite good speed, high damping capacity, very good accuracy, high heat
		Ball-Screw-Fixed both end/or Preloaded	High damping capacity, quite good speed, very high stiffness, very high heating, very good accuracy
		Linear Motor	Very low friction, high positioning, high accuracy, very high speed, very easy or no maintenance
Structure	Frame for the Machining Center's Components	Cast-Iron	Good damping capacity, cheap and easy to build structure
		Special Design or Special Materials Used	Better heat dissipation; it is possible to build structures such as, bridge type, symmetric, double column that provide increased stiffness or thermal stability and damping capacity

4 Application of the Min-Max Approach in Machine Tool Selection Problem

If "Matrix A" and "Matrix B" represent the linkages between machine tool alternatives (*MT*) and machine tool components (*MC*) and machine tool components with performance criteria (*PC*) respectively, one may obtain Matrix C that links machine tool alternatives to performance criteria as follows:

$$A = \begin{bmatrix} & MT_1 & MT_2 & MT_3 & \cdots \\ MC_1 & a_{11} & a_{12} & a_{13} & \cdots \\ MC_2 & a_{21} & a_{22} & a_{23} & \cdots \\ MC_3 & a_{31} & a_{32} & a_{33} & \cdots \\ & \cdot & \cdot & \cdot & \cdot & \cdots \\ & \cdot & \cdot & \cdot & \cdot & \cdots \end{bmatrix} \quad B = \begin{bmatrix} & PC_1 & PC_2 & PC_3 & \cdots \\ MT_1 & b_{11} & b_{12} & b_{13} & \cdots \\ MT_2 & b_{21} & b_{22} & b_{23} & \cdots \\ MT_3 & b_{31} & b_{32} & b_{33} & \cdots \\ & \cdot & \cdot & \cdot & \cdot \\ & \cdot & \cdot & \cdot & \cdot \end{bmatrix} \quad (1)$$

$$C = \begin{bmatrix} & PC_1 & PC_2 & PC_3 & \cdots \\ MC_1 & c_{11} & c_{12} & c_{13} & \cdots \\ MC_2 & c_{21} & c_{22} & c_{23} & \cdots \\ MC_3 & c_{31} & c_{32} & c_{33} & \cdots \\ & \cdot & \cdot & \cdot & \cdot \\ & \cdot & \cdot & \cdot & \cdot \end{bmatrix} \quad (2)$$

$$c_{11} = Best\{Worst(a_{11}; b_{11}); Worst(a_{12}; b_{21}); Worst(a_{13}; b_{31})\} \quad (3)$$

$$c_{21} = Best\{Worst(a_{21}; b_{11}); Worst(a_{22}; b_{21}); Worst(a_{23}; b_{31})\} \quad (4)$$

$$c_{31} = Best\{Worst(a_{31}; b_{11}); Worst(a_{32}; b_{21}); Worst(a_{33}; b_{31})\} \quad (5)$$

$$c_{12} = Best\{Worst(a_{11}; b_{12}); Worst(a_{12}; b_{22}); Worst(a_{13}; b_{32})\} \quad (6)$$

$$c_{22} = Best\{Worst(a_{21}; b_{12}); Worst(a_{22}; b_{22}); Worst(a_{23}; b_{32})\} \quad (7)$$

$$c_{32} = Best\{Worst(a_{31}; b_{12}); Worst(a_{32}; b_{22}); Worst(a_{33}; b_{32})\} \quad (8)$$

Where; a_{ij}; i = 1...9; j = 1...4 (a_{ij}; N or L or M or H or V); b_{ij}; i = 1...4; j = 1...5; (b_{ij}; N or L or M or H or V); c_{ij}; i = 1...9; j = 1...5 (c_{ij}; N or L or M or H or V); MT: Machining Tool, MC: Machine Tool Component, PC: Performance Criteria.

4.1   Calculation of Ranking Scores of Machine Tool Alternatives

In this section alternative machine tools' (MTs) ranking scores are calculated. As a first step, alpha-numeric elements in MT-PC matrix, are converted to numerical numbers. The numerical numbers corresponding to N, L, M, H, V are 0, 3 (3^1), 9 (3^2), 27(3^3) and 81 (3^4) respectively (Table 7). In the second step, row elements of MC-PC matrix are summed to obtain ranking scores of alternative machine tools. The ranking scores are provided in the last column of Table 7.

Table 2. Evaluation and pointing of 9 alternative machine tools

	MAZAK FH 6000	MAZAK FJV120	MAZAK VTC400	OKUMA MA40HA	EXCEL H-360	MILLTR VM15	KAFO HMC 360	FEELER FV1300	EAGLE VMC1300
STIFFNESS									
Guides	0.80	0.80	0.70	0.70	0.70	0.70	0.90	0.70	0.70
Spindle/Bearing type	0.50	0.50	0.40	0.40	1.00	1.00	0.50	0.50	0.60
Feed drive	0.50	0.50	0.50	0.40	0.40	0.40	0.60	0.40	0.50
Structure	0.80	1.00	0.80	0.80	0.60	0.40	0.40	0.40	0.40
DAMPING CAPACITY									
Guides	0.80	0.80	0.70	0.70	0.70	0.70	0.50	0.70	0.70
Spindle/Bearing type	0.50	0.50	0.40	0.40	0.80	0.80	0.50	0.50	0.50
Feed drive	0.70	0.70	0.70	1.00	0.70	0.70	0.40	0.40	0.80
Structure	0.80	1.00	0.80	0.80	0.60	0.40	0.40	0.40	0.40
THERMAL STABILITY									
Guides	0.70	0.70	0.70	0.70	0.70	0.40	0.00	0.70	0.70
Spindle/Bearing type	0.90	0.90	0.70	1.00	0.80	0.80	0.70	0.80	0.70
Feed drive	0.60	0.60	0.40	0.50	0.40	0.40	0.40	0.40	0.40
Structure	0.80	0.80	0.80	1.00	0.40	0.40	0.40	0.30	0.30
ACCURACY									
Guides	0.70	0.70	0.70	0.70	0.70	0.70	0.50	1.00	1.00
Spindle/Bearing type	0.60	0.60	0.40	0.50	0.80	0.80	0.50	0.50	0.50
Feed drive	0.50	0.50	0.40	0.40	0.40	0.40	0.60	0.30	0.40
SPEED CAPACITY									
Guides	1.00	1.00	1.00	1.00	1.00	1.00	0.00	1.00	1.00
Spindle/Bearing type	0.80	0.80	0.70	0.90	0.80	0.80	0.70	0.80	0.70
Feed drive	0.40	0.40	0.40	0.40	0.40	0.40	0.50	0.40	0.40
Average Points									
Guides	0.80	0.80	0.76	0.76	0.76	0.70	0.38	0.82	0.82
Spindle/Bearing type	0.66	0.66	0.52	0.64	0.84	0.84	0.58	0.62	0.60
Feed drive	0.54	0.54	0.48	0.54	0.46	0.46	0.50	0.38	0.50
Structure	0.80	0.93	0.80	0.87	0.53	0.40	0.40	0.37	0.37

Table 3. Alpha numeric symbols and their ranges and descriptions

Range	Alpha-numeric symbol	Description
x < 0.30	N	None important
0.30 <= x < 0.50	L	Low important
0.50 <= x < 0.65	M	Medium
0.65 <= x < 0.80	H	High important
x >= 0.80	V	Very important

Table 4. Comparison matrix for MT alternatives against machine tool components

Machine tool alternatives	Guide	Bearing	Feed drive	Structure
FH6000	V	H	M	V
FJV120	V	H	M	V
VTC300C	H	M	L	V
MA40HA	H	M	M	V
H360	H	V	L	M
VM15	H	V	L	L
HMC360	L	M	M	L
FV1300	V	M	L	L
VMC1300	V	M	M	L

Table 5. Performance values of MT components at performance criteria

	Stiffness	Damping capacity	Thermal stability	Accuracy	Speed capacity
Guide	V	V	M	V	H
Bearing	H	H	V	H	V
Feed drive	V	H	V	V	H
Structure	V	V	V	M	N

Table 6. Peformance values of MT alternatives at the performance criteria

Machine tool alternatives	Stiffness	Damping capacity	Thermal stability	Accuracy	Speed capacity
FH6000	V	V	V	V	H
FJV120	V	V	V	V	H
VTC300C	V	V	V	H	H
MA40HA	V	V	V	H	H
H360	H	H	V	H	V
VM15	H	H	V	H	V
HMC360	M	M	M	M	M
FV1300	V	V	M	V	H
VMC1300	V	V	M	V	H

Table 7. Calculation of the final ranking scores of machine tool alternatives

Machine tool alternatives	Stiffness	Damping capacity	Thermal stability	Accuracy	Speed capacity	Row sum	Rank
FH6000	81	81	81	81	27	351	1
FJV120	81	81	81	81	27	351	1
VTC300C	81	81	81	27	27	297	3
MA40HA	81	81	81	27	27	297	3
H360	27	27	81	27	81	243	7
VM15	27	27	81	27	81	243	7
HMC360	9	9	9	9	9	45	9
FV1300	81	81	9	81	27	279	5
VMC1300	81	81	9	81	27	279	5

5   Conclusion

In this paper, a machine tool selection model has been presented for machine tool purchasing decisions. Additional evaluation for a limited number of best alternatives can be done by using simulation models that can be easily built with simulation languages such as SIMAN, ARENA, PROMODEL. The simulation models can be run for an appropriate manufacturing scenario such as mass production or mass customization for a pre-determined planning horizon. The outcomes of the simulation runs can be converted to cash-flows (revenues and expenses over the years) for an economic analysis.

References

1. Altıntaş, Y.: Manufacturing automation: metal cutting mechanics, machine tool vibrations, and CNC design. Cambridge University Press, Cambridge, UK (2000)
2. Kalpakjian, S., Schmid, S.R.: Manufacturing Engineering and Technology, 4th edn. Prentice-Hall Inc, Upper Saddle River, NJ (2001)
3. Tlusty, G.: Manufacturing processes and equipment. Printice Hall, USA (2000)
4. Tabucanon, M.T., Batanov, D.N., Verma, D.K.: Intelligent decision support system (DSS) for the selection process of alternative machines for Flexible Manufacturing Systems (FMS). Comput. Ind. **25**, 131–143 (1994)
5. Wang, T.Y., Shaw, C.-F., Chen, Y.-L.: Machine selection in flexible manufacturing cell: a fuzzy multiple attribute decision making approach. Int. J. Prod. Res. **38**(9), 2079–2097 (2010)
6. Arslan, M.C., Catay, B., Budak, E.: Decision support system for machine tool selection. In: Baykasoglu A. and Dereli T. (eds.) Proceedings of ICRM-2002, 2nd International Conference on Responsive Manufacturing. University of Gaziantep, Turkey, pp. 752–757 (2002)
7. Lin, Z-C., Yang, C-B.: Evaluation of machine selection by the AHP method. J. Mater. Process. Technol. **57**, 253–258 (1996)

8. Oeltjenbruns, H., Kolarik, W.J., Schnadt-Kirschner, R.: Strategic planning in manufacturing systems- AHP application to an equipment replacement decision. Int. J. Prod. Econ. **38**, 189–197 (1995)
9. Ic, Y.T., Yurdakul, M.: Development of a decision support system For machining center selection. Expert Syst. Appl. **36**(2), 3505–3513 (2009)
10. Yurdakul, M.: AHP as a strategic decision-making tool to justify machine tool selection. J. Mater. Process. Technol. **146**, 365–376 (2004)
11. Beard, T.: A shorter path to machine tool selection. Mod. Mach. Shop **69**(8), 110–111 (1997)
12. Vasilash, G.S.: Machine tool selection made simple. Automot. Manuf. & Prod. **109**(3), 66–67 (1997)
13. Sun, S.: Assessing computer numerical control machines using data envelopment analysis. Int. J. Prod. Res. **40**(9), 2011–2039 (2002)
14. Devedzic, G.B., Pap, E.: Multicriteria-multistage linguistic evaluation and ranking of machine tools. Fuzzy Sets Syst. **102**, 451–461 (1999)
15. Georgakellos, D.A.: Technology selection from alternatives: a scoring model for screening candidates in equipment purchasing. Int. J. Innov. Technol. Manag. **2**(1), 1–18 (2005)
16. Naik, B., Chakravarty, A.K.: Strategic acquisition of new manufacturing technology: a review and research framework. Int. J. Prod. Res. **30**(7), 1575–1601 (1992)
17. Better bearing lead to better parts by Ken Mizuteni, http://www.machinedesign.com/articles
18. Technical information bulletin, Hardinge Inc, http://www.hardinge.com
19. Advanced bearing for high-speed machining by Mark Mc Irath and Anthony Romero, http://www.machinedesign.com/articles
20. High speed machining home page/spindles, http://www.mmsonline.com/articles/hsmgp/spin3.html
21. Scheerer Bearing Corporation, http://www.scheererbearing.com
22. Ball screw or linear motors? Apps, Not Specs, Determine which is Best by Greg Hyatt, http://www.mmsonline.com/articles
23. How a ball screw working, http://www.barnesball-screw.com/ball.html
24. Ball screw components & assemblies, http://www.comptrolinc.com/bs-products.html
25. Accuracy of feed drive by Jan Braasch, http://www.heidenhain.com/techart
26. Irath, MMc, Romero, A.: Advanced bearing for high-speed machining. The Timkon Co. Coventry, UK (2005)
27. Tlusty, G.: Manufacturing processes and equipment, Chapter 7–10. Prentice Hall, USA (2000)
28. İç, Y.T., Yurdakul, M., Eraslan, E.: Development of a component based machining center selection model using AHP. Int. J. Prod. Res. **50**(22), 6489–6498 (2012)

Multiple Service Home Health Care Routing and Scheduling Problem: A Mathematical Model

Asiye Ozge Dengiz[1(✉)], Kumru Didem Atalay[1],
and Fulya Altiparmak[2]

[1] Başkent University, Ankara, Turkey
{aokarahanli,katalay}@baskent.edu.tr
[2] Gazi University, Ankara, Turkey
fulyaal@gazi.edu.tr

Abstract. The home health care routing and scheduling problem (HHCRSP) is an extension of the vehicle routing problem (VRP) that are scheduled and routed to perform a wide range of health care services. Nurses, doctors and/or caregivers provide these services at patients' home. In this study, a mathematical model for HHCRSP is presented. The model is extended to take into account additional characteristics and/or constraints based on specific services, patient needs. In the home health care (HHC) problems, services that must be performed simultaneously or within a convinced time are undoubtedly very important. Thus, we consider several numbers of services, skill requirements for the care workers and time windows. Generally, the main aim of the HHC problems is minimizing the travelling distance as well as maximizing the patients' satisfaction. Thus, the model in this study contains both of these objectives taking into account several measurements.

Keywords: Home health care · Routing and scheduling ·
Mathematical model · Time windows · Multiple services

1 Introduction

Vehicle Routing Problem (VRP) is one of the main problems in logistics and operations research literature. After the first study by Dantzig and Ramsey [7], it has been applied to many fields and many studies have been published by deriving different extensions. Research on multi-featured VRPs, taking into account different features, constraints and systems, has been actively used to elicit realistic determinations and theoretical challenges, and to combine real-life cases and combinations of constraints in practice.

Especially in European countries, increasing the population of elderly people and patients with chronic diseases, having the lower cost compared to the hospital and the developments in health care technologies make the home health care (HHC) more preferable and advantageous. The other advantages of HHC are to cause reduction the patient densities and hospitalization periods in hospitals. In addition, it is allowed for the patients to be treated in their own homes by keeping away from the serious

© Springer Nature Switzerland AG 2020
W. Karwowski et al. (Eds.): AHFE 2019, AISC 971, pp. 289–298, 2020.
https://doi.org/10.1007/978-3-030-20494-5_27

(bedridden and /or chronic) illnesses of hospital environment. Also, HHC, which is the only way to reach some patients and prevents unnecessary occupation of hospitals, draws the attention of the health sector and requires new strategies, policies, arrangements and solution approaches to be developed for arising new needs and problems.

Due to the demands and advantages of the HHC system, many hospitals have HHC departments and give variety of the services to patient their own home through health care staff/team. The team is included doctor, nurse, driver, dietitian, physical therapist, psychologist and the like. They have different qualifications so give variety of services to patient such as bloodletting, examine the patients, take the biological samples, exercise, prepare diet and so on. The staff members/caregivers are typically equipped with cars and other medical equipments and through this car, they make visits between patients' home and hospital. At the patients' side, they need specific treatments and this needs should be met by qualified staff members within a given time window.

Home Health Care Routing and Scheduling Problem (HHCRSP) is generally defined as determining which health care staff should be served in the patients who need to be visited by the health caregiver/staff in their homes. The aim of this problem is to minimize the travelling cost or any other selected criteria or to maximize the service quality with considering different constraints during the planning period.

The difference of HHCRSP from classical VRP can be summarized as follows; the need for continuity of care in the health care staff-patient match, the interdependence, separation of services and care services, and the consideration of health caregiver-patient characteristics.

In the HHC system, there is a connection with two main problems; the first one is which caregiver should be sent to the patient and the second one is routing the patients' visits. These two problems can be handled separately or simultaneously. It is generally considered to be the problem of routing of home health care because of the demand from the patient or due to patients existing in the hospital system and corresponds to the different VRP models in the literature [11].

In this study, a problem definition is made based on real-life case and a mathematical model is proposed that is to route staff members and to schedule single, double and triple service operations for the HCC. This problem is called as HHCRSP. The following sections of the study are as follows. In Sect. 2, literature research for HHCRSP is presented. An extended mathematical model is explained in detail and demonstrated with a descriptive example in Sect. 3. Finally, the conclusion is given in Sect. 4.

2 Literature

In recent years, Home Health Care Routing and Scheduling Problem, which can be considered as an extension of VRP in the health sector, has attracted researchers' attention. The study that is accepted as the first study is published by Begur et al. [2]. In the following years, many other researchers studies and three review articles were published in 2017 [6, 10, 13]. In these studies, in the literature, mathematical models have variety of objective functions and different constraints are considered thus different solution approaches are developed. In these studies, there are several differences among the proposed models and solution approaches. First one is related to the staff members (homogeneous or heterogeneous), second one is related to the services that are provided by the caregivers (number of the services and simultaneous or precedence services), third one is related to the objective function of the model (minimized travelling time or maximized service quality), fourth one is related to the scheduling time period (short or long term) and the last one is related to the methodology (exact methods and/or heuristics). Homogenous staff member are considered in studies [1, 5, 14; heterogeneous are taken into consideration in studies [2, 3, 9, 12, 15]. Interdependent services is included to the model in papers [5, 9, 12, 14]. As an objective function, the studies with minimized traveling time/cost is dominant in the literature [1–3, 5, 9, 11, 12, 14, 15] however, the studies with maximized patient satisfaction/quality is rare [4, 8, 16]. Most of the researchers developed their model for short term period [1, 3, 4, 5, 9, 11, 12, 14, 15], with few of them to consider long term period [2, 8, 16]. Finally, the solution method is generally heuristics or metaheuristics [1, 2, 8, 3, 9], there are also studies that used both exact and heuristics methods together [4, 5, 12, 15]. The contribution of this study is extending a mathematical model for the HHCRSP that included time window, heterogeneous staff members and interdependent three consecutive services. Moreover, in order to meet the all patients' needs, the model is allowed to enable tardiness and the sum of all tardiness is minimized while assigning the patients qualified staff members.

3 Mixed Integer Programming Model for HHCRSP

The problem discussed in this study is a routing and scheduling problem in which the patients are given by four different health care personnel / staff consisting of nurses, nurses and doctors, physiotherapists and dietitians, and there is a interdependency between the services. The objective function of the model is to minimize the total travel time of the staff members while serving the patients as much as possible within the specified time windows. Patients can request single, double or triple services. Within the scope of this study, it is seen that the related services are consecutive and at most three services could be given in consecutive terms.

The information regarding the types of staff members, the services provided is given in Table 1. Also, single, double and triple services are shown in Table 2.

Table 1. The type of staff members and the services they are provided.

Staff members (v)	Services (s)
1 Nurse	1 Bloodletting
2 Nurse + doctor	2 Examine the patients
3 Physiotherapist	3 Exercise
4 Dietitian	4 Prepare diet

Table 2. Services based on the set of patients (C^s, C^d, C^t).

Single service	Double service	Triple service
Bloodletting (1)	Bloodletting, examine the patients (1 + 2)	Bloodletting, examine the patients, exercise (1 + 2 + 3)
Examine the patients (2)		
Exercise (3)	Bloodletting, exercise (1 + 3)	Bloodletting, examine the patients, prepare diet (1 + 2 + 4)
Prepare diet (4)	Bloodletting, prepare diet (1 + 4)	

3.1   Assumptions

For the mathematical model, the following assumptions are made considering the characteristics of the system.

- The planning period is one day.
- All patients should be served during the day.
- Each patient should be visited by staff member who is qualified to provide the requested service.
- There are four different types of staff members.
- The routes of the each staff member start and end in the hospital.
- Each patient is serviced within the specified period of time (time windows).
- Patients may request single or multiple services.
- There is a temporary relationship between services.
- The services requested by the patients can be given consecutively.
- Patients are provided with up to three consecutive services.
- Tardiness is allowed to serve all patients.

3.2   Notation

$C = Set\ of\ all\ patients$

$S = Set\ of\ offered\ service\ types$

$V = Set\ of\ staff\ members$

$C^0 = Set\ of\ all\ locations\,, C^0 = C\ \cup \{0\}$

$C^s = Set\ of\ single\ service\ patients$

$C^d = Set\ of\ double\ service\ patients, C^d = C^{sim}\ \cup\ C^{prec}$

$C^t = Set\ of\ triple\ service\ patients$

C^{sim} = Set of patients requiring simultaneous service

C^{prec} = Set of patients requiring service with precedence

$a_{vs} = \begin{cases} 1, & \textit{iff employee } v \in V \textit{ is qualified to provide service operation } s \in S \\ 0, & \textit{otherwise} \end{cases}$

$r_{is} = \begin{cases} 1, & \textit{iff patient } i \in C \textit{ requires service operation } s \in S \\ 0, & \textit{otherwise} \end{cases}$

δ_{i1}^{min} = Minimal time between 1st and 2nd service start times at patient $i \in C^d, C^t$

δ_{i1}^{max} = Maximal time between 1st and 2nd service start times at patient $i \in C^d, C^t$

δ_{i2}^{min} = Minimal time between 2nd and 3rd service start times at patient $i \in C^t$

δ_{i2}^{max} = Maximal time between 2nd and 3rd service start times at patient $i \in C^t$

$[e_i, l_i]$ = Time window of patient $i \in C$

d_{ij} = Travelling distance between locations i and j; $i, j \in C^0$

p_{is} = Processing time of service operation s at patient $i \in C$

$x_{ijvs} = \begin{cases} 1, & \textit{iff staff member } v \textit{ moves from } i \textit{ to } j \textit{ for providing service operation } s \\ 0, & \textit{otherwise} \end{cases}$

t_{ivs} = Start time of service operation s at patient i provided by staff member v

z_{is} = Tardiness of service operation s at patient i

3.3 Mathematical Model

A mathematical model is extended based on the model proposed by Mankowska et al. [12], considering the characteristic of the pilot hospital. The model is explained in detail as follows.

$$min\, Z = \lambda_1 D + \lambda_2 T + \lambda_3 T^{max}. \tag{1}$$

subject to

$$D = \sum_{v \in V} \sum_{i \in C^0} \sum_{j \in C^0} \sum_{s \in S} d_{ij} \cdot x_{ijvs}. \tag{2}$$

$$T = \sum_{i \in C} \sum_{s \in S} z_{is}. \tag{3}$$

$$T^{max} \geq z_{is} \forall i \in C, s \in S. \tag{4}$$

$$\sum_{i \in C^0} \sum_{s \in S} x_{0ivs} = \sum_{i \in C^0} \sum_{s \in S} x_{i0vs} = 1 \quad \forall v \in V. \tag{5}$$

$$\sum_{j \in C^0} \sum_{s \in S} x_{jivs} = \sum_{j \in C^0} \sum_{s \in S} x_{ijvs} \quad \forall i \in C, v \in V. \tag{6}$$

$$\sum_{v \in V} \sum_{j \in C^0} a_{vs}.x_{jivs} = r_{is} \quad \forall i \in C, s \in S \tag{7}$$

$$t_{ivs_1} + p_{is_1} + d_{ij} \leq t_{jvs_2} + M\left(1 - x_{ijvs_2}\right)$$
$$\forall i \in C^0, j \in C, v \in V, s_1, s_2 \in S. \tag{8}$$

$$t_{ivs} \geq e_i \quad \forall i \in C, v \in V, s \in S. \tag{9}$$

$$t_{ivs} \leq l_i + z_{is} \quad \forall i \in C, v \in V, s \in S. \tag{10}$$

$$t_{iv_2s_2} - t_{iv_1s_1} \geq \delta_{i1}^{min} - M\left(2 - \sum_{j \in C^0} x_{jiv_1s_1} - \sum_{j \in C^0} x_{jiv_2s_2}\right)$$
$$\forall i \in C^d, C^t, v_1, v_2 \in V, s_1, s_2 \in S : s_1 < s_2. \tag{11}$$

$$t_{iv_2s_2} - t_{iv_1s_1} \leq \delta_{i1}^{max} + M\left(2 - \sum_{j \in C^0} x_{jiv_1s_1} - \sum_{j \in C^0} x_{jiv_2s_2}\right)$$
$$\forall i \in C^d, C^t, v_1, v_2 \in V, s_1, s_2 \in S : s_1 < s_2. \tag{12}$$

$$t_{iv_3s_3} - t_{iv_2s_2} \geq \delta_{i2}^{min} - M\left(2 - \sum_{j \in C^0} x_{jiv_2s_2} - \sum_{j \in C^0} x_{jiv_3s_3}\right)$$
$$\forall i \in C^t, v_2, v_3 \in V, s_2, s_3 \in S : s_2 < s_3. \tag{13}$$

$$t_{iv_3s_3} - t_{iv_2s_2} \leq \delta_{i2}^{max} + M\left(2 - \sum_{j \in C^0} x_{jiv_2s_2} - \sum_{j \in C^0} x_{jiv_3s_3}\right)$$
$$\forall i \in C^t, v_2, v_3 \in V, s_2, s_3 \in S : s_2 < s_3. \tag{14}$$

$$x_{ijvs} \in \{0, a_{vs}.r_{js}\} \forall i, j \in C^0, v \in V, s \in S. \tag{15}$$

$$t_{ivs}, z_{is} \geq 0 \quad \forall i \in C^0, v \in V, s \in S. \tag{16}$$

In this study, three different performance measures are considered. The first one is indicated by D which is the total distance traveled by all staff members, second one is denoted by T and it is the total tardiness of services that start after the time windows and the last one is indicated by T_{max} and measures the maximal tardiness noticed all service operations. The objective function shown in Eq. (1) is weighted and reflected the combination of these three performance measures. According to the decision maker, coefficients of the function can be modified. In constraints (2)–(4), we can see the determination of the performance measures. Route of each staff member starts and ends at the central office (hospital) and this condition is ensured by constraints (5). Constraints (6), keeps the network flow even and constant. Constraints (7) guarantee that every service operation s is provided by exactly one qualified staff member. Starting time of service operations is calculated in constraints (8) considering processing time of the service and travelling times between locations. Constraints (9) and (10) are time windows and also calculate the tardiness of the services. Constraints (11)–(14) show the temporal interdependencies for the double and triple service operations. Minimal time distance between the service start times is ensured by constraints (11) and (13); when maximal time distance is ensured by constraints (12) and (14). Constraint (15), determined the decision variable of the route. Lastly, constraint (16) is a non-negativity constraint for the other decision variables.

Table 3. Qualification of staff members.

Staff member (v)	Qualifications (a_{vs})			
	$s = 1$	$s = 2$	$s = 3$	$s = 4$
1	1	0	0	1
2	1	1	0	0
3	0	0	1	0
4	0	0	0	1

Table 4. Service requirements of the patients.

Patient i	Service requirements (r_{is})				Time window		Time distances			
	1	2	3	4	e_i	l_i	δ_{i1}^{min}	δ_{i1}^{max}	δ_{i2}^{min}	δ_{i2}^{max}
1	1	0	0	0	0	30	-	-	-	-
2	1	1	0	0	10	95	100	150	-	-
3	0	1	0	0	95	160	-	-	-	-
4	0	0	1	0	50	300	-	-	-	-
5	1	0	0	0	5	120	-	-	-	-
6	0	0	0	1	20	400	-	-	-	-
7	0	0	0	1	10	250	-	-	-	
8	1	1	1	0	10	300	100	150	0	40
9	1	1	0	1	10	120	100	150	0	80

3.4   Example

To demonstrate the process of the extended model, a problem instance with nine patients, four services and four types of staff members is created. The locations of the hospital and the patients are generated randomly in 50x50 distance units. Travelling time is assumed to be the proportional to the travelling distance and it is Euclidean. Qualifications of the staff members, service requirements of the patients, time windows and time distances between services are displayed in Tables 3 and 4 respectively. There is six single service patients, one double service patients and two triple service patients. Processing time is assumed to be 10-time units for all service types. The weights of the each objective function subgoals are set as equal. The instance problem is solved using IBM ILOG CPLEX Optimization Studio linear programming software package Version 12.6.1 in the computer that is Intel® Core™ i3-3217U CPU @1.80 GHz, 4.00 GB Ram, 64-bit Operating System, x63-based processor. The optimal routing of the instance and the time space diagram of this solution is shown in the Figs. 1 and 2 respectively.

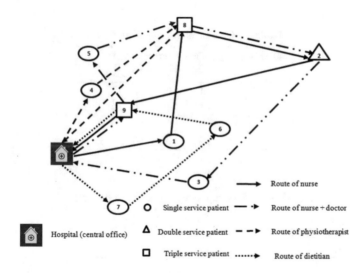

Fig. 1. Optimal routing of the staff members for instance.

4   Conclusion

HHCRSP is an important problem for many countries, especially European countries with high elderly population. In the literature, the number of studies on this subject has increased in recent years. In this study, a mathematical model for the home health care routing and scheduling problem is proposed with interdependent services. In real life, services that provided by health care staff members is diverse because of the patients needs and can be performed by at most three staff members simultaneously or consecutively. Therefore, this situation is considered in the mathematical model and

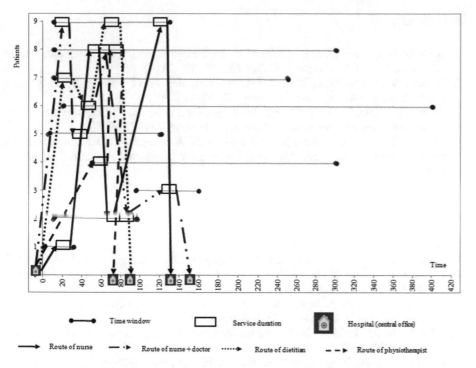

Fig. 2. Time space diagram of the optimal solution for instance.

variables to reflect this real-life case are created. Especially when the increasing demand and advantages of the HHC are taken into consideration, it is seen that many hospitals now have HHC departments and after this stage, many problems has arisen in the system. For these problems, researchers from different disciplines, especially industrial engineers, should make suggestions and develop approaches for the solutions.

For this purpose, in this study first the mathematical models for HHCRSP is considered as an extension of the VRP and the existing literature is investigated. Then an extended model is proposed for this problem. The proposed model is run for a sample problem using the data from the pilot hospital selected for this research. For staff members; acceptable travelling cost, for patients; reasonable waiting times and tardiness or a combination thereof is observed from the sample problem solutions.

To sum up, the problem defined in this study is HHCRSP, which has a time window, heterogeneous staff members and interdependent services. In the problem, different than the studies in the literature, the situation where three consecutive services can be handled is taken into consideration and the mathematical model is extended.

References

1. Akjiratikarl, C., Yenradee, P., Drake, P.R.: PSO-based algorithm for home care worker scheduling in the UK. Comput. Ind. Eng. **53**, 559–583 (2007)
2. Begur, S.V., Miller, D.M., Weaver, J.R.: An integrated spatial DSS for scheduling and routing home health care nurses. Interfaces **27**, 35–48 (1997)
3. Bertels, S., Fahle, T.: A hybrid setup for a hybrid scenario: combining heuristics for the home health care problem. Comput. Oper. Res. **33**, 2866–2890 (2006)
4. Braekers, K., Hartl, R.F., Parragh, S.N., Tricoire, F.: A bi-objective home care scheduling problem: analyzing the trade-off between costs and client inconvenience. Eur. J. Oper. Res. **248**, 428–443 (2016)
5. Bredstrom, D., Ronnqvist, M.: Combined vehicle routing and scheduling with temporal precedence and synchronization constraints. Eur. J. Oper. Res. **191**, 19–31 (2008)
6. Cisse, M., Yalcındag, S., Kergoisen, Y., Sahin, E., Lente, C., Matta, A.: OR problems related to home health care: A review of relevant routing and scheduling problems. Oper. Res. Health Care **13–14**, 1–22 (2017)
7. Dantzig, G., Ramsey, J.H.: The truck dispacthing problem. Manage. Sci. **6**, 80–91 (1959)
8. Duque, P.M., Castro, M., Sorensen, K., Goos, P.: Home care service planning: The case of Landelijke Thuiszorg. Eur. J. Oper. Res. **243**, 292–301 (2015)
9. Eveborn, P., Flisberg, P., Ronnqvist, M.: Laps care an operational system for staff planning of home care. Eur. J. Oper. Res. **171**, 962–976 (2006)
10. Fikar, C., Hirsch, P.: Home health care routing and scheduling: a review. Comput. & Oper. Res. **77**, 86–95 (2017)
11. Liu, R., Yuan, B., Jiang, Z.: Mathematical model and exact algorithm for the home care worker scheduling and routing problem with lunch break requirements. Int. J. Prod. Res. **55**, 558–575 (2017)
12. Mankowska, D.S., Meisel, F., Bierwirth, C.: The home Health care routing and scheduling problem with interdependent services. Health Care Manage. Sci. **17**, 15–30 (2014)
13. Paraskevopoulos, D., Laporte, G., Repoussis, P.P., Trantilis, C.D.: Resource contrained routing and scheduling: review and research prospects. Eur. J. Oper. Res. **263**, 737–754 (2017)
14. Rasmussen, M.S., Justesen, T., Dohn, A., Larsen, J.: The home care crew scheduling problem: preference-based visit clustering and temporal dependencies. Eur. J. Oper. Res. **219**, 598–610 (2012)
15. Trautsamwieser, A., Hirsch, P.: Optimization of daily scheduling for home health care services. J. Appl. Oper. Res. **3**, 124–136 (2011)
16. Wirnitzer, J., Heckmann, I., Meyer, A., Nickel, S.: Patient-based Nurse rostering in home care. Oper. Res. Health Care **8**, 91–102 (2016)

Management Model Logistic for the Use of Planning and Inventory Tools in a Selling Company of the Automotive Sector in Peru

Luis Carazas[1(✉)], Manuel Barrios[1(✉)], Victor Nuñez[1(✉)],
Carlos Raymundo[2(✉)], and Francisco Dominguez[3(✉)]

[1] Escuela de Ingeniería Industrial, Universidad Peruana de Ciencias Aplicadas
(UPC), Lima, Peru
{u201323941, u201413555, Victor.nunez}@upc.edu.pe
[2] Dirección de Investigaciones, Universidad Peruana de Ciencias Aplicadas
(UPC), Lima, Peru
Carlos.raymundo@upc.edu.pe
[3] Escuela Superior de Ingeniería Informática, Universidad Rey Juan Carlos,
Mostoles, Madrid, Spain
francisco.dominguez@urjc.es

Abstract. One of the most important problems affecting companies that assemble and market vehicles is stock depletion of finished products. Therefore, many small and medium-sized enterprises (SMEs) have attempted to manage this situation by using tools such as the Q mode, which continuously reviews inventories but does not indicate when and in what quantity a company must supply its sales outlets but requires a more complex level of supplier development and supply chain management efficiency than SMEs can achieve. Likewise, with our proposal we achieved a 50% decrease in stock depletion, based on an analysis of model results, with a confidence level of 95% and a certainty level of 75.30%, projecting a profit of 59000 soles with a return cost of 1.41 and a recovery period of 2 years and 4 months, ensuring both the sustainability and profitability of the proposal.

Keywords: Stock depletion · Inventory management · MRP · RFID · CRP

1 Introduction

Currently, organizations must be able to correctly organize and manage their inventories in order to meet customers' needs and expectations, while staying in business in tough competition with other companies. However, one of the most significant problems afflicting companies and impeding them from achieving their objectives is stock depletion of their finished products. This problem is compounded due to poor inventory management, which has become a key factor for company success and also a central issue in the failure of some enterprises, especially among SMEs, of which approximately 70%–80% fail within the first 3–4 years due to poor inventory management [1]. This occurs, especially in this country because most SMEs lack strategies and good

© Springer Nature Switzerland AG 2020
W. Karwowski et al. (Eds.): AHFE 2019, AISC 971, pp. 299–309, 2020.
https://doi.org/10.1007/978-3-030-20494-5_28

operational practices due to their empirical and informal development. These figures increased by 5.7% in the last year according to INEI and BCR macroeconomic figures.

Therefore, as a starting point, various efforts and studies were found that consist of developing solution proposals to reduce stock depletion and its financial impact on companies. To this end, the current situation and sector (retail) needs were addressed in researching inventory-related problems, such as inventory inaccuracy, deterioration or expiration, shortages, and excesses, which occur in all types of companies, from SMEs to world-class corporations. Therefore, a proposal to develop an inventory management system for enabling the integration of the materials requirement planning (MRP), radio frequency identification (RFID), and continuous replenishment program (CRP) techniques with an integrative, innovative, and technological vision to improve the optimization and efficiency of processes. It is important to note that in order to enable the proper functioning of this proposal, various components such as electronic data interchange, standardized procedures, maintenance management, change management, risk management, acceptance plans, automated processes, training, and consultancies have been introduced to solve the problem of stock depletion.

Therefore, this research paper is divided into the following six sections that describe the entire research project: introduction, progression, contribution, proposal validation, discussion, conclusions, and respective references.

2   Progression

The focus of this research has been developed based on inventory management topics and the application of technology to techniques aimed at counteracting stock depletion. This has resulted in a proposal for concurrently integrating these techniques to manage an organization's inventory in a more effective manner.

2.1   Material Requirement Planning (MRP)

In this sense, there have been various proposals with tools that focus on inventory management, ranging from planning to replenishing strategies with retailers. Solutions such as MRP in SMEs resulted in a 50% stock reduction and a 40% improvement in customer service, which could have had better results with more collaborative work within a management model [2]. Therefore, many research projects have proposed reducing stock depletion and minimizing operational costs, such as MRP, by applying a particle swarm-based algorithm to solve the problem of multi-level lot size with capacity restriction. This method was followed in a case presented in literature that showed a possibility of demonstrating a cost reduction of 6.31%, in addition to decreasing the generation of poor solutions. This led to the evaluation of the same case study by adding a restriction in the manufacturing of the product to 20%, 40%, and 60% in which it could provide feasible solutions with good performance based on the model and considerably reducing stock depletion. For this reason, this algorithm is ideal for improving company performance and increasing competitiveness [3]. This is why many studies compare MRP methodology with others, concluding that MRP develops properties that extract inventory flow management policies in manufacturing and distribution, which improve company efficiency [4]. Thus, the use of MRP through various approaches, such as multi-objective, diffuse linear integer programming to model a problem of material requirement planning, incorporating the distinct possibility that each product can be produced in optimum delivery time.

2.2 Radio Frequency Identification (RFID)

With respect to inventory control and monitoring, automated systems such as RFID are becoming increasingly relied upon, since acquisition costs are usually more affordable and they manage to increase the visibility of inventories in real time, while reducing inventory inaccuracy by up to 81.25% [5]. Similarly, in another part of the supply chain, CRPs have been used, which, when integrated with RFID systems, resulted in an effective customer response because of streamlining information exchanges through electronic data interchange (EDI) [6]. This resulted in the successful validation of research projects and proposals in the industry. In a competitive environment, the automotive industry's ability to meet demands requires ensuring the visibility and traceability of items for which RFID has been validated in a university environment and through the implementation of a simplified version of the solution model in the logistics chain of an automaker in automotive companies. This witnessed an increase of 0.87% in the level of service compared to the previous month [7]. The development of RFID in the Bosch Car supply chain also demonstrated an increase in material traceability and visibility in the receiving, storing, and confirming of stock phases, resulting in long-term profits, reduced transportation costs, deterioration costs, inventory losses, sales losses, and human errors [8].

3 Contribution

3.1 Proposal Overview

In this manner, the model will work around the synergy of three main axes and uphold with eight supports for its efficient operation (Fig. 1):

Fig. 1. General view of the proposal

As can be seen, the proposal will integrate MRP-RFID-CRP tools to solve the stock depletion problem. This will be accompanied by various components to ensure the proper functioning of the proposal, such as electronic data exchange, standardized procedures, maintenance management, change management, risk management, acceptance plans, automated processes, training, and consulting. Therefore, this proposal is unique and innovative due to its integration and optimization of powerful tools, whose efficiency are increased when combined. It is important to note, that the model is adjusted to the organization's needs and directly solves the problem, thus enabling more reliable decisions from automated processes and the use of emerging technology. Therefore, this mechanism, through integrated support of tools, ensures that the benefits of the proposal are obtained in a direct and efficient manner.

3.2 Detailed View of the Proposal

For a more precise understanding of the model, a more detailed view of the model is presented below (Fig. 2):

Planning
After receiving data from its sales outlet, for example, sales reports and inventory inflow and outflow (RFID), planning will be conducted at the company's headquarters. This information, such as inflow for MRP, is useful in making purchase orders with suppliers, and for the corresponding assemblage production master plan to meet distribution and stock replenishment requirements according to the CRP.

Control and Monitoring of Inventories
Inventory control and monitoring will be conducted upon receiving the raw material at the central headquarters, which is automatically registered by reading the RFID tags and updating stock levels of this category in the company's information system to draft the MRP. Likewise, the finished product inventory is monitored by reading the RFID antennas located at the exit of the central headquarters and at the entrance of the sales outlets in order to close the control cycle when updating inventory levels at the sales outlets. This information is input into the CRP. As a control and inventory tracking technology, RFID manages to interrelate this information through the EDI electronic data exchange system, which facilitates the flow of information as a result of its common language in the integration and operation of the complete model.

Replenishment and Distribution
The replenishment of finished products at sales outlets will be configured through a mathematical algorithm that considers the following two fundamental factors for the CRP: optimal inventory levels and merchandize delivery times. Reliable and up-to-date information is also required due to the electronic exchange of EDI data that the model considers. This makes it possible to complement the integration of inventory planning and control techniques with replenishing sales outlets by placing orders according to the demand of each sales outlet, guaranteeing efficient attention to the consumer and also efficiency of inventory levels throughout the supply chain without putting

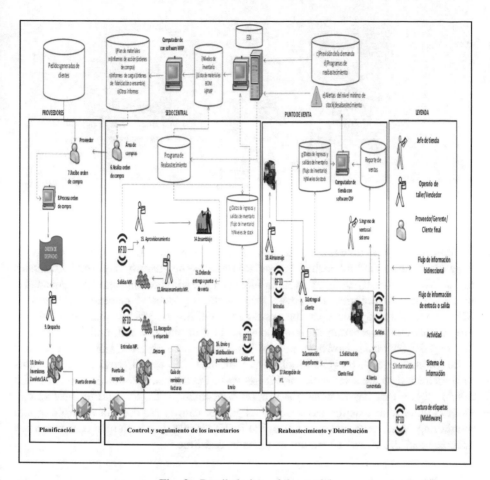

Fig. 2. Detailed view of the model

availability at risk and stock depletion being an issue. It can be seen in the previous graphs that the inventory management model integrates three main tools: CRP–RFID–MRP. This is linked to the first technique, CRP, where it will be used specifically to gain an information system between the headquarters and its sales outlets, thus achieving efficient distribution to sales outlets in order to increase the percentage of completed and quality orders delivered. Therefore, documentation accuracy will also be increased along with an improvement in timely delivery efficiency of products to the customers (Fig. 3).

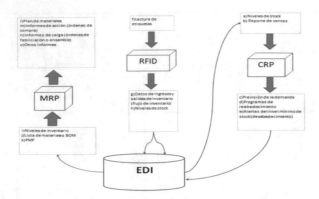

Fig. 3. Integration framework for the model techniques

It is important to note that this technique is also important in improving orders with suppliers because the replenishment program presents the quantity required by the sales outlet, which carefully reduces errors in forecasting, and more appropriately, managing purchases and inventories. Moreover, there is the requirement planning system with MRP; a methodology and planning system that will improve order quality and reduce prognostics and forecasting errors because it is aware what, when, and how much to buy to satisfy demand. However, this is linked not only to objectives but also expected to be a complete and integrated tool along with the CRP, which provides the amounts to be delivered to the sales outlet and therefore to the clients, determining an exact and reliable figure that has to reach the proposed destination. Lastly, due to the fact that work is done with existing stock and thus greater care is required in controlling and monitoring inventories, the model uses the RFID technique, which increases inventory accuracy and the efficient location of items in real time.

3.3   Process View

The project development process consists of seven phases: beginning from project initiation when the project team is formed, and responsibilities are assigned to the analyses of requirements according to the scope of the project by considering the type of organization and the available infrastructure. The planning phase breaks down the project deliverables over time along with the allocation of resources in order to concretely define the proposal in the design phase. Based on this, the implementation and execution phase of the proposal will continue, with the turnkey project including manuals and support plans (consultancy, training, risks, and maintenance). Lastly, phases 6 and 7 will determine the correct operation and sustainability of the proposal over time through feedback from the testing plan phase and the monitoring, adjustment and control phase of the implementation. The implementation phases of the model are as follows (Fig. 4):

3.4   Indicators View

The indicators used in the research are accepted worldwide and obtained from reliable sources. Each indicator is found in the problem and is directly related to the pillars that will help determine the achievement of the objectives. These indicators are as follows:

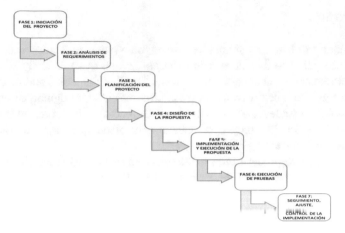

Fig. 4. Implementation phases of the model

Demand Forecast Error

The demand forecast error measures the percentage of error between current and forecasted demand (Supply Chain Council, 2012).

$$\% \text{ Demand Forecast Error} = \frac{\text{Current-Prognostic}}{\text{Current}} \times 10$$

Quality of Generated Orders

This indicator measures the quality of the orders made by the purchasing area through the calculation of the number and percentage of orders generated without delays or without the need for additional information (Mora García, 2015).

$$\% \text{ Quality of Generated Orders} = \frac{\text{Orders Generated Without Problems}}{\text{Total Generated Orders}} \times 100$$

Inventory Registration Accuracy (IRA) or Inventory Exactitude

This measures the precision of inventories through the ratio of correct records over total records (Supply Chain Council, 2012).

$$(3)$$

However, to calculate the variation in accuracy, the following indicator is proposed by the paper, "Measuring Retail Supply Chain Performance: theoretical model using-key performance indicators (KPIs)" (Neeraj Anand, 2015).

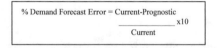

$$= \frac{\text{Inventory in System} - \text{Physical Inventory}}{\text{Inventory in System}} \times 100$$

4 Validation

For functional validation, a simulation based on system dynamics was developed. Dynamics according to Jay Forrest is a language that systematically discovers and describes problems. It refers particularly to the fact that the world and its components are changing in a complex way and in these circumstances, designing an organization, a decision policy, a strategy, or a social system is not easy. Therefore, system dynamics is a methodology for the study and management of complex systems that provides practical management for problem solving.

In this manner, to proceed with validation, the theoretical research model will be used. It employs the following three phases (Fig. 5):

> Conceptualization > Model Formulation > Model Evaluation

Fig. 5. Functional validation phases

Each of the validation phases are detailed below.

4.1 Conceptualization

Conceptualization usually begins with familiarization of the problem to be studied. Once this is done, the aspects of the problem to be solved must be precisely defined. In this study, the identified problem is stock depletion. To this end, the goal is to improve the indicators shown in Fig. 6.

4.2 Model Formulation

The Forrester diagrams were made and equations for the simulation were completed.

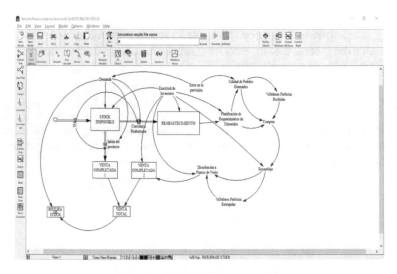

Fig. 6. Forrester diagram model

4.3 Evaluation of the Model

The Forrester diagram previously constructed was now simulated. To accomplish this, the following general equation was determined to execute the simulation in different scenarios:

f (Stock Depletion) = (% Inventory Accuracy * Demand Forecast Error * Quality of Generated Orders * Precise Orders Delivered and Received)

4.4 Results

The results obtained from the simulation are as follows (Table 1):

Table 1. General results from the simulation

RESULTADOS DE SIMULACION PARTE I				RESULTADOS DE SIMULACION PARTE II			
ANO	DEMANDA	VENTAS	ROTURA DE STOCK (%)	ANO	DEMANDA	VENTAS	ROTURA DE STOCK
2021	1000	917	8.3%	2046	3386	3382	0.1%
2022	1050	975	7.1%	2047	3556	3556	0.0%
2023	1103	1030	6.6%	2048	3733	3733	0.0%
2024	1158	1100	5.0%	2049	3920	3920	0.0%
2025	1216	1160	4.6%	2050	4116	4116	0.0%
2026	1276	1226	3.9%	2051	4322	4322	0.0%
2027	1340	1299	3.1%	2052	4538	4538	0.0%
2028	1407	1370	2.6%	2053	4765	4765	0.0%
2029	1477	1440	2.5%	2054	5003	5003	0.0%
2030	1551	1514	2.4%	2055	5253	5253	0.0%
2031	1629	1592	2.3%	2056	5516	5516	0.0%
2032	1710	1674	2.1%	2057	5792	5792	0.0%
2033	1796	1760	2.0%	2058	6081	6081	0.0%
2034	1886	1851	1.8%	2059	6385	6385	0.0%
2035	1980	1946	1.7%	2060	6705	6705	0.0%
2036	2079	2047	1.6%	2061	7040	7040	0.0%
2037	2183	2152	1.4%	2062	7392	7392	0.0%
2038	2292	2263	1.3%	2063	7762	7762	0.0%
2039	2407	2379	1.1%	2064	8150	8150	0.0%
2040	2527	2502	1.0%	2065	8557	8557	0.0%
2041	2653	2631	0.9%	2066	8985	8985	0.0%
2042	2786	2766	0.7%	2067	9434	9434	0.0%
2043	2925	2909	0.6%	2068	9906	9906	0.0%
2044	3072	3058	0.4%	2069	10401	10401	0.0%
2045	3225	3216	0.3%	2070	10921	10921	0.0%

As shown in the simulation results, the interrelation and behavior of each variable is seen over time. Hence, the functionality of the model is verified in the case study, in which stock depletion decreased and sales increased. It can be concluded that the model produces the optimal and desired effect by integrating the tools (MRP–CRP–RFID) that directly impacted the variables included in the model.

5 Conclusions

The proposed model is innovative in solving the primary problem of stock depletion, reducing it by more than 50%, along with its main causes in the supply chain with a certainty level of 75%, based on a results analysis obtained from the 95% confidence level model.

This study shows the inventory management model through a collaborative integration of planning, control, and monitoring tools, as well as continuous replenishment with its scope due to applied technology, enabling it to be integral in the organization's supply chain links. In this manner, savings of 29% in scheduling, 37% in control and monitoring, and 33% in replenishment were achieved.

The proposed model is characterized by its applicability to SMEs that have procurement processes of raw materials, production or assembly, and distribution to sales outlets. This is due to the scalability of the implementation of tools and technology that make them efficient in line with new trends and demands of the market and the customer.

During the development and implementation of the inventory management model, the main indicators of the supply chain are significantly improved, such as the percentage of perfect deliveries, forecast error percentage, quality percentage of generated orders, and inventory accuracy percentage, with reference to best supply chain indicator practices. Furthermore, this improvement is seen and perceived in a tangible way in the results of financial analysis through savings in solving the main causes that causes the problem of stock depletion.

References

1. Nikita, N.S.M., Fitshane, S., Matsoso, M., Juan-Pierré, B.: Inventory management systems used by manufacturing small medium and micro enterprises in Cape Town. Mediterr. J. Soc. Sci., 1–9 (2015)
2. Chung-Yean, C., Chia-Hung, C.: Comportamiento dinámico y estocástico de la incertidumbre del coeficiente de la demanda incorporado con las variables EOQ: Una aplicación en el inventario de productos terminados de los concesionarios de General Motors. Int. J. Prod. Econ. (2015)
3. Rivera, J.M., Ortega, E., Pereyra, J.: Diseño e implementación del sistema MRP. Producción y Gestión, 4855 (2014)
4. Padroncano, J., Coronadohernández, J.R., Caicedotorres, W., Mercadocaruso, N., Ospinovadiris, F.V.: Materials requirements planning through an application based on particle swarm optimization. Espacios **38**, 14 (2017)

5. Miclo, R., Fontanili, F., Lauras, M., Lamothe, J., Milian, B.: An empirical comparison of MRPII and Demand-Driven MRP. IFAC-PapersOnLine **49**(12), 1725–1730 (2016)
6. Díaz-Madroñero, M., Mula, J., Jiménez, M.: Material requirement planning under fuzzy lead times. IFAC-PapersOnLine **48**(3), 242–247 (2015)
7. Sandeep Goyal, B.C., Hardgrave, J.A., DeHoratius, N.: Efectividad del RFID en la gestión de inventario de trastienda y piso de ventas. Int. J. Logist. Manag. (2016)
8. Wei, H.-L. The strategic value of supply chain visibility: Increasing the ability to reconfigure. Strat. Dir. 26(9) (2010)

Order Acceptance and Scheduling Problem: A Proposed Formulation and the Comparison with the Literature

Papatya S. Bıçakcı[1($\boxtimes$)] and İmdat Kara[2]

[1] Faculty of Economics and Administrative Sciences, Department of Management, Başkent University Bağlıca Campus, 06790 Ankara, Turkey
papatyas@baskent.edu.tr
[2] Faculty of Engineering, Department of Industrial Engineering, Başkent University Bağlıca Campus, 06790 Ankara, Turkey

Abstract. In classical scheduling problem, it is assumed that all orders must be processed. In the order acceptance and scheduling (OAS) problem, some orders are rejected due to limited capacity. In make-to-order production environment, in which the OAS problem occurs, accepting all orders may cause overloads, delay in deliveries and unsatisfied customers. Oğuz et al. (2010) introduced the OAS problem with sequence-dependent setup times and release dates. In this paper, we propose a new mixed integer programming formulation with $O(n^2)$ decision variables and $O(n^2)$ constraints for the same problem. We conduct a computational analysis comparing the performance of our formulation with Oğuz et al. (2010) formulation. We use the benchmark instances, which are available in the literature. We observe that our formulation can solve all the instances up to 50 orders in a reasonable time, while Oğuz et al. (2010) formulation can solve only the instances with 10 orders in the same time limit.

Keywords: Scheduling · Order acceptance · Order rejection · Mathematical formulation · Single machine

1 Introduction

In scheduling problem, it is mostly assumed that all orders must be processed. However, in real-life applications this may not be the case. In make-to-order production environment, accepting all orders may cause overloads, delay in deliveries and unsatisfied customers. Therefore, firms tend to reject some orders. Order acceptance and scheduling (OAS) problem consists of deciding which orders to be accepted and determining the schedule of the accepted orders [1].

The OAS problem considered in this study was introduced by Oğuz et al. [2] in 2010 and a mathematical formulation was proposed. The problem is defined as follows. In a single machine environment, there is a set of orders shown as N = {1, 2, …, n}. For each order i∈N, let r_i be the release date, p_i be the processing time, d_i be the due date, $\overline{d_i}$ be the deadline, e_i be the revenue, w_i be the unit penalty weight for the tardiness of the orders and s_{ji} be the sequence-dependent setup time occurring if the order i is processed immediately after order j. The setup operation of order i∈N can only be

© Springer Nature Switzerland AG 2020
W. Karwowski et al. (Eds.): AHFE 2019, AISC 971, pp. 310–316, 2020.
https://doi.org/10.1007/978-3-030-20494-5_29

performed after its release date. The objective is to select and schedule a subset of orders that maximizes the total profit.

To the best of our knowledge, there are 27 studies related to single machine OAS problem in the literature. Only 9 of them include a mathematical formulation. These studies are summarized in Table 1. In [4]'s study, a job is rejected if it is not finished before its due date. This study is considered as the first example of OAS problem in the literature [1]. [5] considered varying prices and customer chosen due dates, while [6] considered inventory costs. They proposed two different formulations. [2] introduced the OAS problem with sequence-dependent setup times and release dates. They proposed a formulation and their study has taken remarkable attention by researches. In [7]'s study, there are obligatory jobs and they proposed two mathematical formulations. [8] addressed the OAS problem considering resource limits and due windows and proposed a formulation. In his study, a job is rejected if it cannot be finished within its due window. [9] studied the OAS problem with sequence-dependent setup times depending on lot sizes. They considered total available time constraint and proposed a formulation. [10] defined an upper limit for the number of accepted jobs and considered sequence-dependent setup times. They proposed a nonlinear formulation. [3] dealt with the problem as in [2]. They proposed a time-indexed formulation with different revenue calculation method. [3]'s time indexed formulation has some structural difficulties, so we did not consider this formulation in detail.

Table 1. Single machine OAS studies with mathematical formulation

Ref. #	Year	Authors	Problem structure	Release date	Setup time
[4]	1990	Stern and Avivi	$1\|rej,pmtn\|\sum P_i$	–	–
[5]	2004	Charnsirisakskul et al.	$1\|rej,pmtn\|\sum P_i-\sum w_j T_j$	–	–
[6]	2006	Charnsirisakskul et al.	$1\|rej,pmtn,prc,r_j\|\sum P_i-\sum w_j T_j$	✓	–
[2]	2010	Oğuz et al.	$1\|rej,\ s_{ij},\overline{d_i},r_j\|\sum R_j$	✓	✓
[7]	2011	Nobibon and Leus	$1\|rej\|\sum P_j-\sum w_j T_j$	–	–
[8]	2016	Garcia	$1\|rej,r_j,D_j^-,D_j^+\|\sum P_j-RC$	✓	–
[9]	2016	Trigos and Lopez	$1\|rej,batch,s_{ij}\|\sum \Pi_j$	–	✓
[10]	2017	Zandieh ve Roumani	$1\|rej,s_{ij}\|\sum P_i-\sum w_j T_j$	–	✓
[3]	2018	Silva et al.	$1\|rej,\ s_{ij},\overline{d_i},r_j\|\sum P_i-\sum w_j T_j$	✓	✓

Note 1: Single machine; rej: rejection; pmtn: preemption; s_{ij}: setup time; prc: pricing; r_i: release date; $\overline{d_i}$: deadline; p_i: processing time; $R_i = \max(0, e_i-w_i T_i)$; $\sum P_i$: total revenue; $\sum T_i$: total tardiness; RC: total rejection cost; $\sum c_i$: total cost; $\sum w_i T_i$: total weighted tardiness; OC: total contract manufacturing cost; F^j: product families; $\sum \Pi_j$: total profit; D_j^-,D_j^+: due windows. ✓ Included; - Not included

In recent years, the developments on computer technology and softwares enable us to solve the combinatorial problems to optimality by well-designed mathematical formulations. Mathematical formulations can be useful in real-life applications and allow post-optimality analysis. In most of the studies on OAS problem, solving methods generally focus on heuristic algorithms. However, as [12] express that mathematical formulations are still useful for the scheduling problems.

In this study, we propose a new formulation for single machine OAS problem with sequence-dependent setup times and release dates. We conduct a detailed computational analysis to compare Oğuz et al. [2] formulation and our proposed formulation.

The remainder of this paper is organized as follows. Section 2 introduces the new integer linear programming formulation for single machine OAS problem with sequence-dependent setup times and release dates. Section 3 presents the results of the computational experiments comparing the performance of our proposed and Oğuz et al. [2] formulation. Finally, Sect. 4 provides the concluding remarks of this work.

2   A New Formulation

In this section we introduce a new mixed integer linear programming formulation for OAS problem with sequence-dependent setup times and release dates. Let "0" and "n +1" be the dummy orders, which indicate the first order of the schedule and the last order of the schedule respectively. Decision variables are given as follows. Let T_i be the tardiness of order i, for i∈N. Let Z_{ij} be the completion time of order j, if order j is processed immediately after order i. Let Y_i be 1, if order i is accepted, 0 otherwise. Let X_{ij} be 1, if order j is processed immediately after order i, 0 otherwise. Our proposed formulation is given below:

$$\max \sum\nolimits_{i=1}^{n} R_i \tag{1}$$

s.t.

$$\sum\nolimits_{i=1}^{n} X_{0i} = 1 \tag{2}$$

$$\sum\nolimits_{i=1}^{n} X_{i,n+1} = 1 \tag{3}$$

$$\sum\nolimits_{j=1,i\neq j}^{n+1} X_{ij} = Y_i \quad \forall_i = 1, \ldots, n \tag{4}$$

$$\sum\nolimits_{j=0,i\neq j}^{n} X_{ji} = Y_i \quad \forall_i = 1, \ldots, n \tag{5}$$

$$\sum\nolimits_{j=1}^{n+1} Z_{ij} - \sum\nolimits_{k=0}^{n} Z_{ki} = \sum\nolimits_{j=1}^{n+1} (r_j + s_{ij} + p_j)X_{ij} \quad i \neq j, \quad \forall_i = 1, \ldots, n, \\ \forall_j = 1, \ldots, n+1 \tag{6}$$

$$Z_{0i} = (r_i + s_{0i} + p_i)X_{0i} \quad \forall_i = 1, \ldots, n \tag{7}$$

$$\sum_{k=0}^{n} Z_{ki} \leq \overline{d_i} Y_i \quad \forall_i = 1, \ldots, n \tag{8}$$

$$Z_{ij} \leq \overline{d_{max}} X_{ij} \quad i \neq j, \quad \forall_i = 0, \ldots, n, \quad \forall_j = 1, \ldots, n+1 \tag{9}$$

$$T_i \geq \sum_{k=0}^{n} Z_{ki} - d_i Y_i \quad \forall_i = 1, \ldots, n \tag{10}$$

$$T_i \leq (\overline{d_i} - d_i) Y_i \quad \forall_i = 1, \ldots, n \tag{11}$$

$$T_i \geq 0 \quad \forall_i = 1, \ldots, n \tag{12}$$

$$R_i = e_i Y_i - T_i w_i \quad \forall_i = 1, \ldots, n \tag{13}$$

$$R_i \geq 0 \quad \forall_i = 1, \ldots, n \tag{14}$$

$$Y_i \in \{0, 1\} \quad \forall_i = 1, \ldots, n \tag{15}$$

$$X_{ij} \in \{0, 1\} \quad i \neq j, \quad \forall_i = 0, \ldots, n, \quad \forall_j = 1, \ldots, n+1 \tag{16}$$

$$Z_{ij} \geq 0 \quad i \neq j, \quad \forall_i = 0, \ldots, n, \quad \forall_j = 1, \ldots, n+1 \tag{17}$$

In the objective function (1), the difference between total revenue of accepted orders and total tardiness penalty of accepted orders is maximized. With constraint (2) and (3), it is guaranteed that "0" is assigned to the first position and "n+1" is assigned to the last position of the schedule. Constraint set (4) ensures that if an order is accepted, another order is processed immediately after this order; and constraint set (5) ensures that if an order is accepted another order is processed immediately before this order. Constraint set (6) calculates the completion time of the orders. Constraint set (7) calculates the completion time of the first processed order in the sequence. Constraint set (8) guarantees that if an order is not completed before its deadline, then the order is not accepted. Constraint set (9), where $\overline{d_{max}} = max_{i=1, \ldots, n} \{\overline{d_i}\}$, provides X_{ij} be zero, when Z_{ij} be zero, and gives an upper bound to Z_{ij}'s. Constraint set (10) calculates the tardiness of the orders. Constraint set (11) gives an upper bound to T_i's, while constraint set (12) provides a lower bound to T_i's. Constraint set (13) calculates the revenue of the orders, while constraint set (14) bounds it. Constraint sets (15) and (16) define the binary variables. Constraint set (17) gives a lower bound to Z_{ij}'s. Our proposed formulation has $2n^2+10n+2$ constraints and n^2+2n binary decision variables.

Constraints (2), (3), (4) and (5) are traditional assignment constraints, therefore they are same as in Oğuz et al. [2] formulation. Constraints (11), (12), (13) are due to the relations between tardiness, targets and revenue, so they are same as in Oğuz et al. [2] formulation. Constraints (14), (15), (16) and (17) are non-negativity and binary constraints. These constraints also are not related with the structure of the formulation directly.

Main difference between our formulation and Oğuz et al. [2] formulation depends upon the decision variables corresponding to completion times of the orders. Oğuz et al. [2] defines completion times by indexing with C_i for i^{th} order and develops main constraints of their formulation as a function of C_i's and other decision variables. We define completely different decision variables for completion time in relation with preceding order i and j as Z_{ij}, thus our main decision variables are arc-based whereas Oğuz et al. [2] formulation's main decision variables are node-based.

In accordance with these above explanations, main body of our formulation which includes constraints (6), (7), (8), (9) and (10) is completely different from Oğuz et al. [2] formulation.

3   Computational Experiments

In this section we summarize the results of computational experiments comparing the performance of Oğuz et al. [2] formulation (OSB) and our proposed formulation (OPF). Mathematical formulations were coded in C++ and benchmark instances were solved by CPLEX 12.4 in an Intel Xeon Phi 7290 with 1.5 GHz and 384 GB of RAM.

Benchmark instances were generated by [11] and they are available on the web address   http://home.ku.edu.tr/ ∼ coguz ∼ /Research/Dataset_OAS.zip.   There are six instance groups which consist of n = 10,15,20,25,50 and 100. Each instance group has 250 instances, and there are totally 1500 benchmark instances. For the instances with 10,15,20,25 and 50 orders the time limit is set 7200 seconds; for the instances with 100 orders the time limit is set 14400 seconds. Run times and LP relaxations of OPF and OSB for n = 10 are showed in Table 2.

Table 2. The results of OSB and OPF for n = 10

τ	R	OPV	CPU		LPR		% DEV.	
0.1	**0.1**		OSB	OPF	OSB	OPF	OSB	OPF
Instance 1	119		90,06	0,49	124	124	0,04	0,04
Instance 2	126		115,9	0,69	131	131	0,04	0,04
Instance 3	90		127,72	0,64	102	102	0,13	0,13
Instance 4	123		40,16	0,29	123	123	0,00	0,00
Instance 5	94		69,26	0,39	96	96	0,02	0,02
Instance 6	111		113,28	0,61	115	115	0,04	0,04
Instance 7	102		117,02	0,51	111	111	0,09	0,09
Instance 8	104		91,76	0,29	114	114	0,10	0,10
Instance 9	117		104,76	0,65	123	123	0,05	0,05
Instance 10	105		76,8	0,31	107	107	0,02	0,02
Avg.			94,67	0,48			0,05	0,05

Note: τ and R are the parameters for using to generate benchmark instances. OPV: optimal value; CPU: run time; LPR: linear programming relaxation; % DEV: percentage deviation between LPR and OPV which is (LPR-OPV)/(OPV). OSB: Oğuz et al. [2] formulation; OPF: our proposed formulation.

From Table 2, we observe that OPF is extremely faster than OSB with the average run times 0,48 and 94,67 respectively. LPR values are not different and the average percentage deviation is 0,05 which indicates the LPR values are very close to the optimal values.

We solved all the instances with n = 10 with each formulation. The average run time and the average percentage deviation are given in Table 3.

Table 3. The results of OPF for n = 10, 15, 20, 25, 50

	OSB		OPF	
	Ave.CPU	Ave.%DEV	Ave.CPU	Ave.%DEV
n = 10	68,70	0,07	0,56	0,07
n = 15	–	–	1,13	0,05
n = 20	–	–	4,18	0,04
n = 25	–	–	10,02	0,05
n = 50	–	–	1977,55	0,03

Ave.CPU: Average run time; Ave.%DEV: Average percentage deviation between LPR and OPV which is (LPR-OPV)/OPV.

OSB cannot solve the instances greater than 10 orders in given time limit. Therefore, we continue computational analysis with OPF. Average run times and average percentage deviations for n = 15, 20, 25 and 50 are given in Table 3. OPF can solve all the instances up to 50 orders in a reasonable time.

There are no benchmark instances between 50 and 100 orders. OPF cannot solve the instances with 100 orders in 14400 seconds time limit.

4 Concluding Remarks

In this study we proposed a new arc-based mathematical formulation for solving a variant of OAS problem that includes sequence-dependent setup times and release dates. Our proposed formulation and Oğuz et al. [2] formulation were tested on instances ranging from 10 to 50 orders. Oğuz et al. [2] formulation can solve only the instances with 10 orders in given time limit. Our proposed formulation can solve the instances up to 50 orders to optimality in the same time limit. Future studies may consider proposing new mathematical formulations capable of solving larger instances in a reasonable time.

References

1. Slotnick, S.A.: Order acceptance and scheduling: a taxonomy and review. Eur. J. Oper. Res. **212**, 1–11 (2011)
2. Oğuz, C., Salman, F.S., Bilgintürk Yalçın, Z.: Order acceptance and scheduling decisions in make-to-order systems. Int. J. Prod. Econ. **125**, 200–211 (2010)

3. Silva, Y.L.T., Subramanian, A., Pessoa, A.A.: Exact and heuristic algorithms for order acceptance and scheduling with sequence-dependent setup times. Comput. Oper. Res. **90**, 142–160 (2018)
4. Stern, H.I., Avivi, Z.: The selection and scheduling of textile orders. Eur. J. Oper. Res. **44**, 11–16 (1990)
5. Charnsirisakskul, K., Griffin, P.M., Keskinocak, P.: Order selection and scheduling with leadtime flexibility. IIE Trans. **36**, 697–707 (2004)
6. Charnsirisakskul, K., Griffin, P.M., Keskinocak, P.: Pricing and scheduling decisions with leadtime flexibility. Eur. J. Oper. Res. **171**, 153–169 (2006)
7. Nobibon, F.T., Leus, R.: Exact algorithms for a generalization of the order acceptance and scheduling problem in a single-machine environment. Comput. Oper. Res. **38**(1), 367–378 (2011)
8. Garcia, C.: Resource-constrained scheduling with hard due windows and rejectionpenalties. Eng. Optim. **48**, 1515–1528 (2016)
9. Trigos, F., López, E.M.: Maximising profit for multiple-product, single-period, single-machine manufacturing under sequential set-up constraints that depend on lot size. Int. J. Prod. Res. **54**, 1134–1151 (2016)
10. Zandieh, M., Roumani, M.: A biogeography-based optimization algorithm for order acceptance and scheduling. J. Ind. Prod. Eng. **34**, 312–321 (2017)
11. Cesaret, B., Oğuz, C., Salman, F.S.: A tabu search algorithm for order acceptance and scheduling. Comput. Oper. Res. **39**, 1197–1205 (2012)
12. Della Croce, F.: MP or not MP: That is the question. J. Sched. **19**(1), 33–42 (2016)

Manufacturing Aspects of Work Improvement

Use of Quality Management Tools to Identify Ergonomic Non-conformities in Human Activities

Adam Górny[(⊠)]

Faculty of Management Engineering, Poznań University of Technology,
ul. Strzelecka 11, 60-967 Poznań, Poland
adam.gorny@put.poznan.pl

Abstract. In order to operate properly, manufacturing enterprises find it increasingly critical to create a proper production environment. To that end, focus is placed on worker health and safety along with other environmental factors that boost efficiency. Organizations need to take measures to ensure that their tasks are performed optimally and that their employees enjoy comfort and well-being. Due to the nature of such requirements and measures, enterprises are compelled to make use of ergonomic criteria [1–3] to select the scope of their response, employ tools to identify issues, analyse any issues discovered and make improvements. The process requires the use of quality management tools associated with quality engineering. This article defines the options for using traditional as well as new quality management tools. It identifies the need to employ such tools and the potential benefits they can bring, including those applicable to broadly-defined ergonomics of the workplace.

Keywords: Quality management tools · Human activities ·
Work environment · Ergonomics

1 Introduction

Any business organization wishing to improve its operating potential needs to identify and address the non-conformities that affect its operating efficiency. Such improvements will be of benefit to all stakeholders. To ensure their adoption is a success, organizations need to employ tools that boost the efficiency of all stages of the improvement process. Such tools must be selected with the specific nature of the relevant non-conformities in mind [4]. This is a necessary prerequisite for properly identifying issues and rectifying them effectively by means of improvement measures that best reflect the profile of the non-conformities faced.

To identify problems and advisable improvement measures, organizations are advised to use traditional quality management tools. Such tools will enable them to examine multiple aspects of the issues at hand, analyse them and select the improvement measures that best reflect the nature of such issues [5].

The aim of this article is to outline the options for the use of quality management tools to analyse the ergonomic non-conformities that result from failures to ensure

© Springer Nature Switzerland AG 2020
W. Karwowski et al. (Eds.): AHFE 2019, AISC 971, pp. 319–329, 2020.
https://doi.org/10.1007/978-3-030-20494-5_30

proper working conditions for workers. Focus is placed on both more and less common tools, as employed for such purposes. These tools are:

– Traditional quality management tools: Ishikawa diagram, Pareto diagram and flowcharts, used for detailed problem analysis,
– New quality management tools: affinity and interrelationship diagrams used to analyse problems as well as systems flow/tree and matrix diagrams used to identify the improvement measures that best reflect the nature of the non-conformities at hand.

The article refers to non-conformities of an ergonomic nature. Such non-conformities prevent employees from working effectively and affect their well-being at work. They can be associated with working environment factors, the scope of the tasks performed, the manner in which work is performed, the conditions that help properly interact with machines and technical devices and the elimination or mitigation of loads encountered in the course of work performance.

2 The Nature of Ergonomic Non-conformities in the Environment for Human Activities Affecting the Choice of Improvement Measures

Working environment factors describe the environment in which work is performed. Such factors allow workers to perform effectively and efficiently in a specific setting. Viewed more broadly, the factors pertain to the ability of humans to operate in specified conditions. They are identified most commonly with reference to their systemic impacts on the environment [1, 6]. Their unambiguous identification is a complex process [7, 8]. Ergonomic conditions include social, psychological, physical and environmental factors. These play a crucial role in ensuring that workers are capable of performing their tasks properly.

The literature abounds with references to such factors, complete with descriptions of working environment components, including factors that are critical for the proper identification of irregularities [9–14]. Irregularities can be associated with vital requirements pertaining to anthropometric, physiological, psycho-physical and hygienic requirements. Such requirements are summarised in Table 1.

In reference to ensure that humans operate properly in a working environment, one needs to eliminate any ergonomic non-conformities. These are the kinds of non-conformities that have an adverse impact on both worker health and safety and their ability to perform work effectively. The occurrence of ergonomic non-conformities depends on the scope of work and the specific manner in which it is performed. Such non-conformities can be viewed in reference to ways in which humans perform tasks, engage in work processes and interact with technical equipment. The nature of irregularities in that respect determines the choice of measures and solutions designed to protect workers' health and life [13]. The means of ensuring a proper environment for the performance of work are incorporated into the functional systems of organizations

Table 1. Characteristics of ergonomic requirements.

Ergonomic requirements	Scope of requirements
Anthropometric requirements	Apply to modifying technical items to best suit the metrics and weight of the human body or parts thereof in either a static or a dynamic system, with a view to, e.g., ensuring a proper posture at work
Physiological requirements	Apply to modifying technical items to best suit the physiological characteristics of man, with proper account taken of optimal loads on the muscular, skeletal, respiratory, circulatory systems as well as joints and limbs
Psycho-physical requirements	Apply to modifying psycho-physical loads in the working environment to reflect human limitations with a view to mitigating adverse impacts on key human senses
Hygienic requirements	Apply to modifying the working environment to best meet human needs and limitations, with a view to mitigating the impacts of deleterious factors and achieving comfort in the working environment

[2, 12, 15, 16]. The lengths to which businesses go in creating such environments reflects the significance of the so-called ergonomic risk factors [17].

Due to their nature, ergonomic non-conformities cannot be examined solely in terms of compliance with laws governing the performance of work and with reference to securing a safe working environment. For the analysis to be reliable, one needs to apply criteria having to do with working comfort and worker well-being.

With respect to the issue referred to in the article's title, proper account should be taken of the factors for worker efficiency and their ability to perform work without undue strain. In creating a working environment, businesses should put workers first, recognizing humans as their most important asset. By satisfying worker needs, an organization gains the ability to meet its goals. Its benefits follow the Maslow's hierarchy of needs and reflect the nature of ergonomic criteria. The potential benefits include:

- Proper relations with the external environment that enable workers to perform properly,
- Workers' ability to operate in the working environment without having to face risks and hazards,
- Having the environment modified to satisfy particular needs of workers,
- Having ambitious tasks that help improve working conditions delegated to employees,
- Safety ensured by eliminating all types of risks and strains,
- Proper amenities and operating comfort ensured for workers.

It is of great importance to ensure that the working conditions reflect interactions among the individuals involved in the work process.

Ergonomic non-conformities require improvements that cover a broad spectrum of impacts. To truly benefit from improvements to their working environment, organizations should first identify irregularities and then respond accordingly to the non-conformities discovered and to environmental characteristics. Once the physical and psychological strain resulting from ergonomic non-conformities is alleviated by ensuring adequate working conditions, enterprises can enjoy reduced losses and higher efficiency. Other benefits include a greater ability to perform tasks and superior operating conditions [6, 10, 17–20]. To make that possible, they need to take the right approach that will enable them to eliminate irregularities [21]. The measures they take must be designed to mitigate loads and ultimately tailor the working environment and working conditions to the identifiable characteristics of workers or, viewed more broadly, the general characteristics of human beings. It should also be noted that operational improvements must not be limited to technical aspects only. Organizations should empower the concerned people to influence decisions about any further improvement steps. Humans are at the heart of organizations and, as such, should be afforded a fair degree of operating comfort [6].

Ergonomic non-conformities should be recognised in assessing the compliance of the manner of work performance with the adopted principles. The organization's awareness of any existing non-conformities is equally important. The bulk of the non-conformities result from working discomfort. Their very nature suggests which improvements need to be emphasised and associated with utilitarian requirement areas. Such areas include [22]:

– Effectiveness, which describes the degree to which any adopted objectives are achieved,
– Efficiency, which refers to cost-to-benefit ratios,
– Satisfaction, which describes the satisfaction of persons engaged in the performance of work.

The primary purpose of improvement measures is to enable organizations to achieve their adopted goals [23]. Their achievement should be seen as an outcome of the improvements. Their specific outcomes include the mitigation of hazards and strains achieved through the adoption of appropriate improvement (corrective and preventive) measures. The key aim in adopting improvement measures in the context of the subject matter of this article is to ensure safe and hygienic working conditions while complying with the ergonomic requirements that are helpful in optimizing loads.

3   Quality Management Tools

3.1   Description of Quality Management Tools

Quality management tools are one of the categories of instruments employed for task management. They can give a company a substantial competitive edge [12, 24]. In their basic form, quality management tools are used to collect and process data on various aspects of quality management. They include tools used to [12, 25]:

- Collect data and information on actual conditions,
- Analyse such data to obtain insights into specific problems,
- Analyse the specifications of problems to ascertain their nature,
- Select advisable measures,
- Identify the order of measures,
- Secure the resources needed to carry out measures.

In most cases, in order to use quality management tools effectively, organizations need to include them in their existing quality management systems.

As quality management tools can only be used effectively to address specific non-conformities, their impact depends on the availability of reliable information, which one uses as the input data and a starting point for the proper application of such tools and for selecting a proper approach to their use [26, 27]. Such tools can be employed to monitor (oversee) and assess the measures taken and verify the successful achievement of intended outcomes. In the process of improving systems, quality management tools enable organizations to gain efficiency, in particular in the areas in which such tools were applied.

Failures to meet requirements should be seen as a critical disruption of company performance [6, 28]. To remedy the problem, business organizations should apply quality management tools to all factors for their success [29]. Quality management tools can be employed to check whether improvements were selected and applied properly.

3.2   Application of Quality Management Tools to Ergonomic Non-conformities

In order to both protect worker health and ensure that a business organization operates efficiently, its management should seek to reduce the impact of non-conformities. The scope and specific nature of improvements must be tailored to the nature of such non-conformities. In an effort to design the working environment, organizations should take measures that best match the tasks performed by their workers [6, 8].

To enhance their workers' ability to complete their work, organizations should improve their working environments. This is done in part by responding to the need to meet such ergonomic criteria as are essential for workers' performance at work and, to that end, using tools that support the relevant improvements.

In selecting the quality management tools that provide the best match for the issues at hand, enterprises should satisfy basic needs, such as the need to handle the key non-conformities encountered in the working environment.

Successful improvements require proper identification of potential non-conformities and an accurate assessment of their nature. The latter is accomplished with tools designed to minimise the impact of issues. Possible applications of the traditional ("old") quality management tools for that purpose are summarised in Table 2. The tools can be used to identify potential non-conformities.

To enhance their improvement capabilities and streamline the improvement process, organizations may choose to employ the so-called "new" quality management tools, as shown in Table 3. Such tools support in-depth analyses of problems and the

Table 2. Possible applications of traditional quality management tools.

Quality management tool	Purpose of using quality management tool	Ergonomic aspects of using quality management tool
Ishikawa diagram	- Enables organizations to identify the impacts of individual factors on process outcomes	- During problem analysis, helps identify non-conformities associated with human operation - Enables organizations to account for the human aspect in the course of non-conformity analyses that are not directly focused on humans
Pareto diagram	- Illustrates the unevenness of cause distributions showing that a relatively small number of causes contribute to a wide range of critical effects	- Enables organizations ascertain the significance of problems resulting from failures to comply with ergonomic requirements, - Enables organizations to focus on key causes of non-conformities and identify the relevant role of ergonomic requirements
Flow-chart	- Enables organizations to capture relationships among process components and identify process vulnerabil-ities, - Represents a starting point for process analysis, for defining the chronology of measures and for identifying existing interrelationships	- Enables organizations to identify the suggested impacts of ergonomic non-conformities on process outcomes, - Enables organizations to identify ergonomic non-conformities resulting from the adopted chronology of measures

selection of best improvement measures. This selection reflects the nature of non-conformities and the prospects of effective implementation. It is made possible through the in-depth analysis of issues and the selection and deployment of improvement measures on the basis of its outcomes.

4   Characteristics and Specificity of the Benefits

The descriptions of the outcomes of the use of quality management tools to analyse ergonomic non-conformities and identify and take reasonable improvement measures, as shown in Tables 2 and 3, suggest a wide range of potential benefits. These are achieved in an effort to ensure that workers function effectively in the working environment. Their outcomes can be observed all across business organizations and specifically in working conditions. A key prerequisite for employing quality management tools to identify irregularities and take proper improvement measures is the ability to use relevant data and information. Specifically, companies need to be able to gain access to accurate, complete, up-to-date, comparable and detailed data efficiently and have a system of making informed fact-based decisions. It is particularly vital for

Table 3. Possible applications of new quality management tools.

Quality management tool	Purpose of using quality management tool	Ergonomic aspects of using quality management tool
Affinity diagram	- Enables organizations to group a large number of ideas thematically by selected affinity criteria (such as the affinity of proposed improvement measures)	- Enables organizations to identify factors that define the issue at hand with proper account taken of the ergonomic aspects of the improvement objective, - Enables organizations to identify the leading thematic issue in the adopted group that can be described as an ergonomic non-conformity, - Enables organizations to use an appropriate graphical format to illustrate the dominant role of e.g. the ergonomic non-conformities that affect the outcome of problem analysis
Interrelationship diagram	- Enables organizations to make a graphical representation of factors affecting the outcome of the measure in question, - Enables organizations to define interrelationships among the causes of non-conformities	- Enables organizations to identify all causes of problems, including ergonomic non-conformities, - Enables organizations to identify interrelationships between the ergonomic causes of non-conformities and causes from other cause categories, - Enables organizations to identify the causes of ergonomic non-conformities, - Enables organizations to identify those ergonomic non-conformities that may be seen as both the causes and the effects of failures to ensure conformity
Systems flow/tree diagram	- Provides a graphical representation of problem causes, - Enables organizations to decompose the main objective and identify the non-conformities whose rectification is the main objective	- Enables organizations to identify logical and temporal links between the causes (such as failures to meet ergonomic requirements) and action outcomes, - Enables organizations to select a course of action that will allow them to eliminate ergonomic non-conformities,

(continued)

Table 3. (*continued*)

Quality management tool	Purpose of using quality management tool	Ergonomic aspects of using quality management tool
		- Enables organizations to clearly present process structures and specify the role and significance of ergonomic requirements, - Enables organizations to define the impact of ergonomic non-conformities on the ability to ensure beneficial outcomes of measures
Matrix diagram (matrix data analysis)	- Enables organizations to identify interrelationships among various sets of requirements and features	- Enables organizations to identify interrelationships among the causes of existing non-conformities by defining the ways and magnitudes of their mutual impacts, - Enables organizations to assess the suggested impacts of ergonomic non-conformities on measure outcomes

their decisions to reflect the nature and frequency of occurrence of non-conformities. The benefits to be gained depend on organization's abilities to meet these prerequisites.

Once enterprises learn to apply tools, they can prevent the occurrence of a large proportion of potential irregularities, select any existing non-conformities and rectify them effectively. Potential non-conformities can be identified as early as the stage of designing the working environment and the conditions in which processes are performed. This is made possible by ensuring that the working environment complies with legal and normative standards. A crucial factor for organization's ability to ensure working comfort is to address the expectations of workers and other concerned parties. The need to address worker expectations applies also to expectations concerning working environment design and supervision as well as human resource management [27]. This is based on the desire that by providing their employees with proper working conditions, businesses will gain the ability to better meet their complex corporate objectives. Their success in doing so hinges on the availability of complete and accurate information necessary for proper problem assessment.

The outcome, in the form of suggested feasible measures, can be described in terms of the nature of benefit-generating factors. The benefits can be associated with ensuring a worker-friendly working environment that will enable them to perform their occupational tasks effectively. Factors of essential importance for the successful implementation of improvement measures can be classified as follows:

- Controllable factors, i.e. factors that can be manipulated to achieve desired effects,
- Uncontrollable factors, i.e. factors whose impacts are beyond the control of decision-makers,
- External distorting factors, i.e. such factors that impede process operation whose impacts are unrelated to the manner in which processes are performed,
- Internal distorting factors, i.e. such factors that impede process operation whose impacts are associated with processes.

When assessing conformity with ergonomic standards, critical success factors include factors that either generate benefits or eliminate human-related distortions to processes. Ultimately, this process will benefit all stakeholders across given organizations.

Satisfaction with work and working conditions comprises the key benefits to be derived. Such satisfaction can generally be described as being happy with the existing status quo. Other than conformity with applicable ergonomic requirements, the achievement of such satisfaction requires the proper treatment of employees and being adequately responsive to their needs. Improvement outcomes are predicated on the availability of information on both the existing status as well as improvement opportunities [9]. In undertaking improvements, one should account for worker needs, any requirements pertaining to the tasks being performed and the assignment of responsibilities and powers. Organizations should also ensure the best possible match between the solutions they employ and employee needs and the approval of their solutions by workers.

5 Summary

Organizations can ensure occupational safety by employing such quality management tools as are helpful in identifying problem areas and reasonable improvement measures.

The primary purpose of an improvement process is to ensure that workers feel well in their organizations and encourage them to be more efficient and creative. Working environment improvements rely on solutions designed to achieve lasting improvement outcomes. To manage and effectively improve their working environment, organizations need to identify factors affecting working conditions.

Worker health protection forms an integral part of adequate working conditions. Human life and health are closely interrelated and warrant special protection and recognition as the supreme value. The capacity to ensure worker effectiveness is increasingly crucial for effective performance.

Notably, the measures should extend beyond such minimum requirements as are necessary to ensure safety. Safety is assessed against standards which, if met, ensure the prevention of harm to worker health. However, one should be clear that compliance with such standards alone may not suffice to secure employee health and well-being. Organizations may also find it necessary to refer to ergonomic criteria and standards. For production to proceed efficiently, the working environment must satisfy worker needs. This may be seen as a precondition for effective performance.

Measures targeted at the ergonomic aspects of the working environment have the potential to improve working safety, hygiene and comfort. These will ultimately increase the satisfaction of workers as well as all other concerned parties [8]. The use of quality management tools is an appropriate response to all kinds of non-conformities as it is helpful in properly identifying the issues at hand and properly assessing the organization's capability to prevent negative effects.

The need to ensure a more comfortable working environment, which is seen as part of the social environment available to a given community, is increasingly viewed as a key challenge to today's economy. To establish such a working environment, one needs solutions that prevent problems from occurring while enabling enterprises to adapt flexibly to existing operating conditions.

References

1. Dobson, K.: Human Factors and Ergonomics in transportation control systems. Procedia Manuf. **3**, 2913–2920 (2015)
2. Górny, A.: Ergonomic requirements in system management of industrial safety. Found. Control. Manag. Sci. **11**, 127–138 (2008)
3. Eklund, J.: Development work for quality and ergonomics. Appl. Ergon. **31**(6), 641–648 (2000)
4. Dul, J., Neuman, W.P.: Ergonomics contribution to company strategies. Appl. Ergon. **40**(4), 745–752 (2009)
5. Gick, A., Tarczyńska M.: Motywowanie pracowników. Polskie Wydawnictwo Ekonomiczne, Warszawa (1999)
6. Wilson, J.R.: Fundamentals of systems ergonomics/human factors. Appl. Ergon. **45**(1), 5–13 (2014)
7. Hamrol, A.: Wybrane myśli o dążeniu do doskonałości w zarządzaniu jakością. In: Sikora, T. (ed.) Zarządzanie jakością. Doskonalenie organizacji, vol. 1, pp. 86–101. Cracow University of Economics, Cracow (2010)
8. Górny, A.: Occupational risk in improving the quality of working conditions. In: Vink, P. (ed.) Advances in Social and Organizational Factors, pp. 267–276, AHFE Conference (2014)
9. Górska, E., Tytyk E.: Ergonomia w projektowaniu stanowisk roboczych. Podstawy teoretycz-ne. Oficyna Wydawnicza Politechniki Warszawskiej, Warszawa (1998)
10. Matczyński, F.: Organizacja pracy na stanowiskach roboczych. Wydawnictwo Naukowo-Techniczne, Warszawa (1978)
11. Górny, A.: Use of quality management principles in the shaping of work environment. In: Stephanidis, C. (ed.) Posters Extended Abstracts: International Conference, HCI International 2015 (part II). CCIS, 529, 136–142 (2015)
12. Górny, A.: The role of safety in ensuring efficient working conditions. In: Fertsch, M., et al. (eds.) Engineering and Technology Research, pp. 348–353. DEStech Publications Inc, Lancaster (2017)
13. Gołaś, H., Mazur, A.: Macroergonomic aspects of a quality management system. Foundation of Control and Management Science **11**, 161–170 (2008)
14. Penc-Pietrzak, I.: Analiza strategiczna w zarzadzaniu firmą. Koncepcje i stosowanie. C.H. Beck, Warszawa (2003)

15. Górny, A.: Total quality management in the improvement of work environment - conditions of ergonomics. In: Goossens, R.H.M. (ed.) Advances in Social & Occupational Ergonomics, AHFE 2017 International Conference. AISC, vol. 605, pp. 91–100. Springer, Cham (2017)
16. Górny, A., Sadłowska-Wrzesińska, J.: Ergonomics aspects in occupational risk management. In: Arezes, P., et al. (eds.) Occupational Safety and Hygiene, SHO 2016, pp. 102–104. Portuguese Society of Occupational Safety and Hygiene (SPOSHO), Guimarães (2016)
17. Jaffar, N., Abdul-Tharim, A.H., Mohd-Kamar, I.F., Lop, N.S.: A literature review of ergonomics risk factors in construction industry. Procedia Eng. **20**, 89–97 (2011)
18. Mazur, A., Gołaś, H., Rosińska, A., Drzewiecka, M.: The identification of potential incompat-ibilities as the most important aspect of quality management. In: Borkowski, S., Selejdak, J. (eds.) Quality Improvement of Products, pp. 117–130. Tripsoft, Trnava (2011)
19. Tytyk, E., Mrugalska, B.: Towards innovation and development in ergonomic design: Insights from a literature review. Procedia - Soc. Behav. Sci. **238**, 167–176 (2018)
20. Butlewski, M., Tytyk, E.: The assessment criteria of the ergonomic quality of anthropotechnical mega-systems. In: Vink, P. (ed.) Advances in Social and Organizational Factors, pp. 298–306. CRC Press, Taylor and Francis Group, Boca Raton (2012)
21. Rothmore, P., Aylward, P., Oakman, J., Tappin, D., Gray, J., Karnon, J.: The stage of change approach for implementing ergonomics advice – Translating research into practice. Appl. Ergon. **59**(A), 225–233 (2017)
22. Hankiewicz, K.: Ergonomic characteristic of software for enterprise management systems. In: Vink, P. (ed.) Advances in Social and Organizational Factors, pp. 279–287. CRC Press, Taylor and Francis Group, Boca Raton (2012)
23. Cooper, M.D.: Towards a model of safety culture. Saf. Sci. **36**(2), 111–136 (2000)
24. Sousa, S., Rodrigues, N., Nunes, E.: Application of SPC and quality tools for process improvement. Procedia Manuf. **11**, 1215–1222 (2017)
25. Hamrol, A.: Zarządzanie jakością z przykładami. Wydawnictwo Naukowe PWN, Waszawa (2005)
26. McCormick, K.: Quality. Butterworth-Heinemann, Elsevier, Amsterdam (2017)
27. Kiran, D.R.: Total Quality Management. Key Concepts and Case Studies. Butterworth-Heinemann, Elsevier, Amsterdam (2017)
28. Boer, J., Blaga, P.: A more efficient production using quality tools and human resources management. Procedia Econ. Financ. **3**, 681–689 (2012)
29. Pater, L.R., Cristea, S.L.: A systemic characterization of organizational marketing. Procedia - Soc. Behav. Sci. **238**, 414–423 (2018)

The Method of Ergonomic Design of Technological Devices

Małgorzata Sławińska$^{(\boxtimes)}$

Faculty of Engineering Management, Poznan University of Technology,
Strzelecka 11, 60-965 Poznan, Poland
malgorzata.slawinska@put.poznan.pl

Abstract. A technological device (**TD**) is an object that facilitates the execution of a given process; often formed by a set of interconnected parts that combine to make an operational whole; used for a specified purpose, such as energy conversion, mechanical work, information processing; and has a specified structure depending on the given work parameters and its assigned purpose.

The aim of the proposed method is to improve the design process of the way technological devices are operated from the point of view of operator fatigue. This method is a tool that will help the engineer design ergonomic conditions for TD operation and a safer workplace. The direction of ergonomic modification of TD operation processes is determined by those ergonomic factors for which relevant system relations were identified in the chain of operation during ergonomic diagnosis. The essence of the proposed design method is the process integration of work system resources.

Keywords: Information load · Ergonomic variables · Work system security

1 Introduction

Performing an evaluation of the ergonomic quality of the technological device (**TD**) operation processes using standard tools of ergonomic diagnosis is ineffective. The results obtained during preliminary studies most often do not facilitate the modification of the system, because the answers to questions that are too general do not yield a way to solve individual problems. Ergonomic engineering of TD at the initial stage requires the identification of the technical and organizational conditions of work processes.

As part of the discussed method a research procedure was proposed, which entails gradually narrowing down the set of analyzed factors contributing to task load. Since the operator's tasks focus on the continuous processing of information, the general requirements regarding the functioning of employees in the work environment were initially limited to the information structure of the work environment. Using adequate information means generates the transmission of information in such portions that none of them will exceed the capacity of its reception, and all of them together guarantee necessary fullness to reflect the reality [13]. The actions of the unit were treated as an autonomous system in which searching, receiving, processing and storing information from the environment as well as making decisions take place The analysis of the demands put on the operator during the process of TD operation was narrowed down to

© Springer Nature Switzerland AG 2020
W. Karwowski et al. (Eds.): AHFE 2019, AISC 971, pp. 330–339, 2020.
https://doi.org/10.1007/978-3-030-20494-5_31

factors related to cognition and decision-making), while the ergonomic modification focused on improving the description of reality on physical media and facilitating the use of the sign system during arithmetic and logical mental operations. In order to take into account the impact of the physical working environment, fulfilling the appropriate ergonomic standards in the area of the remaining ergonomic factors was assumed to be equally important. Due to this distinction, designs of such solutions are created, thanks to which working conditions are improved, fatigue is reduced and safety is increased. The large number of available solutions encourages the researcher to choose those that significantly reduce task load [1]. But in any situation to reduce the number of accidents at work we can additionally monitor the psychomotor performance of operators [3].

Therefore, in order to ensure a high ergonomic quality of the interaction between the operator and the interface, it is necessary to identify the quality of the means ensuring the correct course of information processes, which consists of the following factors: information environment, physical environment and the technical and organizational factors in the operating space of a control device. The large number and diversity of particular categories of factors require that engineers designing control device apply a systemic multi-stage creative process. In these activities is important to taken into account ergonomics requirements that provide compliance with the users' needs [1]. Based on the results of research, it was assumed that the analysis of phenomena defined by the situational context is a source of knowledge about those systemic relations that degrade the ergonomic quality of the interaction [11]. The following phenomena were deemed significant, among others: user experience, distracting stimuli, information noise, excessive motivation and additional tasks.

2 Management of the TD Operation Subsystem

The methodology of ergonomic engineering of technological devices is based on the idea of interdisciplinary data integration in the process of creatively uncovering new facts. It is a type of creative research involving associations and analogies, and has been logically arranged and presented in a structured sequence of activities presented in the process diagram (Fig. 1). In the initial preparatory stage, data on the work system structure are obtained. The characteristics of (1) the devices used in the system, (2) the requirements of the operator working method, and (3) the system relations in the human – TD system are determined. By obtaining this data, a task model of the functional structure of the work system is created. It allows one to accurately present the operator's planned tasks and to estimate the efficiency of TD operation. This involves an analysis of the structural features of the technical elements of the work system and the organizational components of the studied technological process. The interaction model is reproduced at the experimental workstation with the simulation of task load conditions during human interactions with the technical elements of the work system. In order to experience difficult conditions, the information environment of the task context is recreated. For this, the information that the operator utilizes in his work is necessary. This is precise information pertaining to a specific course of an interaction and is used when performing a specific, anticipated task. Since it is necessary to experience the expected load conditions for which the software and interface are designed, the information model is applied here [14]. The information employed during the experiment is obtained from the analysis of safety requirements [10].

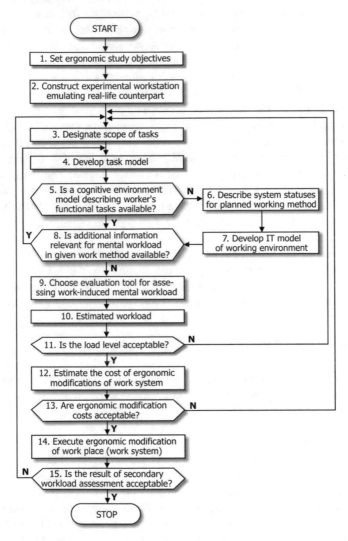

Fig. 1. General model of ergonomic engineering of TD operation. Own elaboration

During work situation modeling, real data relevant to the ergonomic modification are integrated (Fig. 2), and analogies to business process integration may be observed [9]. This is a structure-based design of a set of models that describe and define the structure of the system of TD operation, its functionality and operation. This environment consists of process models, data models, and infrastructure models. It provides a unified vision for the direction of reconfiguration of the chain of TD operation as well as information flow and resource utilization.

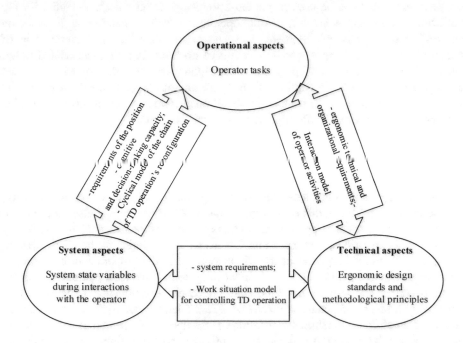

Fig. 2. Diagram of process integration in ergonomic engineering of TD.

In a multi-level analysis of the work system, identified are those system relations that facilitate the comprehensive integration of a variety of data and connections between the levels during modeling. Because ergonomic engineering of TD is based on five basic models, integration includes the following components:

1. General model of ergonomic engineering of technological devices, which is represented by the algorithm presented in Fig. 1,
2. Technical model – represented by signaling and controlling devices as well as technical measures in the workplace at the experimental workstation,
3. Interaction model of operator activities, otherwise known as the information model for operator tasks, Fig. 3,
4. Work situation model for controlling operation, otherwise known as the task model – represented by the functional structure diagram of the work system,
5. Cyclical model of the chain of operation's reconfiguration, supporting ergonomic design of TD (1 - Defining the goals of the created IT system; 2 - Defining the operator tasks; 3 - Identifying ergonomic factors; 4 - Defining the requirements of the IT system; 5 - Determining the indicators and measures of the level of load; 6 - Ensuring the flexibility of the work process requires answering the question what is best for the user in the given moment; 7 - To estimate the costs of the effects of a TD operator's undesirable behavior; 8 - Evaluating alternative IT system solutions in

simulations of feedback conditions; 9 - IT system verification; 10 - Stop or Defining the goals of the created IT system).

Each entry that can be used to present knowledge about what is essential for the performed tasks and what affects the operator's decision-making is a welcome description of the system's elements and system relations [1, 8]. A description of dynamic phenomena is an important element of documenting the facts related to the chain of operation. For this purpose, a description of the chain of events seems appropriate. It's a brief description of the cause-and-effect relationships between the system's components, which resulted in transitions between states of the system. This is basic information for the description of the work situation.

3   Safety Management of the Work System

Systems theory assumes that a relation between particular subsystems is a new quality, a so-called system feature, that is, it assumes that each of the components of the system, in addition to its specific properties, contains system properties that are not easily derived from certain system components, but from the integration of the whole system. Thus, the integrative properties of the whole system are the properties of components that comprehensively make up the whole work system. Therefore, achieving a useful result becomes the goal of the following subsystem components: operators (their personality traits, health status, competence, stress tolerance, motivation, etc.), devices (working conditions, durability, performance standards and compliance with them, reliability, availability, etc.) and management (load normalization, staffing, training, accident analysis systems, etc.).

3.1   Efficiency of TD Operation Processes

Managing the resources of the work system, which are characterized by their relations with the tasks of the TD operator, requires an estimation of the toll, or cost, on the human while performing tasks. This cost includes decreased task performance, decreased wellbeing, and threats to safety or even health [12]. The main goal of human resource management is therefore to plan processes in such a way as to keep work performance at a satisfactory level. It becomes important to prevent the consequences of overload in situations when human information processing capacity becomes insufficient to meet the requirements of the system [6]. If a rational management strategy is adopted, that is, using the available resources to accomplish intended goals and to achieve the best results – then planning should be aimed at achieving the highest efficiency. Economic efficiency is part of a broader concept of social efficiency, which, in addition to economic effects, includes non-economic effects, such as social, political and cultural aspects. The development of technologies useful to the elderly will therefore be a challenge that must be met, while ergonomic modeling will allow to achieve a high quality of the developed solutions in a much faster and cheaper way. However, it is important to develop methods and schemes for evaluation of the technical environment of the elderly [2].

Efficiency is measured as a ratio of the result achieved to the resources used to achieve this result (formula 1).

$$E_f = W/N,\tag{1}$$

where:
E_f – measure of efficiency,
W – achieved result,
N – resources used to achicve the result.

In ergonomic engineering of TD, the elements of the chain of operation, i.e. a complex work system, are reconfigured, therefore an overall assessment should be made during the validation of the final results of this modification. For this reason, the concept of system efficiency is introduced. This is a feature of the system that expresses its overall ability to achieve goals. In this case, efficiency indicators may include the following:

1. The extent to which planned results are achieved – effectiveness,
2. The relation between effects and costs – cost-effectiveness.

Efficiency is a function (formula 2):

$$E_f = fc(W_u, W_R, P_s)\tag{2}$$

where:

W_u – a set of a device's usability features that are a function of its technical, dynamic, supply properties,
W_R – a set of operating conditions, i.e. a set of highlighted environmental influences on the device,
P_s – a set of control stimuli, i.e. stimuli generated by the decision maker, ensuring the execution of tasks by the device.

Knowing the E_f function allows the designer to select such a set of control stimuli that in specified conditions and with a specified set of device features allows one to achieve the required efficiency or at least close to the required level.

In ergonomic redesign of processes, efficiency is evaluated with the costs of modifying system elements and relations with the lowest values of indicators of reliability of the sociotechnical subsystem as the denominator, and as the numerator – the estimated benefits, which are a result of avoiding the potential effects of failure at a given stage of TD operation.

Ergonomic research strives to identify the opportunities for achieving balance in the work system. The evaluation of system efficiency has been adopted as the initial stage of providing fundamental ergonomic standards or defining needs in order to redesign the human – technical object system. The solutions must take into account all the external factors that affect the safety and efficiency of work [4, 5].

3.2 Effectiveness of the TD Chain of Operation

In modern complex technological systems the responsibility resting on the human operator and the cost of committed errors continue to grow. The human's tasks do not end with performing routine operations, but rather with the regulation of huge streams of energy and information [7].

During interactions with a complex sociotechnical system, the human takes on the role of the operator as well as a user of a multiuser information system. This leads the researcher of ergonomic factors to enrich the research subject – a recognition-regulatory system – with aspects of understanding and dialogue on the user's actions. The ergonomic variables of the operator's working conditions are relegated to a secondary role as of greater importance are the limitations related to situations of real dialogue, where the human regulates his actions through thought and understanding and not by one-sided criteria of speed and accuracy.

In the evaluation of the reliability of task execution, the subject of research is the mechanism of interaction in the human – TD system, which can be modelled by an interaction graph (Fig. 3). Pj(t); j = 1, 2, 3, 4, 5, 6 denote the probability that the human – technical object system is in a specified state while performing the given task.

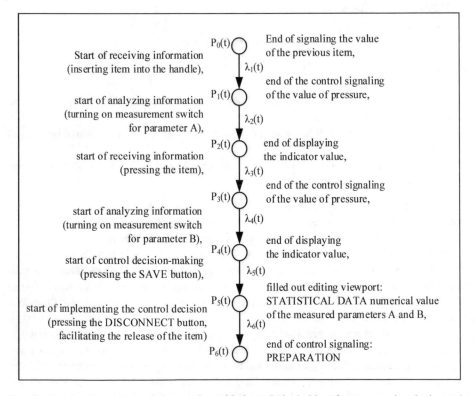

Fig. 3. Graph of the operator's interaction with the technical object for a measuring device at the technical inspection station.

Task execution can be divided into six phases (according to Fig. 3). The corresponding probabilities may be defined as the following:

P0(t) – probability of the appearance of information about the need to perform control actions, P1(t) – probability of receiving information, P2(t) – probability of analyzing information, P3(t) – probability of receiving information, P4(t) – probability of analyzing information, P5(t) – probability of making a control decision, P6(t) – probability of implementing the control decision.

The functions $\lambda j(t)$ are transition intensity functions of the transitions between system states from the $(j - 1)$ state to the j-th state (formula 3). They are defined by the following relationship:

$$\lambda j(t) = \frac{fj(t)}{1 - Fj(t)} \tag{3}$$

where:

fj(t) – probability density function,
Fj(t) – cumulative distribution function of a random variable Tj; j = 1, 2, 3, 4, 5, 6.

The reliability of the human – TD system (system reliability - RS) can generally be denoted as the intersection of independent events, i.e., the reliability of technical measures (RH) and the reliability of the operator (RM): RS = RH RM.

Methods that include documenting the impact of the situational context on human functioning provide a thorough basis for effective TD modification, because it will be focused on the problem of the individual user. At the level of systems analysis design of TD creates an environment that integrates the following elements: a single problem with a posed hypothesis, which is tested using the accepted models of the studied phenomena as well as experiments and recommendations proving the adopted hypothesis.

Monitoring the situational context at the same time as conducting the ergonomic modification is a prerequisite for achieving a higher ergonomic quality during operator tasks. Providing detailed information about the relation between the modified conditions of the interaction and their subjective assessment effectively decreases the length of the process of improving the ergonomic quality of TD.

If self-monitoring mechanisms are applied, i.e., elements of self-design with the TD user's active involvement in design as well as methods of assessing the impact of the situational context on operator tasks, then conditions of efficient ergonomic engineering of TD will be created.

4 Summary

The development of manufacturing forces changes the character and the structure of the target of work of an individual employee. Man is being given automatic devices and he looses the direct contact with the object of his work. Therefore, his tasks concern information processing, scanning the course of events and states of controlled objects, as well as performing steering actions. Man is responsible for the organization of the entire system and achievement of the planned objective; the goal and the structure of his functioning are the central element in the man - to – technique system. Therefore, designing technique that will be economic, reliable and safe, requires a humano-technical approach.

The suggested method of identification of ergonomic parameters, which are essential for the security of the work process enables implementation of knowledge about symptoms of psychical overload to the design area within the organization's structures. Taking under consideration the technical aspect of ergonomic design of a work system enables in the same time reaching following system objectives: (1) providing flexible adjustment of parameters of the employees environment to his individual psycho-physical abilities, (2) creating a dynamic information environment with possibility to adjust it to different course of realization of target objectives, (3) reduction of operating costs of automated process equipment.

The identification of ergonomic parameters important for the security of the work process represents a chance for implementing regulatory mechanisms on operator activities in automated technological devices.

References

1. Berlik, M., Dahlke, G., Sławińska, M.: The idea of modification of work conditions for the reduction of the pilot's workload in a glider, type SZD-30. J. KONBiN **45**, 9–28 (2018). https://doi.org/10.2478/jok-2018-0001
2. Butlewski, M.: Practical approaches in the design of everyday objects for the elderly. In: Slatineanu, L., Merticaru, V., Nagit, G., Coteata, M., Axinte, E., Dusa, P., Ghenghea, L., Negoescu, F., Lupescu, O., Tita, I., Dodun, O., Musca, G. (eds.) Engineering Solutions and Technologies in Manufacturing: Applied Mechanics and Materials, vol. 657, pp. 1061–1065 (2014)
3. Butlewski, M., Hankiewicz, K.: Psychomotor performance monitoring system in the context of fatigue and accident prevention. Procedia Manufact. **3**, 4860 – 4867 (2015). https://doi.org/10.1016/j.promfg.2015.07.603
4. Górny, A.: Ergonomic requirements for the operation of machines and technical equipment. In: Balc, N. (ed.) MATEC Web of Conferences, Modern Technologies in Manufacturing (MTeM 2017 - AMaTUC) **137**, 03005 (2017)
5. Górny, A.: Human factor and ergonomics in essential requirements for the operation of technical equipment. In: Stephanidis, C. (ed.) Posters Extended Abstracts: International Conference, HCI International 2014. CCIS, vol. 435, pp. 449–454 (2014)
6. Kalatpour, O., Farha, S.: The content analysis of emergency scenarios: thematic survey of the context in the process industries. Saf. Sci. **92**, 257–261 (2017)
7. Lasota, A.M., Hankiewicz, K.: Working postures of spot welding machine operators. In: Arezes, P.M., Baptista, J.S., Barroso, M.P., Carneiro, P., Cordeiro, P., Costa, N., Melo, R.B., Miguel, A.S., Perestrelo, G. (eds.) Occupational Safety and Hygiene IV, pp. 261–264. Taylor & Francis Group, London (2016). ISBN 978-1-138-02942-2
8. Nisar, T.M., Prabhakar, G., Strakova, L.: Social media information benefits, knowledge management and smart organizations. J. Bus. Res. **94**, 264–272 (2019)
9. Pacholski, L., Cempel, W., Pawlewski, P.: Reengineering, Reformowanie procesów biznesowych i produkcyjnych w przedsiębiorstwie. In: WPP, pp. 191–230 (2009)

10. Sławińska, M., Mrugalska, B.: Information quality for health and safety management systems: a case study. In: Arezes, P.M., et al. (eds) Occupational Safety and Hygiene III, pp. 29–32. Taylor & Francis Group, London (2015). ISBN 978-1-138-02765-7
11. Sławińska, M.: Modeling ecologic processes of production. Res. Logist. Prod. **6**(3), 217–229 (2016). https://doi.org/10.21008/j.2083-4950.2016.6.3.3
12. Tattersall, A.J.: Obciążenie pracą i przydział (alokacja) zadań. w: Chmiel, N. (red.) Psychologia pracy i organizacji, p. 205. GWP, Gdańsk (2002)
13. Więcek-Janka, E., Sławińska, M.: Improvement of interactive products based on an algorithm minimizing information gap. In: Goossens, R.H.M. (ed.) Advances in Social & Occupational Ergonomics, Part of the Advances in Intelligent Systems and Computing book series (AISC), vol. 605. Proceedings of the AHFE 2017 International Conference on Social & Occupational Ergonomics, pp. 101–109, 17–21 July 2017. https://doi.org/10.1007/978-3-319-60828-0
14. Xu, Y., Reitter, D.: Information density converges in dialogue: towards an information-theoretic model. Cognition **170**, 147–163 (2018)

Cognitive and Emotional-Motivational Aspects of Communication to Improve Work Safety in Production Processes: Case Study

Joanna Sadłowska-Wrzesińska[✉]

Faculty of Engineering Management, Poznan University of Technology,
Strzelecka Street 11, 60-965 Poznan, Poland
joanna.sadlowska-wrzesinska@put.poznan.pl

Abstract. The aim of the publication is to emphasize the role of effective communication practices in shaping and maintaining a high level of occupational safety in production processes. Theoretical considerations were supported by a synthetic review of research results and supplemented by a case study of a manufacturing company. The study exploits the author's questionnaire for the Communication Performance Assessment in the area of Health and Safety (CPA-OHS). The results of the research clearly indicate deficits in vertical communication - between employees and supervisors. The main problems are related to the lack of freedom in daily communication between employees and management and a low level of mutual understanding. The survey showed that the consequences of the current problems in the area of interpersonal communication may be intensified in the near future as a result of a deep reconstruction of the human-machine relationship, based for the first time in the history of civilization on cooperation with non-human intelligence.

Keywords: Human factors · Health and safety ·
Interpersonal and group communication · Safety culture ·
Behavior-based safety (BBS) · Industry 4.0

1 Introduction

In public life, communication in the form of information exchange, discussion and argumentation is the basis of all social, economic and political decision-making processes. The individuals can express their individuality only through communication; they can responsibly act and regulate their relations with other people. In Poland - despite large-scale preventive activities of a promotional and educational character, as well as control activities - the rate of accidents at work still remains unacceptable. The development of research and interests with the accidents subject indicates a high number of threats resulting from human behavior and improper work organization, which confirms the thesis about the need to consider the human factors in safety systems [1, 2]. Hence, responsibility for the worker to take a specific action, eg related to the high risk of accident, is not only on the technical and organizational aspect of the occupational safety culture - the employee's attitude to occupational safety, shaped to the large extant by the employee's knowledge on the occupational risk incurred and the quality of communication in the field of occupational health and safety, will also be a key component.

© Springer Nature Switzerland AG 2020
W. Karwowski et al. (Eds.): AHFE 2019, AISC 971, pp. 340–349, 2020.
https://doi.org/10.1007/978-3-030-20494-5_32

Communicating is one of the most important phenomenas determining the quality of functioning of social groups and individuals in these groups in the same time. Considering the communication in the organizational context, it should be taken into account that it is related to a number of indicators of individual employee functioning as well as their relationship with the organization. For example, Einarsen with the team [3] showed in his research the relationship between the lack of satisfaction with the quantity and quality of tips, instructions and information that employees receive from superiors and the higher scale of abuse in interpersonal relations. In turn, studies by Jo and Shim [4] show that better communication between superiors and subordinates results in a higher level of trust in professional relationships. Dessy [5] on the basis of the research of health care workers indicated that effective communication is a factor protecting against the development of negative consequences of stress and burnout. To a similar extent, Kelly and Berger [6], using the results of their own research, stressed the importance of internal communication processes for the level of safety at work.

The importance of the problem of poor communication is confirmed by empirical research both on a micro scale and in the pan-European approach [7]: in Poland less frequently than in other European Union countries issues of occupational safety and health are regularly discussed at employee meetings or team meetings (in 56% companies compared to 65%); also in the process of occupational risk assessment, Polish employees participate much less frequently than employees of other EU countries (33% compared to 47%). Both examples are strongly related to the need to cooperate and actively participate in the life of the organization, mainly through the aspect of interpersonal and group communication [8]. It should also be emphasized that corporate communication is a key element of organizational identity, alongside determinants such as symbolism, behavior, values, and organizational culture [9]. The complexity of production processes means that modeling of in-house communication systems is usually time-consuming and prone to errors, so additional conditions that may affect job security are sought for. One of the areas increasingly subjected to research exploration are the safe behavior of employees (BBS) and, consequently, the growing interest of industrial enterprises in non-technical methods of shaping the desired health and safety at work [10–13]. It can be noticed that the organizational measures are increasingly focused on building a good atmosphere at work, creating a sense of community, eliminating fear, creating a sense of justice and implementing cooperation at all levels of the hierarchy in the workplace. Such cooperation requires intensive efforts to shape communication skills.

2 Methodology of Communication Capacity Assessment

While *physical capacity* means the ability to exercise, without serious disturbances of homeostasis and in consequence fatigue symptoms, communication capacity should be understood as the organization's ability to communicate and interpret messages among organizational units and between these units and their organizational environment, without significant disturbances destabilizing exchange of information. Just as physical capacity determines the body's potential to exercise, *the communication capacity in the area of occupational health and safety* - according to the author - *will determine the company's potential to organize and implement the communication process in such a way that provides information flow in all directions of the organizational structure, including a dialogue focused on exchange of thoughts (cognitive element) and feelings and expectations (emotional and motivational element) regarding safe work performance.*

The analyzed entity is an international corporate enterprise, having its branch in Poland in the Wielkopolska province, where food products are produced. The factory employs approximately 850 employees. The company's structure includes: director, logistics and planning department, quality department, maintenance department, two production departments, OSH and environmental protection department, innovation implementation department, productivity improvement department as well as invest-ment, HR and financial departments, not directly related to production. An important resource of the factory are production workers who make up the majority of employees. The company, which applies high security standards, depends on convincing employees to participate in occupational health and safety management and encour-aging them to report any dangerous incidents and/or inappropriate working conditions - in line with the idea of continuous improvement. In the company surveyed, employees have the opportunity to communicate on occupational health and safety issues through: personal contact with the health and safety department, sending a message by e-mail to the occupational safety and health department, personal contact with the immediate supervisor, and calling on the so-called secure phone. The problem identified by the company was the unsatisfactory level of involvement of production workers in the implementation of the behavioral safety program (BBS). This is manifested first of all by the lack of expected employee activity in the area of reporting potentially accidental events, improper working conditions and/or other security problems in the workplace.

In order to pre-diagnose deficits in the field of communication and cooperation in the area of occupational health and safety, talks were conducted with production workers and leaders about the safety culture in the workplace. The study used an informal, unclassified version of the interview, taking into account the problem of the study. The respondents confirmed their reluctance to exchange information in the area of occupational health and safety and a low level of participation in diagnosing and reporting weaknesses, arguing with fear of misunderstanding from supervisors, lack of constructive feedback, disrespectful managerial attitude, fear of dismissal and tiring bureaucracy and intricate notification procedures. Most employees paid attention to the danger of being seen as an *informer*. According to the respondents, this situation will be definitely negative in the work environment and may result in the loss of bonuses, unfavorable opinions, reprimands of the superior and deterioration of interpersonal relations in the team.

In the light of the information obtained, a decision was made to conduct an inquiry survey among production workers - this is the next stage of the research process, where the original questionnaire for the Communiaction Capacity Assessment in the area of occupational health and safety (CCA-ohs) was used. The questionnaire is characterized by satisfactory reliability and accuracy coefficients (Cronbach's internal α and II coefficient of parts I and II is 0.87) and enables to get to know employees' opinions on in-house communication regarding work safety problems (Tables 1 and 2).

Table 1. CCA-ohs part I: everyday communication in OHS area.

Nr	Statement
1.	Information on hazards at work are communicated on ongoing basis
2.	I know to whom I should talk on OHS issues
3.	Accidents at work and events potentially dangerous are discussed during meetings, together we try to find their causes
4.	Employees are encouraged to give their comments on potential improvement of OHS in the company
5.	Communication between managers and staff is free and frequent enough
6.	Employees are well informed on OHS issues – they have access to all the information they need to perform their work safely
7.	Managers appreciate the employees caring for OHS issues
8.	Management's visitis at workplaces create uncertainty and tension
9.	I give my comments on hazards I recognize at the workplace
10.	Conflicts between employees representing different departments are frequent
11.	There is plenty of understanding in relations between employees and managers
12.	I feel responsible for safety of my co-employees

Table 2. CCA-ohs part II: barriers in in-house communication.

Nr	Statement
1.	Barriers in communication between departments
2.	Barriers in communication *from the bottom up*
3.	Low quality of managerial communication
4.	Significant role of a gossip in a company
5.	Blocking information by managers
6.	Lack of proper communication tools
7.	Inefficient use of implemented communication tools
8.	Sense of lack of knowledge/skills to communicate effectively with others
9.	Overload with information
10.	Other - which?

The study was conducted among production workers (n = 98), who were asked to give their opinion on twelve statements regarding daily communication in the workplace (part I OWK-bhp), to indicate barriers to internal communication (part II OWK-bhp), to assess own communication skills (part III OWK-bhp) and to provide data on age, sex and seniority (part IV of the General Conditions of Labor Protection). In the part concerning the opinion on the level of communication, a five-point Likert scale was used, and in the part devoted to self-evaluation - the scale of school grades.

3 Results of Reaserch and Discussion

The study revealed clear deficits in vertical communication between employees and supervisors. Based on the respondents' opinions, the main problems are related to the lack of freedom in daily communication between employees and management and a low level of mutual understanding (Fig. 1). Considering the relatively high averaged values, the more problematic the area in point 9 is: *The management's visit at the workplace creates uncertainty and tension.* Together with points 10 and 11 (also below the satisfactory score = 70%), which concern the assessment of conflict among employees and the level of mutual trust - this gives an initial picture of the interpersonal problems of an emotional basis. Figure 2 graphically presents the assessment of *the most satisfied* employees (highest scoring responses on the scale: *definitely yes*). Here, the difference between the three *points of ignition*: 9, 10, 11 and other opinions can be observed even more strongly[1].

In the second part of the CCA-ohs questionnaire (barriers to in-house communication), responses given by respondents are strongly dispersed, which could suggest a strong impact of individual differences, including differences in the perception of specific behaviors and organizational events. Among the barriers in internal communication, employees most often indicated the large role of the rumor in the company (25%), then barriers in the flow of information *from the bottom up* (17%) as well as barriers to the flow of information between departments (15%).

The third part of the CCA-ohs survey is the self-assessment of employees in respect to their communication skills. The rating is very high; the arithmetic means of the results = 4.59 (on a scale of 1–6), and the median = 4. Such high marks may indicate lack of awareness of internal (psychological) communication barriers or attempt to transfer full responsibility for the shape of communication relations to supervisory staff.

The analysis of data on age, sex and seniority (part IV of the CCA-ohs) brought interesting observations. And so, contrary to the existing opinions on *gender in communication*, no significant differences were noticed among the respondents. The group of employees over 50 years old received the lowest result among all age groups in expressing opinions on internal communication. Such a situation requires an in-depth analysis aimed at diagnosing sources of deficits found; it can be assumed that older employees are less able to tolerate changes, including those that relate to forms of communication and related procedures. A similar situation concerns the distribution of results due to seniority - people with long-term work experience (over 20 years) have poorly assessed the level of occupational health and safety communication than people with shorter seniority, which may be a confirmation of a previously suggested cause. It is surprising, however, that the employees - without any real differences due to their age, gender and work experience - rated highly the cooperation with the Health and Safety Department. This is very good information considering the difficult and/or poorly effective relations between employees and occupational health and safety specialists commonly functioning in Polish work conditions. Good contact with the Health and Safety department is an excellent platform for the exchange of information on

---

[1] Questions 9 and 11 are simmetrical questions.

working conditions and the opportunity to participate in conducting occupational risk assessment or during an annual (obligatory) analysis of the occupational health and safety in the company. However, it needs to be emphasized that as part of daily work tasks, the most important role - according to the author - is played by supervisors who provide their employees with information in direct conversations. Such a method of transmission enables quick assessment of the reaction to the provided information and its possible correction.

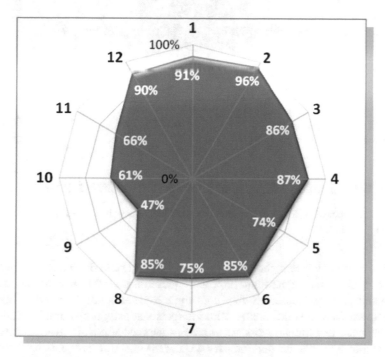

Fig. 1. Assessment of in-house communication in the OHS area – average values.

Lack of communication between employees and supervision or - as in the case of the discussed company - a clear deficit of mutual trust and fear of free expression can effectively disrupt the feedback process in communication and, as a consequence, a harmonious system of in-house relations necessary to build the desired safety culture. And although the organizational identity is a changeable construct, evolving along with the development of the enterprise, it still contains some permanent elements, consisting of widely understood behaviors and corporate practices [14]. In the literature on the subject, one can read that organizational culture is a symbolic re-presentation, shaped during communication [15]. In this context, the safety culture of interpersonal communication is of particular importance - according to the author – and a platform for fundamental assumptions in relation to safety and health protection at work.

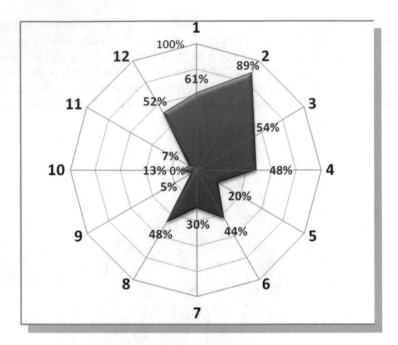

Fig. 2. Assessment of in-house communication in OHS area – responses of *the most satisfied* employees.

The research conducted, though does not entitle author to draw general conclusion, however, results in asking new and important questions on safety at work. Dynamics of changes and growing unpredictibility contribute to developing the need to search for original approaches to issues related with interpersonal and group communication. It is widely believed that human work will undergo unprecedented transformations due to the automation and robotics that Industry 4.0 is carrying. However, among the obvious benefits, there are also problems of socio-economic nature, such as the risk of technological unemployment. McKinsey [16] in a report prepared with the Polish Forbes branch claims that the potential of robotization and automation in Poland is 49 per cent - this is the index on the basis of which working time is devoted to activities that can be automatized. However, it is believed that - although artificial intelligence will radically change the methods of work and its performers - the main impact of this technology should not be to replace human abilities, but to complement and strengthen them. It has been proven so far [17] that enterprises that replace their employees with solutions from the area of IA gain only the benefits of short-term efficiency gains. Organizations achieve the most significant improvement in performance when people and machines work together. For example, cooperation between machines and people allows companies to communicate with employees and clients. A well-known example is the intelligent Microsoft Cortana bot, which facilitates communication between people, eg by creating a transcription of a meeting, providing advice on countless issues, or even being a virtual assistant representing the interests of the system user [18].

Another well-known smart bot - Aida - has access to huge data resources and is capable of conversing with millions of customers in their natural language. Aida is able to ask clients additional questions to solve their problems, and also analyzes the tone of the caller's voice to diagnose whether he is satisfied or frustrated. This is a clear step towards "working out" strongly influencing emotions, which usually have a direct impact on the shape of interpersonal and group communication processes. It can therefore be assumed that both Cortana and Aida could successfully perform the function of a "personal health and safety assistant", which practically eliminates the risk of dangerous occurrences and risky behaviors. The tempting possibility of reducing accidents at work, increasing readiness to respond flexibly to emerging threats and activating pro-health behaviors in practice may turn out to be much more complicated. Even super-bots can not read ambiguous signals, and consequently - they can not differentiate what is good and what is bad. The lack of these skills makes it possible to manipulate students' systems of intelligence, introducing deliberately untrue data and disturbing the cognitive process. Therefore, it is not known what currently poses a bigger problem for the safety of people: the lack of machines of "real" emotional intelligence or the possibility of controlling its components? For the operationalization of the problem, it is legitimate to ask the question: will our security depend on who will be the head of Cortana and Aida?

4   Conclusions

The safety level of modern enterprises is anchored in the quality of management understood as the art of controlling processes, resources and information. In the context of occupational risk prevention, the most advanced solutions function in relation to processes and resources, while the issue of information management is still an area requiring exploration. Effective information management depends on the course of the communication process. This process, as a result of mediation by individual properties of the parties to the transmission, is one of the least-controllable dimension of the functioning of employees and organizations. The survey showed that the participation of the managerial staff in the improvement of interpersonal and group communication in the enterprise may positively affect the involvement and activity of employees in the area of occupational risk prevention. The obtained results confirmed the thesis that in the context of providing broadly understood safety and health at work, communication efficiency will belong to organizational processes of a special nature.

In addition, the described case of the company showed that not only the cognitive aspects of communication - the correct flow of information, transparency of speech, accessibility of communication tools - affect the effectiveness of communication in the organization and safety in production processes. Intra-organizational communication is a complex process and proceeds dynamically. Interacting, for example, two people (entities) involves accepting, understanding and interpreting information. Especially the latter element is marked by the need to both express and receive certain emotions, which enables the emotive communication function to be implemented and allows to satisfy the psychological needs of both sides of the message. Hence, we should be well prepared for changes related to the technological revolution - intelligent machines can help a man both in work processes and in developing his skills, especially those of an analytical and decision-making nature, providing him with the desired information at the right time. It is worth remembering, however, that competences included in emotional intelligence (complementary to rational intelligence) are - for the time being - reserved exclusively for human beings. Cooperation between machines and people, although more and more often enables companies to quickly realize information exchange processes, does not necessarily mean increasing the *communication capacity potential*.

References

1. Górny, A.: Human factor and ergonomics in essential requirements for the operation of technical equipment. In: HCI International 2014 - Posters' Extended Abstracts: HCI International Conference, Heraklion, Crete, Greece, HCI Proceedings, Part II (2014)
2. Hankiewicz, K., Butlewski, M.: Psychomotor performance monitoring system in the context of fatigue and accident prevention. Procedia Manuf. **3**, 4860–4867 (2015)
3. Einarsen, S., Raknes, B.R.I., Matthiesen, S.B.: Bullying and harassment at work and their relationships to work environment quality: an exploratory study. Eur. J. Work. Organ. Psychol. **4**(4), 381–401 (1994)
4. Jo, S., Shim, S.W.: Paradigm shift of employee communication: the effect of management communication on trusting relationships. Public Relat. Rev. **31**(2), 277–280 (2005)
5. Dessy, E.: Effective communication in difficult situations: preventing stress and burnout in the NICU. Early Hum. Dev. **85**(Suppl. 10), S39–S41 (2009)
6. Kelly, B., Berger, S.: Interface management: effective communication to improve process safety. J. Hazard. Mater. **130**(3), 321–325 (2006)
7. ESENER: European Survey of Enterprises on New and Emerging Risks. https://osha.europa.eu/en/surveys-and-statistics-osh/esener
8. Sadłowska-Wrzesińska, J., Mościcka-Teske, A., Stankowiak, A.: Interpersonal and group communication in prevention of potentially hazardous accidents – the behavioral approach. In: Kiełtyka, L., Jedrzejczak, W., Kobis, P. (eds.) Wyzwania współczesnego zarządzania. Kreowanie kapitału intelektualnego organizacji, pp. 71–84, Warszawa (2016)
9. Faber, B.D.: Community Action and Organizational Change: Image, Narrative, Identity. Southern Illinois University, Illinois (2002)
10. DeJoy, D.M.: Behavior change versus culture change: divergent approaches to managing workplace safety. Saf. Sci. **43**(2), 105–129 (2005)
11. Kines, P., Andersen, L.P., Spangenberg, S., Mikkelsen, K.L., Dyreborg, J., Zohar, D.: Improving construction site safety through leader-based verbal safety communication. J. Saf. Res. **41**(5), 399–406 (2010)
12. Li, H., Lu, M., Hsu, S.C., Gray, M., Huang, T.: Proactive behavior-based safety management for construction safety improvement. Saf. Sci. **75**, 107–117 (2015)
13. Ávila, S., Cerqueira, I., Drigo, E.: Cognitive, intuitive and educational intervention strategies for behavior change in high-risk activities – SARS. International Conference on Applied Human Factors and Ergonomics AHFE 2018. In: Advances in Manufacturing, Production Management and Process Control, pp. 367–377. Springer (2018)

14. Carlsen, A.: After James on identity. In: Adler, P.S. (ed.) The Oxford Handbook of Sociology and Organization Studies: Classical Foundations, pp. 421–443. Oxford University Press, Oxford (2009)
15. Keyton, J.: Communication and Organizational Culture: A Key to Understanding Work Experience. Sage Publications, Thousand Oaks (2005)
16. Globalization intransition. Global value chains are being reshaped. McKinsey & Company. https://www.mckinsey.com/
17. Wilson, H.J., Daugherty, P.R.: Jak ludzie i sztuczna inteligencja łączą siły. Harv. Bus. Rev. Pol. **190**, 124–132 (2019)
18. Harari, Y.N.: Homo Deus. A Brief History of Tomorrow. Vintage, London (2018)

Benefits on the Field of Ergonomics and Work Safety from Development of the Information System in the Enterprise

Krzysztof Hankiewicz[1(✉)] and Andrzej Marek Lasota[2]

[1] Poznan University of Technology, Strzelecka 11, 61-845 Poznan, Poland
krzysztof.hankiewicz@put.poznan.pl
[2] University of Zielona Gora, Szafrana 4, 65-516 Zielona Gora, Poland
a.lasota@iibnp.uz.zgora.pl

Abstract. The main objective of this study is to contribute toward a comprehensive understanding of the advantages of modern technology not only in business management but also from the aspect of work safety and ergonomics.

For better use of resources, an enterprise needs to plan and undertake a variety of planning activities like demand planning, production planning, materials planning and budgeting, and financial planning. In this case, employees are a very important resource and should be the main focus of these plans. Their competences, knowledge, skills and needs should be taken into account when designing and implementing management information systems.

The article presents the way an improved information system for business management can affect work safety and ergonomics in the company. The paper includes the procedure for a needs analysis in the field of ergonomics and work safety in order to prepare guidelines for the information system.

Keywords: Ergonomics · Information systems · Enterprise

1 Introduction

It is crucial for business organizations to constantly find new ways to improve their procedures in order to maintain their position on the market. One possible way to achieve greater cost-effectiveness is to develop an information system, as business solution software can bring a lot of value to a company by integrating different applications.

Initially, information systems were very simple to set up as they mostly mirrored the business process that departments had been implementing for years but allowed them to do things faster with less work. However, once the systems were in use for so long, they became very difficult for individual departments to manage and working with these systems was strenuous and generated numerous errors and delays in the company's operations.

In such a situation, the interaction between man and computer became more important, however, this time the ergonomists focused on software, not hardware as it was in the past. Software translates human commands given to a computer into a language that a computer can understand and presents a computer's output in a form comprehensible to the human mind. Excessive complexity of system queries can cause operator errors. The results of such errors can be financial losses – destroyed raw materials, equipment stoppages, delayed deliveries – but they can also reduce work safety. In this situation, it is crucial to consider ergonomic features when designing and improving IT systems.

The aim of ergonomics is to adapt the working conditions and tools to the person who uses them. The process should include not only technical knowledge, but also knowledge from the fields of biology, medicine, and psychology. This principle also applies to adapting the company's information system to the people using it in their work.

International Ergonomics Association distinguish domains of specialization within the discipline of ergonomics as follows [1]:

- Physical Ergonomics - concerned with human anatomical, anthropometric, physiological and biomechanical characteristics as they relate to physical activity.
- Cognitive Ergonomics - concerned with mental processes, such as perception, memory, reasoning, and motor response, as they affect interactions among humans and other elements of a system.
- Organizational Ergonomics - concerned with the optimization of sociotechnical systems, including their organizational structures, policies, and processes.

In the case of human and information system in the enterprise context definitely dominates cognitive ergonomics field which include mental workload, decision-making, skilled performance, human-computer interaction, human reliability, work stress and training. Additionally, it is connected with organizational ergonomics because of communication elements, crew resource management, work design, design of working times, teamwork, participatory design, community ergonomics, cooperative work, new work paradigms, virtual organizations, telework, and quality management.

2   Ergonomics in Software Design

The quality of the software refers to (cf. [2]):

- technical (structural) quality, results from meeting requirements regarding technical characteristics, which are determined by code quality, among others
- ergonomic quality, results from meeting ergonomic requirements, especially in relation to a specific task
- functional quality, results from meeting a user's requirements in relation to a specific task

Software ergonomics deals with user - oriented shaping of human - computer interaction. It concerns the shaping of interactive computer system elements [3]. Ergonomic software is characterized by the relationship between the interactive system and the user and the adjustment of the system to sensory and cognitive human abilities.

The goals of software ergonomics are connected with:

- increasing acceptance
- increasing motivation to work
- increasing competences
- developing personality
- load optimization through the use of new technologies

These goals can be achieved by solving problems related to programming, instal-ling and running programs and user training, maintenance and support systems, as well as program integration, user interface design, dialog style selection, etc. The degree of fulfilment of these goals can be a measure of a software's ergonomic quality.

Scapin and Bastien present in 1997 coherent list of ergonomic criteria for evalu-ating the ergonomic quality of interactive systems [4]:

(1) Guidance

- Prompting
- Grouping and distinguishing items (by location, by format)
- Immediate feed-back Legibility

(2) Workload

- Brevity (Conciseness, Minimal actions)
- Information density

(3) Explicit control

- Explicit user actions
- User control

(4) Adaptability – Flexibility, Users' experience
(5) Error management

- Error protection
- Quality of error messages
- Error correction

(6) Consistency
(7) Significance of codes
(8) Compatibility

The technical and ergonomic quality of the software determines its usability. It is many definitions of usability. Prussak listed many of them [5]. They present various aspects of usability:

- human support at work
- integrating the software with the environment in which it is used
- enabling the user to perform tasks efficiently and effectively
- specifying the user's context
- compliance of the project with the basic rules
- learning support

Usability can be used for characterizing product quality, assuming that other elements of work system are known. Quality in use "is the user's view of the quality of the software product when it is used in a specific environment and a specific context of use" [6]. This Quality characterize product in wide context of real use process for tasks realization in real work environment.

With a focus on user needs, such considerations have a fundamental and critical impact on the final form of an IT product. A large number of subjective and objective factors affects users' satisfaction with using the software (cf. [7]).

3 Development of the Enterprise Information System

In the past, when a company received a client's order, regardless of whether it was a service ordered or a product purchased, the order had to go through a process based on paper documentation, which was forwarded to various departments. In the entire process, the order often had to be re-written as it passed through different departments, increasing the risk of mistakes. There was no exact order status because there was no tracking device that would inform every department. To obtain information about the status of the order, customers were often asked to contact the company's warehouse for on-site verification.

Nowadays, companies have eliminated the uncertainty of paper tracking by implementing enterprise resource planning (ERP) systems. Most companies already use ERP software to eliminate inefficiency in processes such as order tracking. Instead of independent computer systems, ERP uses a unified program that connects various functional departments, such as finance, HR, production, warehouse, planning, purchasing, inventory, sales and marketing. While each department can have its own set of software modules, the software is connected so that information can be shared across the organization. After one department processes and updates an order, it is automatically redirected to the next department so that everyone is notified of the changes.

An enterprise information system is not only a tool, but also a task transfer center. Since users interact with the system's components, they can generate tasks using predefined algorithms. Possible interactions between the tasks, tools and users can bring about work results, organizational results and changes to human well-being.

The user and the computer are connected by a set of actions defined together as communication. The two basic forms of communication are input and dialog. In most applications, input communication has been replaced by dialog, which allows the user to intervene in the processing of previously entered data. In the case of indirect communication, dialog should be developed taking into account the user profile including user skills as well as prior training and experience in using the system.

In many cases the users were skeptical about learning designed rules of IT systems. They can be motivate taking part in system development activities (cf. [8]).

System designers should recognize the needs of each potential user. They should also remember that users who work with the system are also changing. The system should be adapted to the individual needs of the user and facilitate user development. On the other hand, the system should be somewhat "distrustful" of the actions of users, especially those that are irreversible, difficult to undo, or which cause loss of confidential data. Such "distrust" should be built into the system with a choice confirmation requirement and mechanisms of prevention of entering incorrect data. The system should impose informed decisions, i.e., prevent automatic confirmation based on the appearing hints. It should also avoid excessive confirmation requests to not interfere with work and concentration or increase fatigue.

Many people think that working with computers is stressful, although there are rarely similar problems when using other devices, including those whose operation is very complex. The user interface plays a central role in this process. It turns out that the interfaces of many devices are well developed. However, improvements to the user interface can also be observed on computers, especially on personal computers. The graphical user interface (GUI) is still being improved. Progress has significantly contributed to the popularization of personal computers. However, today's products are still not perfect, because the interfaces are still not fully intuitive. On the other hand, many of the tasks performed by the software are very complex – in reality, their complexity increases with the development of systems.

When designing the interface, it is important to distinguish between character-based interfaces and graphical interfaces. The purpose of replacing the character interface with a graphic equivalent is to extend the bandwidth, i.e., the range of useful information visible to the user at a given time and increase the speed of performing the given tasks. This can be achieved by presenting information graphically and additionally using animation. It is also necessary to use the right colors. Even when using plain text, colors make it easy to convey messages. Further improvements result from bold text and changes in font size and text layout.

The interface can be much more intuitive, allowing direct manipulation using a pointing device.

When designing the interface, one should take into account the importance of the layout of the screen's elements, the phrasing of commands, the clarity of the hints, presentation of non-textual information, the highlighting and coding of information.

When arranging screen elements, decisions are made on where to place individual informational elements, forms, and dialog boxes, how to present them graphically, and which color codes and schemes to use. It may be advisable to select part of the screen to display the current status of the program. The screen may need to be divided into sections devoted to displaying information of different types. The division can be used to separate input and output information and system notifications. When deciding on the layout of the user's screen, care should be taken to avoid findings contrary to the user's habits.

The names chosen for the program's items should be easily identifiable; hints and questions should be easy to understand. Unambiguous wording should be implemented in particular with regard to abbreviations. Symbols should be avoided, especially those known only to a limited number of users.

It should also be ensured that the computer messages and feedback are fully understandable to the user. The information units that require special attention from the user should be displayed in a different font (different type, size and/or color). In the best case scenario, such information should be placed in the center of the screen (especially in the case of error messages). Thematically-related and frequently shown information should be displayed in the same place and form.

Important messages that are non-verbal and provided in alphanumeric form, such as telephone numbers or identification numbers, should be provided in a way that facilitates easy entry, retrieval and storage.

Alphanumeric characters should be grouped into sets of two, three or four, separated by spaces. This is especially important for strings with more than six characters. Bigger and often searched datasets are best placed in tables or sorted.

Texts consisting of letters should be aligned to the left, while those that consist only of numbers, especially when they are recognized as numbers, should be aligned to the right. Similar data should be presented in the same way. Traditional methods of data entry should be used – this applies to the order of data and standard formats.

Using adequate information means generates the transmission of information in such portions that none of them will exceed the capacity of its reception, and all of them together guarantee necessary fullness to reflect the reality [9].

Messages, in particular those conveying information, confirmations and notifications about errors, should be formulated in a uniform manner, especially when they refer to similar situations occurring in different parts of the system. Messages should not require the user to search for a related code in the software documentation. Interpretation of verbal messages replacing codes should be clear and unambiguous and easy to understand without referring to the documentation.

Software designers should always take into account operator error. Errors require time to identify and correct their causes. It is therefore desirable for the system to limit the number of errors that can be made. The system should support rapid error detection and should promptly notify the operator.

The design of the dialog box is a key factor in case an error occurs. The impact on the user's mental health should be taken into account. The program should not bore or overwhelm the user. If the user is notified immediately about an error, (s)he will better remember the situation and avoid similar errors in the future.

Some errors are caused by operator overload. Therefore, it is necessary to limit the need to memorize a large amount of information. The ability to repeat the same choices for different types of items greatly facilitates the work of the user. The minimum requirement should be the ability to use the cut, copy and paste functions for different items. In graphical interfaces, it is particularly effective when individual actions can be generalized to be performed by pointing to an icon (e.g. by using a mouse). It is important that the icon be easily associated with the action that it triggers. In addition, the system messages that attract the user's attention and provide error warnings should be carefully worded and contain sufficient information. They should be free of shortcuts and codes that confuse users and force them to refer to the documentation to check their meaning. Forcing the user to do this will result in user stress and further disrupt the workflow. It's best to match the content of the message to the user's skill level. More advanced users should be informed by means of short, concise messages that lead to more detailed information. Beginners require lengthier messages that describe the actions to be taken in the given moment in step-by-step instructions.

It is important that the user receives feedback from the system at the time it is needed. Usually, this time is short, so the user does not have to wait for the system's response. It is difficult to determine the response time acceptable to users, because it depends on the situation. It is certain, however, that if the system response time is short, the computer and the user can exchange more information, which streamlines the workflow. However, despite the progress made in the development of IT equipment, short response times are very expensive. The difficulty often lies in the internal organization of the system and the manner and location of data storage. Users may tolerate longer response times when they realize that large data volumes are being processed or data is retrieved from remote databases. A user convinced that the operation is not complex will not accept a longer response time. Therefore, it is important that the user is informed about the system's status and progress on the task. Users should be aware that the system is working and that they must wait. A message informing of this can be given graphically or verbally on the status bar or in a separate dialog box. In order to extend the waiting time, it is advisable to display a progress bar.

Finally, it should be noted that too short response times are not always desirable since they can make operators feel that they cannot keep up with the system. To prevent this, the system's reactions may be deliberately delayed.

If a problem occurs, the operator can use the help of a more experienced user as a source of information. This is usually possible for systems that have been used for a longer period of time, because by then a relatively large number of previous users is available. It may be advisable to find an advanced user – trainer for new system users. It is particularly valuable for advanced users to pay attention to problems that have proven difficult when they started working with the system. The disadvantage of using such help is that it involves people who may have other duties to perform.

In any case, complete information about the system should be available in the system documentation. Such documentation should be helpful in solving operational problems. System documentation has various forms. The four most popular are:

- user manual - instructions for the user; explains how to perform each task using the software; the user manual should list all the available system functions; users who want to rely on such documentation themselves need some experience;
- quick reference guide - contains basic information about the system; this documentation is intended for advanced users who previously worked with similar systems;
- full user manual - provides detailed information on each function of the system; a full user guide is intended for advanced users who want to get acquainted with and use all of the system's functions;
- tutorial - contains step-by-step instructions for using the system; tutorials are designed for inexperienced users who want to learn how to use the system.

The type of documentation included with the system depends on the type of system, the target users and, in some cases, marketing reasons. It is important that the text of the documentation is supplemented with graphic symbols, tables, charts and screenshots, and even short videos to make it easier to track individual operations. It is also important to provide enough examples to make the content of the documentation easier to understand, especially for novice users.

Quicker access to information needed to solve a given problem is offered in the help program available directly in the system. An additional advantage of software documentation is the ease of updating by changing files. Extended software help should contain context-sensitive help that sends one directly to the help topic related to the given task.

The program's help section should include a keyword index for easy searching. When frequent help updates are needed, it is worth using a program connected to the online version.

An effective way to introduce the user to operating the system is to use the training program. Training programs resemble system guides. They explain the main functions of the system using demonstrations and exercises. They help the user learn how to use the system. Ideally, the training program should include tests to check the user's knowledge.

In summary, the help system should offer users the choice of using the help section of the program or the system documentation. In both cases, the help should be complete enough, and the search option should be clear enough, for the user to not have to seek help from more advanced operators. It is best to provide users with the ability to view the basic functions of the system using a training program.

4   Ergonomic Features of an IT System

In the past, when examining the workstation of a computer operator in terms of meeting ergonomic requirements, the hardware and user interface of the supported device were usually analyzed. Mostly it was an analysis of the human-computer system. Generally, the simultaneous cooperation of many people using one information system, as well as the possibility of interpreting data collected in database systems and automatically generating reports, were not taken into account.

Butlewski and Tytyk define such system as anthropotechnical mega-system where relations between "the human subsystem," consisting of many interacting people, and the "technical subsystem" consisting of many interacting machines [10].

Replacing direct contact between employees during the execution of tasks with interactions through and in cooperation with the IT system creates a different type of nuisance and requires additional analysis. First of all, the way of entering, viewing, modifying and deleting documents and data, as well as approving documents, is different from typical computer use in office work. The point of reference here are computer stations used independently of each other, where direct interaction between users is present and where for the most part the allocation of tasks and reporting on their implementation happens directly. The principle of the modern IT system is to provide access to all documents and tasks. Direct interaction between employees, including supervisors, is replaced by messages generated by the system. These messages can be generated based on a message sent by another employee/supervisor or based on data or documents appearing in the system. Therefore, the system should be equipped with mechanisms that notify the employee about the appearance of new documents and tasks. System notifications should not be burdensome for the employee. If necessary, the notification method can be diversified, adjusting it to the level of significance of messages. It should also be considered whether employees should have an influence on the method of notification. Certainly, the system should follow the stages of document development, its subsequent modifications and approvals, especially when many people participate in the process and the document is of significant importance to the company.

The IT system imposes procedures whose accuracy, consistency, intuitiveness and relevance of the interface have an impact on the following basic assessment categories: ease of use, usefulness, comprehension, speed of use, accessibility, self-descriptiveness, adequacy, error tolerance, ease of learning to use, integrity. In many cases it is emphasized that making a positive impression on the user is also a factor conducive to making correct decisions and minimizing fatigue. Therefore, the above list can be supplemented with another category: aesthetics (cf. [11, 12]). Such a combination of ergonomic categories considered when assessing the usability of software or websites also applies to IT systems, but differs in depth of analysis.

A typical assumption of ergonomics is the replacement of the human in monotonous repetitive activities. In IT systems this applies to entering data into the system or generating repeatable reports, among others. Already at the initial stage of development of IT systems, the aim was to automatically enter all electronic documents. Documents generated in the company could be entered directly into the system. Difficulties arose when introducing paper correspondence and entering data from verbal or telephone conversations. However, nowadays, the replacement of paper documents with electronic versions has mostly eliminated tedious scanning. The problem of recording conversations is solved in various ways – depending on the needs: note from the conversation, record of the conversation (only for archiving) or recording based on voice recognition. There are also new challenges, for example, as a result of active use of social media. In many cases this requires more employee involvement and an individual approach. Taking care of the company's image or building relationships with clients does not involve new possibilities that these services allow, and the IT system gives the opportunity to document these activities and process the information collected from customers.

For many companies, media in the form of pictures, sound or video materials belong to key data. These media are in many cases collected in a high quality, which determines not only the larger capacity of the disks, but also a larger transfer when they are saved and downloaded. In this case, the speed of the network and the time of access to resources are crucial to reduce cumbersome waiting for documents. However, it turns out that organizational activities can be very helpful here. In addition to database replication, or mirroring, in some situations one can access an initial preview of lower quality files. This can significantly increase system performance and reduce nuisance. Obtaining this effect should not involve any additional effort during the introduction, only the mechanisms of automatic conversion. Such data preparation can be treated as a kind of indexing prepared in order to make it easier to search.

The advantages of changing paper documents to electronic ones are obvious, but only introducing them to a system that allows them to be searched based on complex criteria brings significant benefits. Focusing on the benefits for employees – it not only means reduced time spent searching for documents. Employees should feel the need for their intellectual involvement in the work process. This is facilitated by the creation of documents, making creative analyses and working on concepts of development in the enterprise. However, searching through subsequent documents and checking if they do not contain data related to the task currently performed can be frustrating and discouraging. Also, if it is necessary to communicate with other employees in order to obtain information when the necessary mechanisms for finding information in the IT system could be provided, it can be demoralizing. Especially if the time devoted to searching for information and communication with other employees in order to collect all the necessary data is comparable with the time needed to prepare documents.

Working under time pressure is typical for people who operate IT systems. The need to prepare documents with short deadlines is particularly onerous. However, such situations are common when analyses and reports have to be prepared for management. In such instances, the systems that automatically generate summaries and reports, leaving the operator with a decision regarding its content and form, are even more valuable. The advantage of being able to generate complex reports while ensuring the most up-to-date information can ensure that one stays on a competitive market that requires a rapid response to customer needs.

Managers spend an equally large amount of time when looking for information. However, in their case a significant part of the burden is passed on to various assistants and other subordinates who are responsible for answering questions and obtaining information. This increases the proportion of time spent on interpersonal communication in the total working time. This communication time also includes meetings during which employees exchange information in direct contacts. Time spent on performing such activities can also be significantly reduced by using an IT system that generates summaries and reports based on information provided in databases.

The benefits of adopting an IT system go beyond the usual time savings for employees and management, and the associated decreased stress and effort and increased productivity. Such systems additionally offer a number of other advantages. Immediate entry of payment documents into the system helps to manage cash flows rationally and to plan payments in advance, for example due to the availability of invoices in the system immediately after their introduction, and even before their approval.

Another benefit is access to cross-sectional data in a way that is not supported in traditional systems. When first establishing contact with a client (customer, supplier or seller), the contact topic is referred to as a case (or task, in some systems). In further contacts with the client or with any other company, the case is accompanied by a name or reference number. As a result, obtaining information about a given case is not only possible, but also very simple, even if such information may apply to many different companies. An important part of the system is a module that informs employees about the appearance of new documents, informs about the need to approve them, enter them into accounting books, prepare offers, the phase of order fulfillment, etc. A very important advantage is informing employees about any unfinished tasks.

The company's document and information management system can be a module of a comprehensive company IT management system or an independent system that communicates with other enterprise systems. Regardless of the degree of integration, such systems should cover various areas of management, such as document management, document and information flow management, case management and workflow management.

A disadvantage of such extended and automated systems is that the user's activities are constantly monitored, which can lead to a feeling of restriction. It is important to provide employees with breaks during which their behavior will not be monitored. Another solution is the establishment of periodic employee meetings, where direct contact will balance continuous isolation. At the same time, these meetings may concern the improvement of the enterprise information system.

Due to the complexity of each enterprise information system, it seems beneficial if user participate in system development. Such suggestions can be found in the recommendations for designers. The basic element are user requirements. They can also be presented in a different form. For example user requirements can be expressed as scenarios or stories and the user prioritizes these for development. The development team assesses each scenario and breaks it down into tasks (cf. [13]).

5 Conclusions

The use of information technology is an important element of the company's strategy to succeed in a rapidly changing world. Modern information technologies enable enterprises to create information systems that facilitate effective communication and mutual understanding among employees, as well as support the decision-making process at all levels of management.

In the future, IT solutions should be even more efficient at the integration of people, information and business processes into an IT system that is one complex of technological solutions. Currently, this trend manifests as establishing feedback with clients, mainly thanks to social media, among others.

The designers of IT systems, when considering the role of the human in the system, focus on the elements usually associated with the human-computer interaction, i.e., the user's communication with the system, the user interface, system feedback and prompts, system response time and system help. In order for the management information systems to respond better to human capabilities, this scope should be extended to additional activities. They include studies on the methods of entering documents and information as well as document and information flows in the system. It should also be noted that many of today's business management information systems act as intermediaries in interpersonal communication. This applies to both the interactions within the company in which the system is installed, as well as external contacts with business partners and clients. The ability to respond quickly to customer needs and market changes can also improve business performance.

It is often assumed that well-being at work is the higher degree of implementation of ergonomics principles in the workplace. However, well-being is related to reducing the effort of employees to comfort level. In the case of a company's decision-makers, who are at the same time active users of the information system of the company, the definition of well-being as related to their workload is not entirely appropriate. For them, self-fulfillment, job satisfaction and new challenges may be more important than limiting their effort at work. On the other hand, such people may have a lower level of tolerance for repetitive and monotonous actions imposed by the system.

In summary, it can be concluded that the term "adaptation to user needs" that is often used in software ergonomics should be understood more broadly. The implementation and development of the enterprise information system should take place with the participation of its users. The term "made to measure" seems more fitting in this case, although the complexity of modern IT systems precludes the profitability of creating dedicated systems. The pursuit of creating universal systems that include the possibility of adapting them to the recipient during implementation is the currently emerging trend in this industry. Therefore, it is important that in these systems the elements connected with adapting them to the experience of users are configurable to the greatest extent.

References

1. International Ergonomics Association. http://www.iea.cc/whats. Accessed 12 Dec 2018
2. Sikorski, M.: Managing Usability in IT Projects. Wyd, Politechniki Gdańskiej, Poland (2000)
3. Eberleh, E., Oberquelle, H., Oppermann R. (eds.): Einführung in die Software-Ergonomie. In: Gestaltung graphisch-interaktiver Systeme; Prinzipien, Werkzeuge, Lösungen. de Guyter, Berlin (1994)
4. Scapin, D., Bastien, J.: Ergonomic criteria for evaluating the ergonomic quality of interactive systems. Behav. Inf. Technol. 16(4/5), 220–231 (1997)
5. Prussak, W.: Ergonomiczne zasady projektowania oprogramowania komputerowego. In: Jabłoński, J. (ed.) Ergonomia produktu Ergonomiczne zasady projektowania produktów, pp. 303–342. Wydawnictwo Politechniki Poznańskiej, Poznań (2006)
6. ISO 9126-1: Software engineering. In: Product Quality. Part 1. Quality Model. International Organization for Standardization, Geneva (2001)
7. Hankiewicz, K.: Ergonomic characteristic of software for enterprise management systems. In: Vink, P. (ed.) Advances in Social and Organizational Factors, pp. 279–287. CRC Press, Boca Raton (2012)
8. Sikorski, M.: Evolution of end-user participation in IT projects. In: Pańkowska, M. (ed.) Frameworks of IT Prosumption for Business Systems Development, pp. 48–63. IGI Global Hershley, New York (2013)
9. Więcek-Janka, E., Sławińska, M.: Improvement of interactive products based on an algorithm minimizing information gap. In: Goossens, R.H.M. (ed.) Advances in Intelligent Systems & Computing, Proceedings of the AHFE 2017 International Conference on Social & Occupational Ergonomics, AISC, vol. 605, pp. 101–109 (2017)
10. Butlewski, M., Tytyk, E.: The assessment criteria of the ergonomic quality of anthro-potechnical mega-systems. In: Vink, P. (ed.) Advances in Social and Organizational Factors, pp. 298–306. CRC Press, Taylor and Francis Group, Boca Raton, London, New York (2012)
11. Nielsen, J.: Usability 101: Introduction to Usability (2012). www.nngroup.com/articles/usability-101-introduction-to-usability. Accessed 19 Dec 2018
12. Hankiewicz, K., Prussak, W.: Usability estimation of quality management system software. In: Salvendy, G. (ed.) HCI International, 11th International Conference on Human-Computer Interaction. Theories, Models and Processes in HCI, vol. 4 (2005)
13. Sommerville, I.: Software Engineering. Addison-Wesley, Boston (2011)

Production Management and Process Control

Six Sigma-Based Optimization Model in Hauling Cut and Fill Exploitation Activities to Reduce Downtime in Underground Mines in Peru

Kevin Rojas[1]([✉]), Vidal Aramburú[1]([✉]), Edgar Ramos[2]([✉]),
Carlos Raymundo[3]([✉]), and Javier M. Moguerza[4]([✉])

[1] Escuela de Ingeniería de Gestión Minera,
Universidad Peruana de Ciencias Aplicadas (UPC), Lima, Peru
{u201319297, pcgmvara}@upc.edu.pe
[2] Escuela de Ingeniería Industrial,
Universidad Peruana de Ciencias Aplicadas (UPC), Lima, Peru
pcineram@upc.edu.pe
[3] Dirección de Investigaciones,
Universidad Peruana de Ciencias Aplicadas (UPC), Lima, Peru
carlosraymundo@upc.edu.pe
[4] Escuela Superior de Ingeniería Informática,
Universidad Rey Juan Carlos, Mostoles, Madrid, Spain
javier.moguerza@urjc.es

Abstract. In the mining industry, the mining cycle is a very important part in the operating stage of every mining unit. Through the exploitation method used, the mining cycle provides mineral ore, which subsequently undergoes various metallurgical processes and its commercialization and thereby generates profits to the mining company. Currently, within this cycle, the hauling and transportation stages are those that have a lower efficiency with respect to the drilling and blasting stages. Therefore, the sector is always seeking effective ways to optimize these processes, reduce downtime and increase productivity. Six Sigma is a technique that allows for the continuous process improvement. In this study, the factors that generate inefficiency in the hauling cycle are determined, and alternatives are implemented to solve the main problem and improve the operations cycle.

Keywords: Six Sigma · Hauling · Overhand Cut and Fill · Productivity · Downtimes · Dumper

1 Introduction

In Peru, according to the Ministry of Energy and Mines, 38% of mining units in operation use underground mining methods. All these operations conduct the process of drilling, blasting, cleaning, sustaining, hauling, and transportation to obtain economically profitable mineral ore in an efficient manner. In underground mining, the process that hinders mining operations the most is hauling and transportation. Within this stage, high downtime levels generate inefficiency in the hauling cycle. This is due to a low

© Springer Nature Switzerland AG 2020
W. Karwowski et al. (Eds.): AHFE 2019, AISC 971, pp. 365–375, 2020.
https://doi.org/10.1007/978-3-030-20494-5_34

capacity of the operators to effectively handle dumpers, either due to lack of training or inexperience. In addition, an adequate design of ramps and tracks is important for the development of mining processes, since it facilitates the transit of vehicles without resorting to difficult maneuvers. On the other hand, technical vehicles failures hinder dumper conditions for efficient cycle performance. All these factors result in low productivity during mineral ore transportation, causing economic losses. Likewise, dumpers generate excessive consumption of supplies such as tires, lubricants, and fuel due to the length of the cycle, which consequently reduces their operational use.

Faced with this problem, several innovative solutions were developed, such as software and methods to reduce downtime in the Peruvian mining cycle. Various efforts have been undertaken in recent years to find an effective solution to this problem. One of them is the GPSS/H software, which is a programming language that performs discrete event simulations to improve transportations flows within the mine.

In addition, stochastic techniques were implemented to simulate vehicle routes through equations and thus obtain greater transportation control throughout the process. In addition, the AlgoI software is used to adequately select the ideal number of vehicles needed for mining operations. Further, management programs such as Dispatch and Mining Star are currently used to control hauling and transportation vehicles, since these programs monitor these vehicles to decide the best route vehicles should take based on fixed or dynamic assignments.

In this study, the Six Sigma technique will be used to fulfill this purpose. Six Sigma is a data-driven method that examines repetitive business processes and aims to bring quality to levels close to perfection. The great advantage of the Six Sigma methodology is that it may be implemented in all mining operations, from prospecting to commercialization of final products. In addition, Six Sigma also provides the DMAIC Quality Management model: Define, Measure, Analyze, Improve, and Control.

Therefore, the purpose of this study is to determine the critical activities that generate inefficiency in the hauling cycle and decrease productivity. In view of this, we will first define what times are part of the cycle to subsequently make a measurement and finally implement an improvement to solve the problem.

This study attempts to improve operations processes in underground mines and, in particular, reduce downtimes to improve the hauling cycle and generate greater productivity [1].

This paper is divided into 5 parts. Section 2 contains information about previous articles related to the subject matter addressed in this document. Section 3 analyzes and discusses the proposal developed. Then, Sect. 4 presents the validation results for the proposal. Finally, conclusions are presented in Sect. 5.

2 State of the Art

In every mining operation, adequate vehicle selection is paramount. For this reason, various contributions have been made to optimize the vehicle fleet. First, a discrete event simulation is prepared using the SimMine Software developed by Greberg et al. [2, 3]. The selection of hauling and transportation vehicles in underground mines is a challenge due to their impact on production rates and costs. Despite the availability of several transportation systems, diesel trucks have been widely used to transport materials both in open pit and in underground mines [4]. Sub-optimal selections can lead to a high production costs; however, finding an ideal solution is not an easy task either, due to the number of variables involved in system design, such as production requirements, transportation routes, infrastructure designs, material fragmentation, and capital and operating costs. Therefore, in order to make an optimal selection, this technique is used. In the applied situation, the TH660 vehicle exhibits greater production efficiency than the TH430. However, the improvement does not generate significant impact; therefore, the TH660 vehicle is not selected. However, a disadvantage of this technique is its high implementation costs. In addition, the time required to develop the simulations is a bit high and may generate disagreement in senior management.

Second, a part of proper hauling optimization is to appropriately select the vehicles that will operate in the mine, since work assignments are designed based on those vehicles. Santelices et al. [5] develop a stochastic model to reduce vehicle selection and replacement problems, including determining the correct number of trucks and shovels to optimize the fleet. This technique consists in determining the random variables (times, cost, productivity, availability). Then, the stochastic model was applied, in which the aforementioned variables are indexed in a probabilistic space. Next, restrictions are aligned, such as vehicle features, work conditions and operator experience). In this case reported by the authors, the optimum fleet for the loading and transportation cycle comprises 25 type 1 trucks, 16 type 2 trucks, 6 type 1 shovels, and 2 type 3 blades.

Finally, another technique used to optimize the hauling feet is the scheduling model proposed by Córdova et al. [11] using the ALGO1 algorithm. This method consists of assigning "1" if the vehicle starts executing a given task and "0" if not. Then, the target function and the ALGO1 algorithm are defined to establish the Decision-Making process. The main results obtained when comparing the proposed model with the one used in the mine under study is that the variable (operator experience) is the important factor in increasing the efficiency of the hauling cycle.

The Six Sigma methodology is used in different companies around the world to optimize processes. Keley et al., Arango, and Gallardo evidence how this technique works in various areas within organizations [6]. The first article describes how Six Sigma contributes to the mining industry in Africa. For these purposes, a longitudinal study is implemented, using a set of indicators and success factors for the deployment of Six Sigma at the Lonmin PLC mining company in South Africa. Through this methodology, operating costs were reduced by 10%. In the second article, the author focuses on the process of admission at a community college. In this scenario, the specialists found that Six Sigma may also be used successfully in other areas within the same institution. Finally, the last article evaluates a small company in the automotive sector, where Six Sigma implementation successfully reduced 13% of the products in the automotive painting section. This optimized the process by moving the rating from 2.4 to 3.6 on the Six Sigma scale. Thus, process alternation was reduced, providing a more reliable and secure process experience to customers. These three studies evidence how Six Sigma is guiding everyone to be more effective and efficient.

Regarding the optimization process, an important focus in the operations cycle of the mine is the improvement of vehicle management, particularly in the hauling cycle. In the underground mines studied in the articles, deficient fleet management and downtime in the hauling and transportation cycle have been identified as critical problems, generating both vehicular congestion and inefficiencies when transporting mineral ore or providing other transportation services. The limited and reduced space, the one-way roads in which trucks and personnel circulate, vehicles failures, and limited operator experience are some of the causes for the problems identified at the mines studied. To solve these problems, various approaches have been proposed. For example, Haviland and Marshall [7] perform simulations in the GPSS/H programming language as a method to efficiently coordinate strategies of two different traffic-management-policy models. The GPSS/H methodology comprises developing a conceptual or theoretical event based on a real event. Initially, the established data are validated. Then, the conceptual model is validated in order to simulate it in the software. Finally, the results of the simulation are evaluated and compared with the actual situation. The authors analyze 2 models. In the first model, the passage of a single vehicle was allowed while blocking the access of others. The second model contains prioritizing vehicles that travel upwards. In this way, the most productive transportation route was designed for the ramp.

Greberg [8] uses the discrete event simulation method to achieve optimal results in terms of production compliance by determining the appropriate number of trucks for each activity. This method was applied to a ramp of an underground mine. Here, parameters such as geotechnical design, mine geometry, and vehicles used were identified. Then, the behavior patterns of the vehicles are analyzed and understood. In this light, vehicles must be identical, and the ramps must only be used for transporting ore. Subsequently, linear parameters are defined for the ramp. Next, an algorithm is executed wherein an important conditional question, "System Blocked?", is implemented: if the answer is positive, the vehicles may advance since other vehicles have been restricted from accessing the ramp; if the answer is negative, the vehicles may not advance. In both situations, the delays experienced in ore hauling and transportation are assessed to improve the process.

Regarding this methodology, seven cases were analyzed by accounting for variations in truck and production area availability as well as disturbances caused by other mining vehicles on their way to and from hauling production areas [9].

Motivation. In relation to the topics discussed above, the hauling process exhibits shortcomings that directly affect production rates and costs. Although efforts have been made through the application of different methodologies or programs to optimize the hauling cycle, successful results are yet to be reported. However, from the literature review, it can be seen that the Six Sigma methodology has had very good results in reducing costs and improving productivity in various industries such as the automotive, services, and manufacturing industries. Therefore, by implementing Six Sigma in hauling or other mining activities, the cycle may become more efficient.

3 Collaborations

3.1 Rationale

Various authors have improved mining cycle stages. For example, Algo1 has improved transit flows in limited areas such as ramps. In addition, Mining Star has been used to record hauling and transportation times and determine vehicles' efficiency. In addition, the GPSS/H software has improved vehicle transit. All of these contributions are certainly important. From analyzing other models presented, we identified a new opportunity of improvement through the Six Sigma methodology. This methodology may improve hauling vehicle management and cycle times by selecting the ideal vehicles for these activities. Six Sigma is a methodology with proven successful results in various industries (such as automotive and services). However, it has not yet become relevant in the mining industry. Therefore, our objective is to use this technique to improve the hauling cycle, since mining companies are always looking for ways to increase efficiency in their activities. One of the few Six Sigma contributions to the mining industry took place in the Escondida mine in Chile. Here, after its application in the mining operations area, the Six Sigma technique indicated that lack of motivation was a consequence of the repetitive and monotonous activities performed by the workers over the years, directly affecting their performance. A rebound effect was also identified regarding the acceptability that these operators exhibit to new processes and job reconversion. At the end, this study optimized operations processes based on the assessment of critical points. On the other hand, by implementing this technique, this research study attempts to improve traffic flows on hauling roadways.

3.2 Overview

Define
It is performed within the mine, where the problem must be defined by planning, including customer expectations and needs (concentrator plant). The following questions are asked:

- Which cycle times are going to be considered?
- Which formulas will be used in the hauling process?
- What are the most important KPIs?
- How will time information be obtained?

Based on these questions, data are collected from the corresponding mining pit. In this case, from pit 6636 to ore deposit 6704.

Analyze
In this stage, collected data are studied to reduce the gap between the current actual hauling time and the desired time. The objective of this phase is to understand why downtimes are generated. For this, a cause–effect diagram is used. Later, loading times, loaded travel times, and empty travel times are calculated. With this information at hand, statistical tools such as Excel and Minitab are used to process data and prepare statistical tables such as histograms and bar charts. Consequently, each activity is assessed and compared against its mean times. Likewise, the mine layout is assessed through plans and blueprints to better understand the location of the loading chamber at pit 6636 and ore deposit 6704 within the mine.

Improve
Determines the cause–effect relationship and identifies possible process aspects that could be improved, proposing solutions to mitigate or eliminate the causes that

originate process problems, thus fulfilling customer expectations and needs. Finally, the operational range of the input parameters or variables of the process are determined. This phase identifies improvement tools, designs experiments for process optimization, and implements improvements to project processes.

During data collection, it was observed that road traffic is a permanent problem in this mine. Therefore, building a bypass ramp was decided, which will reduce traffic and improve hauling cycle efficiencies.

Control

Finally, when the improvement is implemented to optimize the hauling cycle, the information will be collected once more through periodic reviews to pit 6636 with the purpose of performing a comparative time analysis using Office tools such as Excel. This phase answers the following questions: How much has the process improved after implementation? How are the changes maintained? How much time and money is saved? How are processes monitored?

3.3   Process View

Figure 1 presents an overview of the Optimized Overhand Cut and Fill Exploitation Hauling Process.

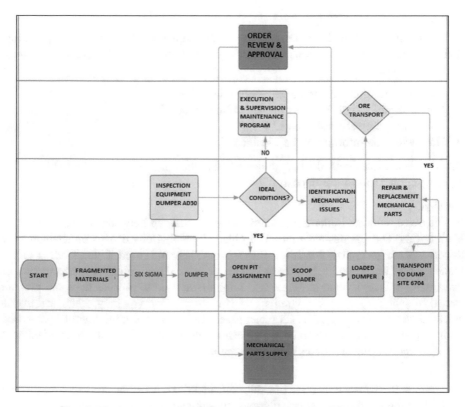

Fig. 1. Optimized Overhand Cut and Fill exploitation hauling process

The general indicators are the monthly mineral ore production and hauling times.

- Mineral Ore Production (ton/month)
- Hauling cycle (hours)

4 Validation

4.1 Case Studies

The case subject was the Socorro mine at the Uchucchacua Mining Unit, located in the department of Lima, province of Oyón, district of Oyón, at an altitude between 4,300 and 4,600 meters above sea level. The objective is to reduce downtimes generated during the transportation of mineral ore to the 6704 ore deposit by applying the Six Sigma methodology. To obtain favorable results, the 5 steps established by Six Sigma must be followed.

4.2 Current Hauling Situation

At the mining unit, hauling is currently performed by MT2010 dumper vehicles and a locomotive. In the studied pit, hauling is performed exclusively by dumpers. The hauling cycle records very high times due to factors such as waiting times at scooping, deposit queues, and heavy road traffic. These factors cause cycle times to rise, thus affecting mining productivity. The amount of ore transported from pit 6636 to ore deposit 6704 is 2,736 tons/month with an average 48-min hauling cycle.

Define
Fieldwork is performed and the times that affect the theoretical cycle and the practical cycle are determined. This way, downtimes are determined in terms of hours, minutes, or seconds. Only 4 times were considered for the theoretical cycle, since they are constant hauling variables. On the other hand, for the field cycle, the non-constant variables were considered—that is, variables that depend on vehicle operators. Likewise, KPIs were defined to make comparisons that may allow us to reduce downtimes in the hauling process. In this first phase, the process or processes that will be subject to evaluation are defined by Management. In addition, the work team that will execute the project is also defined. Finally, the improvement objectives are defined.

Measure
Hauling cycle times are collected for 30 days. A hauling cycle is performed by MT200 dumpers with a nominal capacity of 20 tons. The average time of the current hauling cycle is 0.8067 h (48.40 min). Subsequently, all hauling cycle times are assessed. In this third phase, the dumper times were analyzed based on the hauling data collected per day. According to these charts, we can observe that empty and loaded travel times are variable, which means that they can be standardized to optimize the hauling cycle.

Improve

Through these assessments, downtimes may be reduced if adjustments are made in the following variables: loaded travel time and empty travel time. The improvement proposal is the implementation of a bypass road. The project will finance the construction of this bypass road. A bypass road costs table will be developed. This bypass road will be considered as a 4 m × 4 m section ramp. In this stage, the mesh selected has 49 bores, of which 4 are relief, square, drag. In the drilling and blasting activities, an efficiency of 92% will be considered. According to geomechanics, the quality of the rock is medium. Therefore, for rock support, a mesh and helical bolts with resin will be used. This will cost $170/m of progress. Once implemented, average hauling cycle times were reduced from 0.8067 h to 0.7159 h.

Control

The results obtained evidenced a reduction in transportation cycle times. This favors a greater number of cycles per shift. In addition, loaded travel times and empty travel times have decreased, when compared to the initial times without improvement. Therefore, a substantial reduction was achieved in empty travel times and loaded travel time.

5 Results

The results of the study can be observed in the previous charts. When optimizing the cycle, the hauling cycle from pit 6636 to ore deposit 6704 was improved in terms of reduced production times and costs. Regarding loaded travel times, an average decrease of 13% can be evidenced, as shown in Fig. 2.

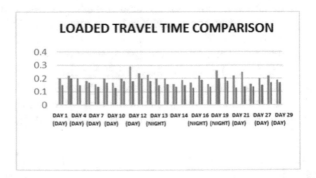

Fig. 2. Comparative variation of the loaded travel time per day

Regarding empty travel times, there is evidence of a 12% decrease once the bypass was implemented, as shown in Fig. 3.

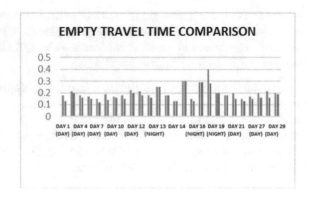

Fig. 3. Comparative variation of the empty travel time per day

With respect to waiting times at the loading chamber in the hauling cycle from pit 66336 to ore deposit 6704, Fig. 4 depicts an average time improvement of 5% against current hauling cycle times.

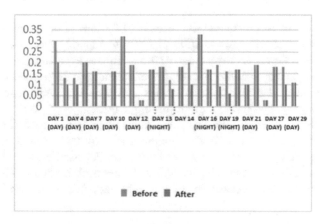

Fig. 4. Comparison of variation of loading waiting times in the hauling cycle

6 Discussion

In the results of this study, it can be observed that the hauling cycle had a substantial improvement when applying the Six Sigma methodology. A 13% improvement was reported for loaded travel times; a 12% reduction was reported for empty travel times; and a decrease by 7% in total loading waiting times. In turn, these reductions also decrease overall hauling cycle times. As can be seen in the following statistical chart, hauling cycle times decreased by 12% when compared to the current cycle times.

On average, the hauling cycle after implementing the bypass was 0.7159 h. Currently, the cycle lasts 0.8067 h. Considering these results, it can be determined that since the productivity and the hauling time are related variables, 12% more mineral ore will be hauled on average. Currently 3,534 tons/month are exploited. With the increase in productivity reported, the amount of mineral ore increases to 4,025.28 tons/month (Tonnage Improvement = 3,534 ton × 1.12 = 3,958 tons).

- Converting this new productivity rate into financial terms, a profit margin of USD $79,768.713 is generated.
- Profit = 4,025.28 tons − 3,534 tons = 491.28 ton ore
- Tonnage × Grade × Price × Metallurgical Recovery × Settlement Factor Profit ($) = 491.28 ton × 14 ounce/ton × 14.49 $/ounce × 0.92 × 0.87 = USD $79,768.713/month.
- Comparing these profits against the bypass investment.

Return on Investment time = 178,146/79,768.71 = 2.23 (2.5 months).

7 Conclusion

In short, this study optimized hauling system performance, substantially decreasing times from 0.80 h to 0.71 h, which represents a 12% increase in production. This was achieved through the application of a bypass road, which improved traffic flows for dumper vehicles.

Mine productivity has increased to process 491.28 tons of ore more than before. In addition, monthly profits amount to USD $79,768. Comparing this amount against the investment made, a Return on Investment of 2.5 months is expected. On the other hand, this study also proved that the Six Sigma methodology is a very useful tool in underground mining. Through further studies and proper variable assessment, great progress can be generated to the current mining activities.

Based on this hauling system analysis, future work is proposed for the different mining stages using Six Sigma. This technique is still not widely used in mining even when it has provided good results in other industries. For examples, studies should be conducted in the drilling and blasting system to supplement this study. These processes should also be optimized because, in mining, efficiency is critical for all exploitation process stages.

References

1. Haviland, D., Marshall, J.: Fundamental behaviours of production traffic in underground mine haulage ramps. Int. J. Min. Sci. Technol. **25**, 7–14 (2015)
2. Greberg, J., Salama, A., Gustafson, A., Skawina, B.: Alternative process flow for underground mining operations: analysis of conceptual transport methods using discrete event simulation. Int. J. Min. **1**, 6–65 (2016)
3. Carrasco, J., Francisco, A.: Asignación dinámica de operadores de lhd para operación a distancia en minería subterránea (tesis para optar el título profesional de ingeniero minero) (2016). http://repositorio.uchile.cl/handle/2250/144483
4. Gustafson, A., Schunnesson, H., Galar, D., Kumar, U.: The influence of the operating environment on manual and automatic load-haul-dump machines: a fault tree analysis. Int. J. Min. Reclam. Environ. **27**, 75–87 (2013)

5. Pérez, E., Edward, W.: Mejoras en el 2011 en la unidad uchucchacua de la compañía de minas buenaventura s.a.a (tesis para optar el título profesional de ingeniero de minas) (2012). http:// cybertesis.uni.edu.pe/bitstream/uni/1299/1/perez_ee.pdf
6. Jang, G., Jeon, J.H.: A Six Sigma methodology using data mining: a case study on Six Sigma project for heat efficiency improvement of a hot stove system in a Korean steel manufacturing company. In: Communications in Computer and Information Science, pp. 72–80 (2009)
7. Vásquez, R., Juan, G.: Elección y aplicación del método tajeo por subniveles con taladros largos para mejorar la producción en la veta Gina Socorro Tajo 6675 - 2 de la U.E.A. Uchucchacua de la Compañía de Minas Buenaventura S.A.A. (Tesis para optar el título de INGENIERO DE MINAS) (2015). http://repositorio.uncp.edu.pe/handle/UNCP/3858
8. Salama, A., Greberg, J., Schunnesson, H.: The use of discrete event simulation for underground haulage mining equipment selection. Int. J. Min. Min. Eng. 5, 256–271 (2014)
9. Gustafson, A., Lipsett, M., Schunnesson, H., Galar, D., Kumar, U.: Development of a Markov model for production performance optimization. Application for semi-automatic and manual LHD machines in underground mines. Int. J. Min. Reclam. Environ. 28, 342–355 (2014)

Analysis of the Technological Capability of Linking SMEs in the Electronic Sector to Integrate into the Maquiladora Industry Electronic Sector in Tijuana, Baja California, Mexico

Maria Marcela Solis-Quinteros[✉], Luis Alfredo Avila-Lopez,
Carolina Zayas-Márquez, Teresa Carrillo-Gutierrez,
and Karina Cecilia Arredondo-Soto

Universidad Autónoma de Baja California, Campus Tijuana,
Calzada Universidad, 14418, Parque Industrial Internacional, 22300 Tijuana,
Baja California, Mexico
{marcela.solis, avila.luis, carolina.zayas, tcarrillo,
karina.arredondo}@uabc.edu.mx

Abstract. The interest of the present investigation is to identify information related to the requirements that Mexican local companies must have in order to be considered as suppliers of the Electronic Sector Maquiladora Industry. For the effects of this research, the Bell and Pavitt model with the matrix of technological capabilities adapted by Dutrenit was chosen. Only the bonding dimension will be analyzed; which is defined as those necessary support capabilities to receive and transmit information, experience and technology of the suppliers of components and raw materials of subcontractors, consultants, service companies and technological institutions. Therefore, the objective is to identify if the technological capability of linkage is an important factor for the Mexican local pro-monitoring to increase its integration in the Decree for the Promotion of the Manufacturing, Maquila and Export Service Industry (IMMEX) electronic sector. The results of the research will allow identifying the technological capabilities of linking the local Mexican supplier of the electronic sector, and thus identify the type and class of suppliers that are required to boost their development.

Keywords: Technological capability · SMEs · Manufacturing sector

1 Introduction

There are several reasons why IMMEX has not developed an important network of local suppliers, but mainly because the plants installed in Mexico already had local suppliers in their place of origin, with whom they had already established a relationship of trust in terms of quality and time of delivery. It is relevant that the local supplier has technological capabilities, so it must be on a par with the needs of transnational firms to respond in terms of quality, service, cost and delivery times, and thereby seeking

© Springer Nature Switzerland AG 2020
W. Karwowski et al. (Eds.): AHFE 2019, AISC 971, pp. 376–387, 2020.
https://doi.org/10.1007/978-3-030-20494-5_35

customer satisfaction. Therefore, the linkage of local supplier companies with the export maquiladora industry represents an important factor for the development of technological capabilities, as well as access to global markets.

The model proposed by Bell and Pavitt turned out to be the most complete and adequate for this investigation; because the proposed framework was created based on the characteristics of the manufacturing sector and does not include other types of capabilities such as organizational ones, however, they have been suggestive in some case studies. Another reason for selecting this model was because it is the most widely adopted as a frame of reference for analyzing developing countries. Mexico has used this model to address empirical studies in manufacturing companies, since it distinguishes more clearly the analysis of technological capability dimensions, which tend to be very varied, ranging from very routine and operational activities to the most advanced ones, taxonomy that Bell and Pavitt classify in the following way: the routine ones as production activities and the advanced ones as innovation activities.

2 Literature Review

In the last thirty years Baja California, Mexico, has become one of the most dynamic regions of the northern border of Mexico for the manufacture of electronic products. This dynamism is based on the fact that it has reached high levels of international competitiveness in comparison with other regions of the country and the world [1]. According to Amburto [2], the World Trade Organization [3] the exportations of electronic products, represented 15% of the total value of merchandise traded in the world, equivalent to 1129 billion dollars; almost twice as many other traditional sectors as textiles and clothing, automotive and chemical. It is estimated that in the current decade the different branches of the electronic industry will maintain high rates of growth worldwide, and that their development will continue to drive other sectors. This impulse will be greater as more accurate are the strategies of each country to integrate the productive chains that intervene in the manufacture and development of new electronic products.

In Baja California, a great variety of products are manufactured, to name a few: printed circuit boards, harnesses, marine sonars, inductors, connectors, cell phones, electronic boards, microchips, semiconductors and mainly televisions [4]. The trigger for the local supplier to the electronics industry in the city of Tijuana was given the occasion of the entry of the North American Free Trade Agreement (NAFTA), as a result of the low relative cost of labor and the proximity to the American consumer market [5]. According to García and Paredes [6] small and medium enterprises have had to face great challenges to be able to compete in an increasingly globalized market, allowing and giving way to productive groupings as a way to increase competitiveness. Bracamontes and Contreras [7] make reference that the traditional entrepreneurs have not been able to have a relevant role in the new productive scheme woven around the maquiladora industry and its large suppliers. The technological capabilities necessary to participate are very low and represent an important line of analysis.

2.1   Technological Capability

Ariffin and Figueiredo [8] conceptualize technological capabilities as: "those resources needed to generate and manage process improvements, as well as the organization of production, products, equipment and engineering projects" highlighting that said technological capabilities can be given both at the individual level (skills, knowledge and experience) and at the organizational level (systems implemented in the organization).

Tapias [9] affirms that technological capabilities are not only composed of the knowledge and skills that individuals in an organization have, but also their organization and purpose. Some are more complex than others, but all require learning and assimilation.

Kim and Nelson [10], Lall [11], Torres [12] converge on the fact that technological capabilities have always been a fundamental component of the competitiveness, growth and economic well-being of the countries. Lugones [13] emphasizes that the concept of technological capability has a very close relationship with elements of technological management, which are the guide for sustained growth and development; Involving knowledge, techniques and skills to acquire, use, absorb, adapt, improve and generate new technologies. In the construction of technological capabilities there are factors that are specific to the company and others that are specific to a country, in such a way that the development of capabilities is the result of the complex interaction between the incentive structure provided by the government the human resources available, the technological efforts made and the incidence of different institutional factors.

The accumulation of technological capabilities is not enough to generate a sustained development, these must be complemented with new combinations of resources, skills, seeking different ideas and new combinations of existing factors (internal to the organization and its environment), the result This is known as innovation capability. This implies that firms must learn to monitor the progress of other players in the market, seeking new more complex or sophisticated innovations [14].

2.2   Measurement of Technological Capability

Matrix of Bell and Pavitt [15] Based on the work of Lall [16] they constructed a taxonomy represented by a matrix, which allows the classification of technological capabilities, in relation to the most important technical functions carried out by a company. These functions will vary or acquire greater relevance over each other depending on the sector in which the company is inserted. The matrix includes four levels of accumulation: one of technological capabilities of routine production and three of innovative technological capabilities: basic, intermediate and advanced. The technological capabilities of routine production are those that use and operate the existing technology. Innovators are capabilities to generate and manage technical change. Basic innovative capabilities allow only a relatively small and incremental contribution to change; but at the intermediate and advanced levels the technological capabilities can have a more considerable, novel and ambitious contribution to change. In relation to technological capabilities, Lall defines three categories: (a) technological investment capabilities; (b) technological capabilities of production, and (c) technological capabilities of linkage. Binding abilities: are those skills that allow firms to receive and transmit information, knowledge, experience and technology from agents located in the external environment such as: suppliers, customers, partners, competitors, technology fairs, specialized magazines, patents, subcontractors, technology consultancies, technical schools, public and private university institutions.

This model clearly distinguishes the technological capabilities to imitate/use/operate technology of technological capabilities to change/create technology, in addition to maintaining the three levels of technological capability that Lall proposes, from least to greatest complexity: basic, intermediate and advanced, what allows to identify the different stages of transition where a company is located. The adjustments made to the Bell and Pavitt model are based on the fact that companies differ in their nature and sources of technological knowledge, which makes it difficult to make generalizations about the correct and appropriate sequence in the process of accumulation of technological capabilities; due to this Dutrénit [17], Arias [18] and Vera-Cruz [19] made adjustments based on the evidence of a study on the characteristics in the processes of accumulation of technological capabilities in the export maquiladora industry in Mexico, identifying three functions within the support functions: external linkage, internal linkage and equipment modification.

3 Methods

In order to achieve the objective of the investigation, the following activities were carried out: (1) In-depth interviews with responsible managers of 20 maquiladora companies of the electronic sector located in the city of Tijuana, BC, Mexico, which are considered as more successful (2) Analyze the technological capacity of linkage that currently has the local SME electronic sector and (3) Propose a scheme of the key indicators of the technological capacity of linkage that affect its limitation as potential suppliers of large companies. The objective of the research is to identify information related to the technological capacity of linkage that SMEs currently have in the electronics sector in Tijuana, B.C., Mexico. As a universe of study, all were considered to be the (SMEs) of the electronic sector and they are registered in the Mexican Business Information System (SIEM), as well as in the directories of the National Chamber of the Industry of the Transformation (CANACINTRA), the National Chamber of the Electronic Industry of Telecommunications and Information Technologies (CANIETI), the National Statistical Directory of Economic Units (DENUE) INEGI 2014 and the Directory of the Maquiladora Industry of Baja California 2014 After debugging the listings of the different directories, the size of the population is very small; In this way, the criterion used is to apply the instrument to 12 companies, which represents 100%.

4 Results

4.1 Qualitative Results

According to the in-depth interviews carried out with those responsible for the commissions of 20 maquiladora companies considered as the most important transnational companies established in Tijuana, B.C. Mexico (because they represent the best practices in the world and therefore the highest level of competitiveness to which a business organization should aspire), the following findings and obstacles were found to maintain relations with local Mexican suppliers: the quality of their inputs, the fulfillment of deliveries on time, the technological capability and that the suppliers do not manufacture what the company needs. In relation to the support provided by IMMEX for the development of local Mexican suppliers (with the exception of 2 companies), they are not interested in providing support in technological capability. However, if there is communication with the SMEs regarding their future requirements of the demand, in such a way that their agreements are mainly related to indirect inputs, such as packaging material, labels and cleaning material. Of the measures recommended by the government or associations of the industrial sector to support the Mexican local suppliers highlighted the following points: (a) Greater financial support, to establish inventory capabilities, (b) The establishment of subsidies for the creation of local suppliers that can satisfy large companies; (c) The formation of alliances between suppliers to manage competitive prices, since Asian companies tend to maintain a differentiation due to their cost and price strategy; (d) A greater stimulus to SMEs as it is done in other countries, but as long as they are redistributive companies so as not to fall into paternalism; (e) To train personnel in the continuous improvement of their processes; (f) The development of distribution chains appropriate to the requirements of the environment, (g) Opening of a greater number of technology fairs, as well as greater support to local suppliers, negotiating raw materials in a large volume, thus uniting the production capabilities of local suppliers.

4.2 Quantitative Results

By capturing data in the SPSS program and determining the Pearson correlation coefficient in each of the variables, the following results are obtained (Table 1).

Table 1. Correlation linking/integration

		Integration	Linking
Integration	Pearson correlation	1	.858(**)
	Sig. (bilateral)		.000
	N	12	12
Linking	Pearson correlation	.858(**)	1
	Sig. (bilateral)	.000	
	N	12	12

According to the results presented and the manifestation of a positive correlation, with a result of 0.858.

Applying the previously described Bell and Pavitt model to the analyzed SMEs, the most relevant results are shown in each of the stages of the technological linkage capabilities (basic, intermediate and advanced) with the clients, universities/public centers of research, chambers and associations and other local suppliers; which can serve as key indicators that affect the competitiveness of the SME electronic sector to be able to venture into large companies (see Tables 2, 3, 4, 5, 6, 7, 8, 9, 10, 11, 12 and 13).

Table 2. Relationship maintained with clients (workers have been trained)

		Frequency	Percentage	Valid percentage	Accumulated percentage
Valid	Eventually	8	66.7	66.7	66.7
	Regularly	3	25.0	25.0	91.7
	Always	1	8.3	8.3	100.0
	Total	12	100.0	100.0	

As shown in this table, only 8.3% of SMEs in the electronics sector in Tijuana, B.C. Mexico has received frequent training of its workers by its clients.

Table 3. Relationship maintained with clients (they support the incorporation of technologies)

		Frequency	Percentage	Valid percentage	Accumulated percentage
Valid	Never	1	8.3	8.3	8.3
	Eventually	4	33.3	33.3	41.7
	Regularly	5	41.7	41.7	83.3
	Always	2	16.7	16.7	100.0
	Total	12	100.0	100.0	

The results shown in this table indicate that only 16.7% of SMEs in the electronics sector in Tijuana, B.C. Mexico, its clients have always supported the incorporation of technology.

Table 4. Relationship with universities/public research centers (design and development of new products/processes)

		Frequency	Percentage	Valid percentage	Accumulated percentage
Valid	Eventually	9	75.0	75.0	75.0
	Regularly	3	25.0	25.0	100.0
	Total	12	100.0	100.0	

It is observed that 25% of SMEs electronic sector in Tijuana, B.C. Mexico, regularly has a link with universities/research centers for the design and development of new products or processes.

Table 5. Relationship with universities/public research centers (support incorporation of their technologies)

		Frequency	Percentage	Valid percentage	Accumulated percentage
Valid	Eventually	9	75.0	75.0	75.0
	Regularly	3	25.0	25.0	100.0
	Total	12	100.0	100.0	

As indicated in this table, 25% of SMEs in the electronics sector in Tijuana, B.C. Mexico, universities/public research centers regularly support them in the incorporation of technologies.

Table 6. Relationship with universities/public research centers (training of their workers)

		Frequency	Percentage	Valid percentage	Accumulated percentage
Valid	Eventually	8	66.7	66.7	66.7
	Regularly	4	33.3	33.3	100.0
	Total	12	100.0	100.0	

The interest of universities/public research centers to train SMEs in the electronics sector in Tijuana, B.C. Mexico on a regular basis is 33%.

Table 7. Relationship with chambers and associations (identification of sources and ways to obtain financing)

		Frequency	Percentage	Valid percentage	Accumulated percentage
Valid	Never	2	16.7	16.7	16.7
	Eventually	2	16.7	16.7	33.3
	Regularly	7	58.3	58.3	91.7
	Always	1	8.3	8.3	100.0
	Total	12	100.0	100.0	

As you can see, only 8.3% of SMEs in the electronics sector in Tijuana, B.C. Mexico frequently receives support from chambers and associations in identifying sources and ways to obtain financing.

Table 8. Relationship with cameras and associations (link with potential customers)

		Frequency	Percentage	Valid percentage	Accumulated percentage
Valid	Never	3	25	25	25
	Eventually	4	33.3	33.3	58.3
	Regularly	5	41.7	41.7	100
	Total	12	100.0	100.0	

The table shows that regularly 41.7% of SMEs in the electronics sector in Tijuana, B.C. Mexico, links with potential clients as part of the support of the chambers and associations have.

Table 9. Relationship with cameras and associations (training)

		Frequency	Percentage	Valid percentage	Accumulated percentage
Valid	Never	1	8.3	8.3	8.3
	Eventually	5	41.7	41.7	50.0
	Regularly	6	50.0	50.0	100
	Total	12	100.0	100.0	

It is observed that 50% of SMEs electronic sector in Tijuana, B.C. Mexico, frequently receive training by chambers and associations.

Table 10. Relationship with chambers and associations (dissemination of SMEs in the sector)

		Frequency	Percentage	Valid percentage	Accumulated percentage
Valid	Never	1	8.3	8.3	8.3
	Eventually	1	8.3	8.3	16.7
	Regularly	3	25.0	25.0	41.7
	Always	7	58.3	58.3	100.0
	Total	12	100.0	100.0	

In this table it is indicated that 58.3% of SMEs in the electronics sector in Tijuana, B.C. Mexico, frequently maintains a relationship with chambers and associations for its dissemination in the sector.

Table 11. Relationship with other local suppliers to the maquiladora industry electronic sector (joint design actions)

		Frequency	Percentage	Valid percentage	Accumulated percentage
Valid	Never	4	33.3	33.3	33.3
	Eventually	4	33.3	33.3	66.7
	Regularly	3	8.3	8.3	75.0
	Always	3	25.0	25.0	100.0
	Total	12	100.0	100.0	

According to the results, frequently 25% of the electronic sector SMEs in Tijuana, B.C. Mexico, carries out joint actions in design with other local suppliers.

Table 12. Relationship with other local suppliers to the maquila industry electronic sector (development or improvement of products and processes)

		Frequency	Percentage	Valid percentage	Accumulated percentage
Valid	Never	3	25.0	25.0	25.0
	Eventually	3	25.0	25.0	50.0
	Regularly	5	41.7	41.7	91.7
	Always	1	8.3	8.3	100.0
	Total	12	100.0	100.0	

As shown in this table, only 8.3% of SMEs in the electronics sector in Tijuana, B.C. Mexico has a frequent relationship with other local suppliers for the development or improvement of products and processes.

It is observed that only 8.3% of SMEs in the electronics sector in Tijuana, B.C. Mexico, frequently share training with other local supply companies

After analyzing the different indicators of the Bell and Pavitt model, the key indicators that can directly affect the different levels of technological capability for linking SMEs in the electronics sector in Tijuana, BC, Mexico (Table 14).

This table presents the technological capabilities of linking the electronic sector SMEs in Tijuana, B.C., Mexico, which are necessary to be able to enter the large companies, according to the analysis of the present investigation.

Table 13. Relationship with other local supply companies to the maquiladora industry electronic sector (share training)

		Frequency	Percentage	Valid percentage	Accumulated percentage
Valid	Never	4	33.3	33.3	33.3
	Eventually	4	33.3	33.3	66.7
	Regularly	3	25.0	25.0	91.7
	Always	1	8.3	8.3	100.0
	Total	12	100.0	100.0	

Table 14. Matrix of linking technological capabilities

Matrix of linking technological capabilities	
Capability levels	Technical support functions (external link)
Routine production capabilities: capabilities to use and operate existing technology	
Basic operational capabilities	Information on raw materials, equipment and technical assistance. Sale of products to existing customers
Innovative technological capabilities: capabilities to generate and manage technical change	
Basic innovative capabilities.	Search for support for certifications and sources, as well as new ways to obtain financing. Negotiations for the purchase of raw

(*continued*)

Table 14. (*continued*)

Matrix of linking technological capabilities	
Capability levels	Technical support functions (external link)
	material. Knowledge to export. Link with new potential customers
Intermediate innovative capabilities	They share machinery with some local suppliers and exchange information. Training of its workers by institutions. They share production capabilities with their clients. Customer support in the set-up of the plants. Joint marketing actions
Advanced innovative capabilities	Customer support for technology incorporation. Share capabilities of new designs

Source: Self-made

5 Conclusions

The indicators that have a strong correlation within the first type of relationship were the support of their clients in the incorporation of technologies, the support of their clients in the set-up of the plant, the cooperation activities to share knowledge for export, the cooperative activities to train their workers; the support of its clients to provide equipment, and the support for opening to modifications and/or recommendations in the design of its parts. In the second type, due to their strong correlation, activities that support the incorporation of technologies, activities that support the training of workers and activities to share design capabilities stood out. Regarding the third type of relationship, the most important ones are listed, the support in the definition of common objectives for the locality, the activities of stimulation in the perception of future visions for strategic actions; the identification of sources and ways to obtain financing; the creation of forums and environments for discussion; collective negotiations for the purchase of raw materials and equipment; linking with potential customers; the support to obtain certifications; the diffusion of SMEs in the sector; the organization of events, technical and commercial fairs; and market and technological studies. Finally, the indicators linked to the fourth type of relationship, which specify the relationships that the maquiladora industry has with other local supply companies, showed a moderate correlation. In this group marginally highlighted the joint actions in design; development or improvements of products and processes; the cooperation actions related to the support to incorporate new technologies, joint marketing actions, and the joint sale of products.

According to the results obtained, the type of relationships maintained by the local Mexican suppliers with some of their clients has allowed them to improve their learning abilities. However, actions must be implemented that help improve their administrative and business structure and allow them to become more competitive; For example, in relation to universities and public research centers it is necessary to develop a greater link that allows accompaniment to achieve the objectives. With regard to the chambers and associations, even though they have assumed a leading role in supporting local procurement companies so that they can improve their position with IMMEX, support has been limited and it is not enough to cover the demands and needs of SMEs. Finally, in relation to activities related to other local suppliers, greater empathy is required for joint work, strategic alliances that allow them to coordinate efforts and resources to strengthen their technological capabilities as a provider group and specialist in the industrial sector.

References

1. Martínez, R.: Quinta Hélice Sistémica (QHS), Un método para evaluar la competitividad internacional del sector electrónico en Baja California, México. Investigación Administrativa, año 4(110), 34–48 (2012)
2. Amburto, L.: La importancia de la industria electrónica en Jalisco. Revista académica electrónica e-scholarum. In: División de apoyo para el aprendizaje, vol. 2, pp. 46–60. Universidad Autónoma de Guadalajara (2008)
3. Organización Mundial del Comercio: Reporte de las exportaciones mundiales (2010). http://www.wto.org/indexsp.htm. Recuperado el 22 de febrero 2011 de
4. Secretaría de Economía: Monografía Industria electrónica en México (2012). http://www.economia.gob.mx/files/comunidadnegocios/industria_comercio/monografia_industria_electronica_Oct2012.pdf. Recuperado el 25 de mayo 2012 de
5. Carrillo, J., Hualde, A.: Existe un tercer clúster en la maquiladora electrónica en Tijuana? Revista Mexicana de Sociología, México 2(2), 9–12 (2001)
6. García, G., Paredes, V.: Programas de apoyo a la micro, pequeñas y medianas empresas en México, 1995–2000. Revista CEPAL Serie Desarrollo Productivo 115, 1–27 (2001)
7. Bracamontes, A., Contreras, F.: Redes globales de producción y proveedores locales: los empresarios sonorenses frente a la expansión de la industria automotriz. Estudios fronterizos 9(18), 161–194 (2008)
8. Ariffin, N., Figueiredo, P.: Internacionalizacão de competências tecnológicas. Editora FGV, Rio de Janeiro (2003)
9. Tapias, H.: Capacidades tecnológicas: elemento estratégico de la competitividad. Revista Facultad de Ingeniería Universidad de Antioquia 33, 97–119 (2005)
10. Kim, L., Nelson, R.: Technology, Learning and Innovation: The Experience of the Asian NIEs. Cambridge University Press, Cambridge (2000)
11. Lall, S.: Las capacidades tecnológicas. In: Salomon, J.J., Sagasti, F., Sachs, C. (eds.) Una búsqueda incierta. Ciencia, tecnología y desarrollo, pp. 301–342. Fondo de Cultura Económica, México (1993)
12. Torres, A.: Aprendizaje y construcción de capacidades tecnológicas. J. Technol. Manag. Innov. 1(5), 12–24 (2006)
13. Lugones, G.: Indicadores de capacidades tecnológicas en América Latina. Revista CEPAL Serie Estudios y Perspectivas. México 89 (2007)
14. Fagerberg, J.: Innovation: a guide to the literature. ponencia presentada en el taller "The Many Guises of Innovation: What we have learnt and where we are heading", Ottawa, 23 y 24 de octubre (2003)
15. Bell, M., Pavitt, K.: The development of technological capabilities. In: Haque, I.U. (ed.) Trade, Technology and International Competitiveness, pp. 69–101. World Bank, Washington (1995)
16. Lall, S.: Technological capabilities and industrialization. World Dev. 20(2), 165–186 (1992)

17. Dutrenit, G., Arias, A., Vera-Cruz, A.: Diferencias en el perfil de acumulación de capacidades tecnológicas en tres empresas mexicanas. El trimestre económico **70**(277), 109–165 (2003)
18. Arias, A.: Acumulación de capacidades tecnológicas: el caso de la empresa curtidora ALFA. Investigación Económica **LXIII**(249), 101–123 (2004)
19. Vera-Cruz, A.: Cultura de la empresa y comportamiento tecnológico: Como aprenden las cerveceras mexicanas. UAM-Miguel Ángel Porrúa (2004)

Discussion on the Iterative Process in Robust Algorithm A

Yue Zhang[3], Fan Zhang[1,2(✉)], Jing Zhao[1,2], Chao Zhao[1,2],
Gang Wu[1,2], Xinyu Cao[1,2], and Haitao Wang[1,2]

[1] Center for Quality and Statistics, China National Institute of Standardization,
Beijing, China
{zhangfan, zhaoj, zhaochao, wugang,
caoxy, wanght}@cnis.gov.cn
[2] AQSIQ Key Laboratory of Human Factors and Ergonomics (CNIS),
Beijing, China
[3] Capital Normal University, Beijing, China
15652668478@163.com

Abstract. ISO 13528-2005 "Statistical methods for use in proficiency testing by interlaboratory comparisons" gives a robust method for calculating mean and standard deviation—Algorithm A. The algorithm gives a robust estimate of population mean and population standard deviation through continuous iterative calculations. In each step, the original value compared with the predetermined range ends. But in practical applications, some experimenters use the values from last iteration instead of the original value, and believe that it brings better results to the experiment. Through simulation, this paper shows that it is unreasonable to use the last iteration results, and the original data should be used for calculation.

Keywords: Robust statistics · Outliers · Iterative calculations · Algorithm A

1   Introduction

In practice, most proficiency testing data sets include some results that are far from most of the data, which may be due to inaccurate measurement methods or less experience of participant. The results that deviate from most of the data may be variable and make traditional statistical techniques unavailable, the numbers that are larger than other data are called outlier values. When there are outliers in the data, it is common to remove them directly from data. However, in most cases, we cannot accurately determine which are outliers. Direct deletion may remove correct values, so it is not accurate to delete the outliers directly. Robust statistical technique is considered in this situation. Robust statistics can get high efficiency of estimating of population mean and population standard deviation from a data set which contains some outliers.

The following is a brief introduction to the theory of robust estimation. If we refer to the results of correct measurement as "good" results, the incorrect results in the experiment are called "bad" results. μ is the mean of "good" measurement, σ is standard deviation of the measurement results of "good", and ε is the proportion of

© Springer Nature Switzerland AG 2020
W. Karwowski et al. (Eds.): AHFE 2019, AISC 971, pp. 388–394, 2020.
https://doi.org/10.1007/978-3-030-20494-5_36

"bad" measurement results, the statistical distribution produced by the overall measured values ("good" and "bad" measurements) is a contaminated normal distribution

$$F(x) = (1 - \varepsilon)\Phi\left(\frac{x - \mu}{\sigma}\right) + \varepsilon H(x), \quad \text{where } \Phi(t)$$

is standard normal distribution, $H(x)$ is an unknown pollution distribution and $0 \leq \varepsilon \leq 1$. Huber [1] proposed a new theory of robust estimation, discussed the asymptotic theory for estimating location parameters of contaminated normal distribution, and formulated three heuristic methods for the solution of the problem of the simultaneous determination of μ and σ for a given ε. Algorithm A [2] is "Proposal 2" from Huber, a process of iterative calculations. In each iteration, the raw data values are chosen as variables to calculate. However, many people use the values of the last iteration in the experiment and bring it into Algorithm A for calculation. They think that this result gives it a better test effect, but that is not true.

This article will explain it through simulation. Firstly, the basic idea and calculation process of Algorithm A are introduced. Secondly, the Algorithm that uses the values of the last iteration to calculate is recorded as the Algorithm A', and the calculation process is introduced. Finally, through data simulation, the irrationality of the operation using the numbers after the last iteration is illustrated.

2 Algorithm

2.1 Algorithm A

Firstly, sort the data in increasing order, so that the outliers are mainly concentrated at both ends. Set a range, then the numbers outside the range are replaced by the end value of the range, and the original data inside the range remains unchanged. Robust estimates of population mean and population standard deviation are obtained through continuous iterative calculations. The calculation process of Algorithm A will be specifically described below.

Denote the n items of data, sorted into increasing order, by:

$$x_1, x_2, \ldots, x_n$$

Denote the robust average and robust standard deviation of these data by x^* and s^*. Calculate the initial values of x^* and s^* as:

$$x^* = medx_i \quad i = 1, 2, \ldots, n \tag{1}$$

$$s^* = 1.483 \times med|x_i - x^*| \quad i = 1, 2, \ldots, n \tag{2}$$

Update the values of x^* and s^* according to the following steps.

For each x_i $i = 1, 2, \ldots, n$, calculate:

$$x_i^* = \begin{cases} x^* - cs^* & x_i < x^* - cs^* \\ x_i & |x_i - x^*| \leq cs^* \\ x^* + cs^* & x_i > x^* + cs^* \end{cases} \tag{3}$$

Then calculate the new values of x^* and s^* by the following formula:

$$x^* = \sum_{i=1}^{n} x_i^* / n \tag{4}$$

$$s^* = \sqrt{\frac{1}{\beta(c)(n-1)} \sum_{i=1}^{n} (x_i^* - x^*)^2} \tag{5}$$

For the parameters ε, c, β have the following relationship:

$$\frac{1}{1 - \varepsilon(c)} = \theta(c) + 2\frac{\varphi(c)}{c} \tag{6}$$

$$\beta(c) = \theta(c) + c^2(1 - \theta(c)) - 2\varphi(c) \tag{7}$$

where $\theta(c) = \int_{-c}^{c} \varphi(t)dt$, $\varphi(t)$ is the density function of standard normal distribution.

2.2   Algorithm A′

In proficiency testing, some experimenters changed Algorithm A and used the last transformed data to calculate the mean and standard deviation estimates. Here, the algorithm is referred to as Algorithm A′.

Algorithm A′ is similar to Algorithm A. First, it is still necessary to sort the original data from small to large and record it as $x_1, x_2, \ldots, x_n$. The robust estimates of population mean μ and population standard deviation σ are denoted as x^* and s^*, respectively.

Calculate the initial values of x^* and s^*:

$$x^{j*} = medx_i^j \quad i = 1, 2, \ldots, n; j = 0 \tag{8}$$

$$s^{j*} = 1.483 \times med|x_i^j - x^{j*}| \quad i = 1, 2, \ldots, n; j = 0 \tag{9}$$

where j represents the number of iterations, and when $j = 0$, x_i^0 represents the original data values, and x^{0*} and s^{0*} are the initial values of the mean and standard deviation.

Update the values of x^{0*} and s^{0*} according to the following steps.

For each $x_i^{j+1} (i = 1, 2, \ldots, n; j = 0, 1, 2, \ldots)$ calculate:

$$x_i^{j+1} = \begin{cases} x^{j*} - cs^{j*} & \text{if } x_i^j < x^{j*} - cs^{j*} \\ x_i^j & \text{if } |x_i^j - x^{j*}| \le cs^{j*} \\ x^{j*} + cs^{j*} & \text{if } x_i^j > x^{j*} + cs^{j*} \end{cases} \tag{10}$$

Then calculate the new values of $x^{(j+1)*}$ and $s^{(j+1)*}$ by the following formula:

$$x^{(j+1)*} = \sum_{i=1}^{n} x_i^{j+1}/n \tag{11}$$

$$s^{(j+1)} = \sqrt{\frac{1}{\beta(c)(n-1)} \sum_{i=1}^{n} \left(x_i^{j+1} - x^{(j+1)*} \right)^2} \tag{12}$$

For the parameters ε, c, β have the following relationship:

$$\frac{1}{1 - \varepsilon(c)} = \theta(c) + 2\frac{\varphi(c)}{c} \tag{13}$$

$$\beta(c) = \theta(c) + c^2(1 - \theta(c)) - 2\varphi(c) \tag{14}$$

where $\theta(c) = \int_{-c}^{c} \varphi(t)dt$, ɸ(t) is the density function of standard normal distribution.

Comparing Algorithm A' with Algorithm A, the main difference between them is in Eqs. (3) and (10).

In the following article, x_i in Algorithm A and x_i^j in Algorithm A' are referred to as independent variables.

3 Simulation

Firstly, the pollution distribution H(x) is taken as the chi-square distribution, t-distribution and uniform distribution, the actual data is generated by the contaminated distribution F(x). Then respectively use Algorithm A and Algorithm A' to estimate the mean and standard deviation, and compare them. Finally, explain the results of the simulation. The following is a detailed simulation process.

Let the contaminated normal distribution be

$$F(x) = (1 - \varepsilon)\Phi\left(\frac{x - \mu}{\sigma}\right) + \varepsilon H(x) \tag{15}$$

where φ(t) is the density function of standard normal distribution and H(x) is an unknown pollution distribution and $0 \le \varepsilon \le 1$.

In the simulation, let pollution distribution H(x) take the chi-square distribution ($\chi^2(4)$), uniform distribution (U(1,3)) and t-distribution (t(8)), respectively, and take c, 0.5, 1, 1.3, 1.4, 1.5, 1.7, 2, 2.5, 3, 3.5, 4. Then, the corresponding ε is calculated by the formula (13), brings it into Eq. (15) and randomly generates 1000 numbers as the original data. Iteratively using algorithm A' and algorithm A, the estimates of each mean and standard deviation are calculated.

Tables 1, 2, and 3 are chi-square distribution, uniform distribution, and t-distribution, using Algorithm A' and Algorithm A to calculate the top three population mean and standard deviation estimates, and compare them.

Table 1. Algorithm A' and Algorithm A mean and standard deviation estimates of Chi-square distribution $\chi^2(4)$

Chi-square distribution $\chi^2(4)$								
c	ε	Algorithm	First calculation		Second calculation		Third calculation	
			Mean estimation	Standard deviation estimate	Mean estimation	Standard deviation estimate	Mean estimation	Standard deviation estimate
0.5	0.4417	Algorithm A	1.5777	1.1515	1.5722	1.1477	1.5688	1.1443
		Algorithm A'	1.5777	1.1515	1.5743	1.1421	1.5718	1.1352
0.7	0.2899	Algorithm A	1.0763	0.9851	1.0796	0.9835	1.0812	0.9825
		Algorithm A'	1.0763	0.9851	1.0783	0.9806	1.0795	0.9779
1	0.1428	Algorithm A	0.5476	0.9241	0.5553	0.9219	0.5577	0.9207
		Algorithm A'	0.5476	0.9241	0.5519	0.9154	0.5539	0.9116
1.3	0.0655	Algorithm A	0.2586	0.9304	0.2630	0.9348	0.2639	0.9370
		Algorithm A'	0.2586	0.9304	0.2596	0.9285	0.2600	0.9278
1.4	0.0498	Algorithm A	0.1982	0.9431	0.2014	0.9470	0.2019	0.9488
		Algorithm A'	0.1982	0.9431	0.1987	0.9421	0.1989	0.9418
1.5	0.0376	Algorithm A	0.1537	0.9520	0.1544	0.9567	0.1545	0.9587
		Algorithm A'	0.1537	0.9520	0.1537	0.9520	0.1537	0.9520
1.7	0.0211	Algorithm A	0.0914	0.9671	0.0916	0.9723	0.0917	0.9739
		Algorithm A'	0.0914	0.9671	0.0914	0.9671	0.0914	0.9671
2	0.0084	Algorithm A	0.0453	0.9810	0.0450	0.9874	0.0450	0.9887
		Algorithm A'	0.0453	0.9810	0.0453	0.9810	0.0453	0.9810
2.5	0.0016	Algorithm A	0.0217	0.9927	0.0219	0.9954	0.0219	0.9956
		Algorithm A'	0.0217	0.9927	0.0217	0.9927	0.0217	0.9927
3	0.0003	Algorithm A	0.0168	0.9929	0.0169	0.9932	0.0169	0.9932
		Algorithm A'	0.0168	0.9929	0.0168	0.9929	0.0168	0.9929
3.5	$3.34 \times e^{-5}$	Algorithm A	0.0163	0.9921	0.0163	0.9921	0.0163	0.9921
		Algorithm A'	0.0163	0.9921	0.0163	0.9921	0.0163	0.9921
4	$3.57 \times e^{-6}$	Algorithm A	0.0161	0.9918	0.0161	0.9918	0.0161	0.9918
		Algorithm A'	0.0161	0.9918	0.0161	0.9918	0.0161	0.9918

Table 2. Algorithm A' and Algorithm A mean and standard deviation estimates of uniform distribution U(1,3)

Uniform distribution U(1,3)								
c	ε	Algorithm	First calculation		Second calculation		Third calculation	
			Mean estimation	Standard deviation estimate	Mean estimation	Standard deviation estimate	Mean estimation	Standard deviation estimate
0.5	0.4417	Algorithm A	0.8849	0.6329	0.8860	0.6333	0.8866	0.6336
		Algorithm A'	0.8849	0.6329	0.8853	0.6316	0.8857	0.6307
0.7	0.2899	Algorithm A	0.5921	0.7171	0.5879	0.7185	0.5859	0.7195
		Algorithm A'	0.5921	0.7171	0.5903	0.7132	0.5892	0.7108
1	0.1428	Algorithm A	0.2952	0.8392	0.2951	0.8384	0.2951	0.8379
		Algorithm A'	0.2952	0.8392	0.2951	0.8384	0.2951	0.8379
1.3	0.0655	Algorithm A	0.1400	0.9107	0.1394	0.9151	0.1393	0.9172
		Algorithm A'	0.1400	0.9107	0.1400	0.9107	0.1400	0.9107
1.4	0.0498	Algorithm A	0.1081	0.9284	0.1077	0.9324	0.1076	0.9343
		Algorithm A'	0.1081	0.9284	0.1081	0.9284	0.1081	0.9284
1.5	0.0376	Algorithm A	0.0836	0.9438	0.0834	0.9480	0.0834	0.9498
		Algorithm A'	0.0836	0.9438	0.0836	0.9438	0.0836	0.9438
1.7	0.0211	Algorithm A	0.0516	0.9632	0.0519	0.9690	0.0519	0.9709
		Algorithm A'	0.0516	0.9632	0.0516	0.9632	0.0516	0.9632
2	0.0084	Algorithm A	0.0284	0.9802	0.0288	0.9870	0.0290	0.9884
		Algorithm A'	0.0284	0.9802	0.0284	0.9802	0.0284	0.9802
2.5	0.0016	Algorithm A	0.0187	0.9929	0.0189	0.9955	0.0189	0.9957
		Algorithm A'	0.0187	0.9929	0.0187	0.9929	0.0187	0.9929
3	0.0003	Algorithm A	0.0163	0.9929	0.0164	0.9932	0.0164	0.9932
		Algorithm A'	0.0163	0.9929	0.0163	0.9929	0.0163	0.9929
3.5	$3.34 \times e^{-5}$	Algorithm A	0.0162	0.9921	0.0162	0.9921	0.0162	0.9921
		Algorithm A'	0.0162	0.9921	0.0162	0.9921	0.0162	0.9921
4	$3.57 \times e^{-6}$	Algorithm A	0.0161	0.9918	0.0161	0.9918	0.0161	0.9918
		Algorithm A'	0.0161	0.9918	0.0161	0.9918	0.0161	0.9918

Observing the above tables, since the independent variables used in the first calculation are all raw data values, the first calculation results of Algorithm A' are the same as Algorithm A. After the second calculation, the standard deviation in Algorithm A' is smaller than the standard deviation of Algorithm A, that is, the data fluctuation range is small and concentrated. Note that when c exceeds 1.5 in Table 1, when c exceeds 1.3 in Table 2, and when c exceeds 1 in Table 3, the results of each iteration of Algorithm A' do not change, which is due to that the original data is restrained in the interval $(x^{0*} - cs^{0*}, x^{0*} + cs^{0*})$ after the first iteration step. And from the second iteration $x^{j*} - cs^{j*}$ (j = 1, 2, 3...) is less than $x^{0*} - cs^{0*}$, $x^{j*} + cs^{j*}$ (j = 1, 2, 3...) is bigger than $x^{0*} + cs^{0*}$. Therefore, for different contaminated distributions, when c exceeds a certain value, the independent variables of each iteration process both in Algorithm A and Algorithm A' are the same set of values. As the number of iterations increasing, the intervals of the independent variables are getting smaller and smaller, that is, the data used for the following calculation is more and more concentrated. Although the estimates of the standard deviation are smaller than that of Algorithm A, more raw information is lost.

Table 3. Algorithm A′ and Algorithm A mean and standard deviation estimates of t-distribution t(8)

t-distribution t(8)

c	ε	Algorithm	First calculation		Second calculation		Third calculation	
			Mean estimation	Standard deviation estimate	Mean estimation	Standard deviation estimate	Mean estimation	Standard deviation estimate
0.5	0.4417	Algorithm A	−0.0275	0.7673	−0.0229	0.7673	−0.0201	0.7672
		Algorithm A′	−0.0275	0.7673	−0.0253	0.7610	−0.0236	0.7565
0.7	0.2899	Algorithm A	−0.0146	0.7877	−0.0120	0.7872	−0.0107	0.7867
		Algorithm A′	−0.0146	0.7877	−0.0132	0.7845	−0.0123	0.7825
1	0.1428	Algorithm A	0.0024	0.8456	0.0010	0.8511	0.0005	0.8547
		Algorithm A′	0.0024	0.8456	0.0024	0.8456	0.0024	0.8456
1.3	0.0655	Algorithm A	0.0022	0.9132	0.0040	0.9177	0.0044	0.9200
		Algorithm A′	0.0022	0.9132	0.0022	0.9132	0.0022	0.9132
1.4	0.0498	Algorithm A	0.0036	0.9284	0.0054	0.9337	0.0057	0.9363
		Algorithm A′	0.0036	0.9284	0.0036	0.9284	0.0036	0.9284
1.5	0.0376	Algorithm A	0.0052	0.9425	0.0066	0.9484	0.0069	0.9510
		Algorithm A′	0.0052	0.9425	0.0052	0.9425	0.0052	0.9425
1.7	0.0211	Algorithm A	0.0082	0.9640	0.0090	0.9700	0.0091	0.9719
		Algorithm A′	0.0082	0.9640	0.0082	0.9640	0.0082	0.9640
2	0.0084	Algorithm A	0.0115	0.9806	0.0117	0.9871	0.0118	0.9885
		Algorithm A′	0.0115	0.9806	0.0115	0.9806	0.0115	0.9806
2.5	0.0016	Algorithm A	0.0154	0.9926	0.0156	0.9955	0.0156	0.9957
		Algorithm A′	0.0154	0.9926	0.0154	0.9926	0.0154	0.9926
3	0.0003	Algorithm A	0.0158	0.9929	0.0159	0.9932	0.0159	0.9932
		Algorithm A′	0.0158	0.9929	0.0158	0.9929	0.0158	0.9929
3.5	$3.34 \times e^{-5}$	Algorithm A	0.0161	0.9921	0.0161	0.9921	0.0161	0.9921
		Algorithm A′	0.0161	0.9921	0.0161	0.9921	0.0161	0.9921
4	$3.57 \times e^{-6}$	Algorithm A	0.0161	0.9918	0.0161	0.9918	0.0161	0.9918
		Algorithm A′	0.0161	0.9918	0.0161	0.9918	0.0161	0.9918

4   Conclusion

Algorithm A′ can obviously improve the iteration speed and get smaller standard deviation than Algorithm A. These can bring high efficiency to the experimenter in the process of proficiency testing. However due to the small control range of the independent variables, more data is removed, making the data set more concentrated and more information lost. In practical applications, such results are unreasonable.

Acknowledgement. This research was supported by National Key Technology R&D Program (2017YFF0206503, 2017YFF0209004, 2016YFF0204205) and China National Institute of Standardization through the "special funds for the basic R&D undertakings by welfare research institutions" (522018Y-5941, 522018Y-5948, 522019Y-6771).

References

1. Huber, P.J.: Robust estimation of a location parameter. Ann. Math. Stat. **35**(1), 73–101 1964
2. ISO 13528-2005 Statistical Methods for Use in Proficiency Testing by Interlaboratory Comparisons

Model for Monitoring Socioenvironmental Conflicts in Relation to the Emission of Particulate Matter in the Prehauling Phase of a Surface Mine in Peru

Marcio Filomeno[1]([✉]), Josemaria Heracles[1], Vidal Aramburu[1],
Carlos Raymundo[2], and Javier M. Moguerza[3]

[1] Escuela de Ingeniería de Gestión Minera, Universidad Peruana de Ciencias
Aplicadas (UPC), Lima, Peru
{u811273, u201314985, pcgmvara}@upc.edu.pe
[2] Dirección de Investigación, Universidad Peruana de Ciencas Aplicadas (UPC),
Lima, Peru
carlos.raymundo@upc.edu.pe
[3] Escuela Superior de Ingeniería Informática, Universidad Rey Juan Carlos,
Mostoles, Madrid, Spain
javier.moguerza@urjc.es

Abstract. This research will focus on proposing a model based on surveys conducted among people of the affected area. The questions were classified by indicators and variables selected to generate solutions to reduce social conflicts, which arise due to the emission of the particulate matter generated in the area before hauling tasks. Particulate matter is produced by hydraulic shovels, which load mineralized material and discharge it to a dump truck in mining operations. This survey was conducted among people who are specifically located in the Huari region. A study of the Social Conflict Monitoring Model (MMCS) tool was executed. It is used for recording, monitoring, and controlling this type of social conflicts, so that mining operations will not be affected in the short or long term. In addition, this model will help in discovering the opinions and/or comments when they are informed on a new method, which decreases generated particulate matter.

Keywords: MMCS · Dust · Social conflict · Particulate matter · Mining · Loading · Surface mining · Emission · Monitoring model

1 Introduction

The exploitation of mineral ore in Peru is one of its main activities, since it is a country overwhelmingly dominated by mining. As a direct consequence of this activity and of mineral processing, the environment has been affected.

In Peru, communities reject mining activities because, for many years, mining has been affecting the environment. There were environmental consequences that produced

© Springer Nature Switzerland AG 2020
W. Karwowski et al. (Eds.): AHFE 2019, AISC 971, pp. 395–406, 2020.
https://doi.org/10.1007/978-3-030-20494-5_37

an impact on lifestyle of the people, whose main activities were livestock and agriculture.

This research project focused on the community of Huari. According to INEC (2016), 35% of the population of this region resides in urban areas and 64% in rural areas.

Therefore, there are different models to assess social conflicts. Among them, "MMCS" (Social Conflict Monitoring Model) has already been applied. To develop this model a comparative study was made with the main existing early warning systems. They were the FAST Early Warning System, developed by the Institute for Environmental Security of the Swiss Agency for Development and Cooperation (SDC) and the Conflict Early Warning and Response Mechanism (CEWARN), launched in 2002 by the seven member countries of Intergovernmental Authority on Development (IGAD). Consequently, it is necessary to consider different social impacts generated by mining companies, because this could have an impact on stoppages and profits of the organization.

One of the most important problems in surface mining is the control of the particulate matter produced in the area where blade loading to trucks is done post-blasting. In most cases, different mining stages have been considered more important due to socioenvironmental conflicts that may arise when there is dust. Over time, the control of particulate matter or its mitigation has become more relevant for mining companies. [1]

The need to control or mitigate particulate matter at a mining site is not an easy task. Slopes, crushers, hillsides, conveyor belts, blasting, drilling, industrial roads in the pre-hauling stage, and the transportation of segregated and production material generate a large amount of fine particulate matter whose elimination system is quite complex when related to dust inventory [2].

Therefore, a study is proposed to be based on the community close to where the mine operates with regard to the particulate matter that is produced in the loading area, which may adversely affect inhabitants. The purpose of the study is to reduce social conflicts of the community close to the area where the Antamina mining company operates, by using a tool (MMCS) and finding out the community's opinions on the implementation of a new suppressant that mitigates the amount of particulate matter. The suppressant will be applied in that area, so that the environment where they live is not negatively impacted.

In addition, the MMCS method showed that the IGCEW method helps to identify the main tensions among interest groups, such as the community of the Huari province. These questions will contribute to identifying if the additive, which is an innovative idea of the mining company, would generate some type of discomfort and/or negative impact to communities close to the mining area.

This combination was based on studied indicators, such as environmental quality. With these surveys, the main problems of the community close to the mine will be discovered, from the perspective of its dwellers.

2 State of the Art

2.1 Mining in Peru

Since ancient times, the activity of mining has been carried out to sell mineral ore. However, as occurs with all production activities, it produces negative consequences or impacts due to inefficient management or the absence of an entity to control the processes for mineral ore exploitation. These conflicts can be considered a series of events that occur due to disagreements of local actors and mining companies. Over the years, mineral ore extraction has grown significantly across the world and Peru has been one of its main expansion areas.

There is probably no other business sector in Peru that has made comparable efforts with miners to minimize social conflicts of the surrounding communities where companies operate. Even so, the mining sector remains one of the most conflictive. The problem is that it is not possible to generate trust in the population or reverse this image of a "black past" left by traditional mining [3].

2.2 Evaluation as Related to Mining Activity

It is important to make an assessment of social conflicts related to mining activity because it helps to identify key issues from the perspective of people who might be affected by projects, to minimize and to provide solution proposals, and to enter this information into systems and strategies in progress to respond proactively to the consequences of development.

On the other hand, Da Cunha Rodovalho, Edmo, and De Tomi, Giorgio [4], conducted a research that required a subnational evaluation to understand the spatial variation of factors that contribute to maintaining livelihoods better. In this research, Nepal was chosen as the place where the case study would be conducted. This is a country that is consistently ranked as one of the poorest in the world. To understand how sustainable development can effectively promote the diversification of livelihoods, we advocate that a multidimensional spatial approach is essential to monitor social and environmental change to help decision-making processes.

Antabe and Atuoye [5] stated that knowledge about the social identity approach might be useful to understand the causes of dysfunctional conflicts in Environmental and Natural Resource Management (ENRM). These conflicts tend to be created by multiple factors, including governance agreements that are in effect and how the discussions are conducted; the behavior and interactions of stakeholders and the general public; and the legacy of conflict, which can perpetuate a "culture of conflict" regarding specific issues. This research presents a comprehensive conceptual model of the sociopolitical landscape of the ENRM conflict, which brings together these multiple factors. The social identity approach is then introduced as an adequate lens through which to continue questioning people in charge of driving the ENRM conflict. Key social identity mechanisms are discussed along with their contribution to the proliferation of dysfunctional ENRM conflicts. Based on this analysis, it was determined that the social identity approach presents a way of understanding the subtle and, at times, invisible social structures that underlie ENRM; and these ENRM issues should be seen as a series of conflicting episodes connected through time and contexts by their legacy. The conceptual model and its interpretation based on the social identity approach pose several implications for the current theory, practice, and institutions involved in the nefarious ENRM sociopolitical context. These implications are examined, followed by a discussion of several options to address the social identity impact on dysfunctional conflicts derived from published Australian and international empirical examples.

In addition, Haslam and Tanimoune [3] developed a model that analyzed social conflicts between mining companies and nearby communities with mines in operation, which have proliferated in recent decades. Therefore, several researchers have tried to understand the causes and consequences of these scenarios in the local environment. They compared research on the curse of national resources, which has relied heavily on quantitative analysis to test hypotheses and to establish causal inference, such as the new literature on the "curse of local resources." The main thing for Haslam and Tanimoune was the quantitative evaluation of the principal causal statements regarding various social conflict factors in the affected communities caused by mining. The main determinants of social conflicts were classified into three central hypotheses:

(1) Company and the property characteristics (which affect the behavior of the company).
(2) Socioeconomic characteristics of nearby population (which affect distribution concerns)
(3) Socioenvironmental characteristics (which affect the concern for livelihoods)

2.3   Most Important Paper for the Case Study

In their research, Vázquez and Riofrío [6] analyzed the case of mining companies located in the border areas between Peru and Ecuador. The main objective of their study was to determine how the population perceives different socioenvironmental conflicts in the area and to characterize problems faced by miners in their efforts to organize and regulate activities in accordance with the new legislation. The main results confirmed the need to redefine the mining legal framework and to promote training processes for workers to reduce socioenvironmental conflicts.

Finally, Vázquez and Riofrío [6], conducted research regarding the expansion of an Ecuadorian mining company, which will double the amount of waste generated in a humid tropical area. This would produce a new social conflict in the surrounding community. The arrival of ECSA, a mining company, affected the use and exploitation of essential natural resources such as land and water and generated serious social conflicts, which have led to a strong resistance movement. The surrounding community is the Shuar indigenous nationality, an ethnic group dating back thousands of years. Also known as *jíbaros*, the Shuars have traditionally lived in the region, and therefore, they are considered the legitimate "owners" of that territory. This community rejected the mining activities of the company. To establish an order of priorities on the conflicts that seemed most important to the indigenous community, the authors used the Social Conflict Monitoring Model (MMCS). It showed several approaches with regard to social conflicts generated by the mining company in the indigenous community. If some of these variables are not satisfied with the population, the possibilities that conflicts will be violent will clearly increase. The opposite will occur, if one of these factors is maintained and strengthened.

2.4 Motivation

Thanks to all the authors mentioned above, we undertook a project in the Huari province, which is located in the same province as one of the most prominent companies in Peru, Antamina. In addition, based on the methods applied, MMCS will be able to identify discomforts and annoyances before a conflict occurs where there might be violent actions, when the company uses a new dust suppressant. The questions were evaluated according to the indicators studied by Vázquez and Riofrío in Ecuador.

Our motivation is that MMCS could help us assess the tolerance level of the community with regard to the mining company. If any problem arises, which might adversely affect the environment, due to the new dust suppression application at the mining unit, it would lead to more social problems and would prevent ongoing projects and those that are waiting for approval from achieving their expected production.

Regarding the IGCEW method applied in Peru, its main objective is the determination of how the population perceives different environmental impacts of the area. A second round of questions was included to evaluate how people perceive environmental impacts in their community.

3   Socioenvironmental Risk Evaluation Model

3.1   Background:

After presenting the models that were studied, Table 1 was drawn up to summarize each component of the models that were included:

Table 1. Higher education

Authors	Components
Caroline Donohue and Eloise Biigs (2015)	• Subnationality level • This country is classified as one of the poorest • The gray grouping method is used to quantify the information
Alexi Delgado and Romero (2014)	• Subnational economic evaluation as one of the poorest in the world
Caroline Donohue and Eloise Biigs (2015)	• Organize and regulate their activities in accordance with the new legislation. Contact quality • Fair procedures
Sánchez Vázquez Luis and Eguiguren Riofrío (2016)	• Importance • Risk • Justice • Collaboration

3.2   Design

3.2.1   Social Conflict Monitoring Model (MMCS)

Its aim is to improve the relationship between communities and the company. These models are focused on collecting information through surveys.

The MMCS model is applicable only to the people living in the Huari province. The evaluation will be made to discover the best indicator for the case study. It will help us to know the opinions of the community regarding the implementation of a new suppressant, and to find out whether communities are accepting of the application of a new additive used as suppressant of particulate matter (Table 2).

Table 2. Indicators

Thematic Variables	Indicators
Environmental quality	
Contamination	Water quality
	Ecosystem quality
Access to natural resources	Changes to soil usage
	Increase in number of mining concessions
Health	Mortality rate
	Disease causes
Sociocultural quality	
Socioeconomic conditions	Evolution of the unemployment rate
	Evolution of the purchasing power
	Changes to income sources
Cultural conditions	Different cultural expressions
	Use of different languages
Security	Crime rate
Social cohesion	
Demography	Population growth
	Migration
Social structure	Usage of common spaces and common interest events
	Number of associations
Transparency and political participation	
Governance	Participation of population in political advisory activities
Major change factor: mining	
Mining and related activities	Advisory and RSE exercise processes
	Advisory and RSE exercise processes
	Number of new employees from the province

3.2.2   Integrated Gray Clustering and Entropy-Weight Method, IGCEW

With the IGCEW method, it will be possible to analyze, from the perspective of the community member, the different impacts of mining activities on the environment. Based on a second round of questions, and thanks to this evaluation, several actions could be taken to avoid some conflicts between the community and the company (Fig. 1).

Probability of Using Violence: X

If Y1 decreases	The probability of using	If Y1 increases	The probability of using
If Y2 decreases	violence increases	If Y2 increases	violence decreases
If Y3 decreases		If Y3 increases	
If Y4 decreases		If Y4 increases	
If Y5 negative repercusions increase, positive repercusions decrease		If Y5 negative repercusions decrease, positive repercusions increase	

Fig. 1. Probability of the model. Note: Y1, Y2, Y3, Y4 and Y5 are any cases where the mining company does not meet the needs of people in their community, such as basic services (power, water), community works, and environmental quality (contamination).

4 Proposal

4.1 Socio-Environmental Risk Evaluation Model

Background Table 3 shows the partial results obtained from model implementation, taking into account the limitations of presenting analyses focused on the baseline, since the model is conceived as a continuous process susceptible to periodic adjustments. Considering the values of the indicator baseline, it shows the optimal values to prevent some type of violence and/or annoyance of communities when facing mining activities. These indicators of perception and satisfaction were defined as shown in Table 3, in %. With the previous Table, the following questions were proposed that cover each of the indicators, but only regarding environmental quality issues (Table 4).

Table 3. Statistical Indicators of MMCS

Thematic variables	Indicator	Value (baseline)
Environmental quality		
Pollution	Water quality (BMWP index)	Good *
	Ecosystem quality (Simpson index)	Good *
Access to	Changes to soil usage (% natural forests)	67
natural resources	Increase in number of mining concessions	16
Health	Mortality rate	47
	Causes of diseases (number of different diseases)	7

Table 4. Questions of communities' opinions

1. What do you think of the pollution level generated by the mining company in your community?
2. Are you satisfied with the state of the environment of your community?
3. Are you satisfied with the environmental management carried out by authorities regarding pollution in your community?
4. Do you think that mining in the area is a threat to nature?
5. Do you consider that health problems caused by dust have increased in your community?
6. Are there people, institutions, or companies using land and natural resources that have been negatively affected?
7. Do you think there are new diseases caused by pollution in your community?

A second group of questions was also created and were focused on finding out if the community knows about the additive called bischofite (magnesium chloride), which is used in the mining sector as a suppressant of particulate matter, to control pollution in mining operations (Table 5).

Table 5. Questions about the bischofite posed to the community

1. To what extent do you agree on the use of a new technique to reduce dust?
2. Have you heard about the bischofite? (magnesium chloride)
3. Do you know the benefits of bischofite?
4. Do you think that bischofite would negatively affect the health of community dwellers?
5. Do you think that bischofite would affect the lifestyle of your community? Would it affect livestock and/or agriculture?
6. If you are aware about bischofite, would you support disseminating its advantages?
7. How do you think the solution will affect the environment?

With the results of the surveys of the 10 families of the Huari community, it was possible to analyze each one in the corresponding indicator as shown in Table 6. The following data were obtained:

Table 6. Results

Thematic Variables	Indicators	Best Scenario (in %)	Reality (in %)
Environmental quality			
Pollution	Water Quality (BMWP index)	80	37.7
	Quality of the Ecosystem	80	23.5
Access to Natural Resources	Changes to soil usage	67	34
	Increase in number of mining concessions	16	26.4
Health	Mortality rate	47	53
	Causes of diseases	7	24.6

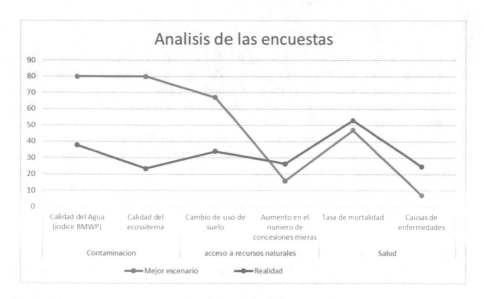

Translations:

Analysis of Surveys

Water quality   Ecosystem quality   Changes to soil usage Increase of number   Mortality rate Causes
(BMWP index)   of mining concessions   of diseases
Pollution   Access to Natural Resources   Health
[Blue] Best Scenario   [Orange]   Actual

Fig. 2. Comparative results

The data were placed as percentages (%) as part of the optimistic scenario. The results from surveys were included as "reality." The following figure was drawn up.

In Fig. 2, a comparison is shown: "80%" was placed in the indicator of water quality and ecosystem quality, as a high average. The chart shows the level of perception and satisfaction. Initially, the first two indicators show the satisfaction level of people regarding water and ecosystem quality. Lower values show a lower level of satisfaction of surveyed people.

Additionally, in the indicator "changes in soil usage," the greater value of this indicator of the real scenario means that people observed soil usage changes caused by persons or companies, such as the mining unit.

Finally, the last points show the following:

(1) An increase in the number of mining concessions that refers to the portion of land that was granted by the state to a mining company.
(2) The mortality rate: indicates if communities have linked the mortality rate to the mining project.
(3) The causes of diseases: This value indicates if communities believe that a greater quantity of diseases was caused by contamination produced by the mining unit.

5 Results Analysis

The results in Fig. 2 show that people believe that activities of mining companies have affected both water and ecosystem quality. This has made them reject the mining company, generating problems for their operations. Additionally, people perceive that there is soil deterioration and its contamination and have been able to observe it (from their perspective).

Although it is true that the community earns its livelihood from livestock and agriculture, the slightest change that the mining activity can generate might make people feel threatened, fearing that their main activity might be adversely affected. That is why, from the perspective of community dwellers, the mining activity is a threat to them, and because of this, the results shown in Fig. 2 show how the community regards mortality rate and diseases to the pollution generated by the nearby mining unit.

In the second round of questions, the additive called bischofite was presented. Community dwellers were unaware about this product. Despite not knowing that this additive is used as a suppressant of particulate matter to prevent pollution, and to decrease dust, they are afraid of the unknown.

As shown in Fig. 3, the question is whether people are aware of the existence bischofite. Many community dwellers are unaware of its existence. Additionally, as can be seen in Fig. 4, in spite of not knowing it, people still think that this product can negatively affect their lifestyle. They feel afraid and reject new methods that could be useful to them. The bischofite is used as a dust suppressant in mining operations.

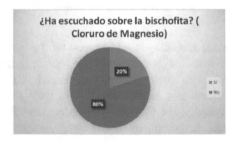

[Have you heard of bischofite? (magnesium chloride)
Yes   No]

Fig. 3. Bischofite question

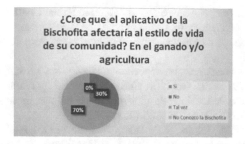

[Do you think that the use of bischofite would affect the lifestyle of you community? Or livestock and/or agriculture?

Yes   No   Maybe   I do not know what bischofite is.]

Fig. 4. Bischofite question (Would this affect them?)

6   Conclusion

The application of the function of these two models in the Huari community allowed questions to be posed to families regarding the environment and how it is affected by mining operations. Additionally, it was possible to find out the perspective of the people who are ill-informed about the use of suppressants and their benefits. They did not know of their existence and thought that it would be a risk to apply brine to the environment. In reality, this is not true, as different companies, at the international level have proven that the bischofite does not cause adverse risks to people, soil, ore, or crops.

Due to the results and their respective analyses, the issues that must be reinforced in the community were disseminated. The main one is misinformation. Therefore, informative and participatory workshops can be conducted to make the community feel included in mining activities and to prevent future social conflicts.

Finally, the fusion of the MMCS model and the IGCEW method turned out to be viable because either their questions did not have technical words or the surveys were conducted in an authoritative manner. Therefore, other mining companies should adopt this new model in case the community dwellers are unaware of the new techniques that will be used in the future to mitigate the adverse environmental impact on their community produced by mining company activities.

References

1. Rodovalho, E.C., de Tomi, G.: Simulation of the impact of mine face geometry on the energy efficiency of short-distance haulage mining operations. Min. Technol. **125**(4), 226–232 (2016)
2. Costanza, J.N.: Mining conflicts and the politics of obtaining a social license: Insight from Guatemala (Minería El conflicto y la política de obtención de una Licencia Social: Conocimiento de Guatemala). World Dev. **79**, 97–113 (2016)

3. Haslam, P.A., Tanimoune, N.A.: The Determinants of Social Conflict in the Latin American Mining Sector: New Evidence with Quantitative Data (Los determinantes del conflicto social en América Latina Sector de Minería: Nueva evidencia con datos cuantitativos). World Dev. **78**, 401–419 (2015)
4. Jackson, S.L.: Dusty roads and disconnections: Perceptions of dust from unpaved mining roads in Mongolia's South Gobi province (Carreteras polvorientas y desconexiones: percepciones de polvo de caminos de minería sin pavimentar en la provincia de Gobi del sur de Mongolia). Geoforum **66**, 94–105 (2015)
5. Antabe, R., Atuoye, K.N., Kuuire, V.Z., Sano, Y., Arku, G., Luginaah, I.: Community health impacts of surface mining in the Upper West Region of Ghana: The roles of mining odors and dust. Hum. Ecol. Risk Assess. **23**(4), 798–813 (2017)
6. Vázquez, L.S., Riofrío, M.B.E.: Aportes teórico-metodológicos para un Sistema de Alerta Temprana de conflictos socioambientales. Experiencias en torno al Proyecto Mirador, Ecuador. Inst. Geogr. **93**, 63–75 (2016)
7. Ranängen, H., Lindman, Å.: Exploring corporate social responsibility practice versus stakeholder interests in Nordic mining (Explorando la práctica de la responsabilidad social corporativa frente a los intereses de los stakeholder en la minería nórdica Ecuador). J. Clean. Prod. **197**, 668–677 (2018)

Research on Sampling Inspection Procedures for Bank Service Time Based on Bayes Method

Jingjing Wang[1,2], Haitao Wang[1], Fan Zhang[1], Chao Zhao[1],
Gang Wu[1], and Jing Zhao[1(✉)]

[1] China National Institute of Standardization, Beijing, China
1513457957@qq.com, {wanght, zhangfan, zhaochao, wugang,
zhaoj}@cnis.gov.cn
[2] School of Mathematical Sciences, Capital Normal University, Beijing, China

Abstract. Using the previous data and experience of the service time provided by bank tellers to customers, this paper constructs a bank service time sampling inspection program based on Bayes method, and re-formulate the sampling plan (n, λ) for the mean value. In this paper, we use nonlinear programming theory to determine (n, λ) in the "λ"-type sampling plan. Then, compared with the conventional sampling inspection program, it is found that the sampling inspection plan for banking service time based on Bayes method requires fewer samples, which is valuable for reducing the sampling cost and strength.

Keywords: Banking service time · Hypothesis testing · Bayes theory · Nonlinear program

1 Foreword

Service is the eternal theme of the banking business, and the quality of service [1] is the lifeline of the bank. Quality of service is a comprehensive evaluation that reflects the extent to which service itself meets regulatory or potential requirements and reflects the customer's perception of the service process. Service time is an indicator to measure the quality of bank teller work. This paper mainly refers to the length of working hours to indicate the quality of work.

Currently, in the process of bank tellers' work quality inspection, it is usually based on the conventional product quality inspection sampling plan specified in the relevant standard specifications to achieve the purpose of understanding and judging the quality of bank tellers' work.

However, in the conventional sampling inspection program of product quality, the excessive sample size of product often aggravates the workload of inspection and increases the time and cost of sampling inspection of product quality. Therefore, it is necessary to improve the current sampling inspection scheme of bank tellers' work quality with the help of new techniques and methods. Bayes statistical analysis is a method to obtain better parameter estimation based on historical data or prior knowledge [2] of parameters to be estimated on the premise of small sample or no sample information. In recent years, it has been applied in product structure optimization design, reliability assessment, quality safety monitoring and other aspects. Hua and Di [3], working in national water-saving irrigation engineering technology research center in Beijing, have used the Bayes method to improve the sampling quality inspection program for discrete products. However, in the existing research literature, the Bayes method has not been found to improve the results of the continuous product quality sampling inspection program. Based on the prior information provided by the manufacturer or the quality inspection organization' data and experience, the Bayes method is used to make necessary improvements to the conventional product quality sampling inspection plan specified in the existing product standard specification, which can effectively reduce the number of samples and reduce the inspection cost while satisfying the requirements of product quality sampling and inspection.

This paper, by using a bank teller service time sampling inspection with the first test data, builds a bank teller service quality sampling inspection plan based on the Bayes method. Then the new sampling calculation of bank teller service time is carried out by using the improved product quality sampling inspection program, and the product sampling quantity in the prescribed conventional product quality sampling inspection plan is analyzed and compared to explore the feasibility of applying the improved program.

2 Hypothesis Test Theory

In the Neyman-Pearson theory, J. Neyman and E.S. Pearson believe that two types of errors [4] can be made when testing a hypothesis H_0. Assuming the null hypothesis $H_0{:}\theta \in \Theta_0$, alternative hypothesis $H_1{:}\theta \in \Theta_1$. Here Θ_0 and Θ_1 are the sets of θ that the hypothesis H_0 is true or not, respectively, obviously $\Theta = \Theta_0 + \Theta_1$.

Type I error is that when H_0 is true (i.e., $\theta \in \Theta_0$), it is judged that H_0 is rejected; Type II error is that when H_1 is true (i.e., $\theta \in \Theta_1$), it is judged that H_0 is accepted (see Table 1). When $\theta \in \Theta_0$, the probability of $P_\theta(X \in W)$ is the probability α of the type I error. When $\theta \in \Theta_1$, the probability of $P_\theta(X \in \bar{W})$ is the probability β of the type I error. Here, sample X is the vector consisting of $X_1, X_2, \ldots, X_n$, and W is the rejected domain. In order to describe the quality of the test, the function $P_\theta(X \in W)$ of θ is called the power function of the test. The power function of this test is shown in Fig. 1, where p_0, p_1, α, and β are selected as needed. This graph clearly describes the probability of making two types of errors.

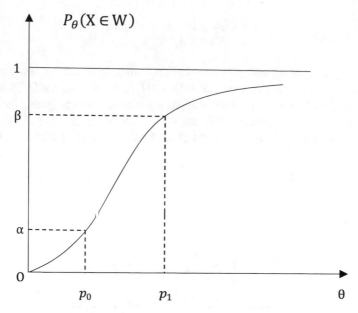

$P_\theta(X \in W)$

Fig. 1. The power function of Hypothesis test

Table 1. The situation of Hypothesis test

Judgment conclusion	Real situation of distribution	
	$\theta \in \Theta_0$ (H_0 is true)	$\theta \in \Theta_1$ (H_1 is true)
Accept H_0	True	Accept H_0
Reject H_0	Type I error	Reject H_0

3   Sampling Procedures for Inspection by Variables

The single upper specification limit U is taken as an example (the single lower specification limit L is the same). Here we use the hypothesis testing theory for analysis. Assuming the null hypothesis $H_0 : \mu \leq \mu_0$, the alternative hypothesis $H_1 : \mu > \mu_0$. Assume that the quality characteristics $X \sim N(\mu, \sigma^2)$, μ_0 and μ_1 represent the mean value of the product, μ_0 is the limit of the high-quality lot, and μ_1 is the limit of the inferior lot. μ_0 and μ_1 corresponds to p_0 and p_1 in Fig. 1, respectively. For the upper specification limit, we want to receive at a high probability of not less than $1 - \alpha$ when the overall mean $\mu \leq \mu_0$, and a low probability of not higher than β when $\mu \geq \mu_1$. That is, μ_0 is the supplier quality level (PQL), α is the supplier risk; μ_1 is the buyer's quality level (CQL), and β is the buyer's risk. For a given μ_0, μ_1, α, β, we hope that the solution satisfies:

$$\begin{cases} P(\mu) \geq 1 - \alpha, & \mu \leq \mu_0 \\ P(\mu) \leq \beta, & \mu \geq \mu_1 \end{cases} \qquad (1)$$

where $\mu_0 < \mu_1$, $P(\mu)$ is the acceptance probability when the product quality mean is μ.

Due to the quality characteristics $X \sim N(\mu, \sigma^2)$, we use u-statistics. Here, we use the sample mean $\bar{x}$ as a measure of the acceptance criteria of the program. For the upper specification limit, the acceptance criterion of the scheme is $\bar{x} \leq \lambda$, and the sampling scheme is (n, λ). Figure 2 shows the acceptance criteria for the scheme (n, λ).

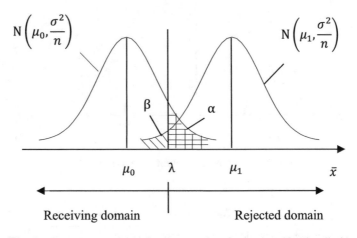

Fig. 2. Acceptance criteria for the program (upper specification limit)

According to the hypothesis test, the following conclusions may exist during the sampling process of product quality, as shown in the following table.

Table 2. Judgment conclusions that may exist in the process of product quality sampling inspection

The true quality of the product lot	Sampling data status	Judgment conclusion	Conclusion evaluation
$\mu \leq \mu_0$	$\bar{x} \leq \lambda$	Receive	True
$\mu \leq \mu_0$	$\bar{x} > \lambda$	Not receiving	The first type of error
$\mu > \mu_1$	$\bar{x} \leq \lambda$	Receiving	The second type of error
$\mu > \mu_1$	$\bar{x} > \lambda$	Not receiving	True

Since $X \sim N(\mu, \sigma^2)$, $\bar{X} \sim N\left(\mu, \frac{\sigma^2}{n}\right)$. For the sampling plan (n, λ), the probability of acception is

$$P(\mu) = P(\bar{X} \le \lambda) = \Phi\left(\frac{\lambda - \mu}{\sigma/\sqrt{n}}\right), \tag{2}$$

where (2) is the decreasing function of μ, and the monotonicity of $P(\mu)$, it is known that the Eq. (1) is equivalent to

$$\begin{cases} P(\mu_0) = 1 - \alpha \\ \quad P(\mu_1) = \beta \end{cases} (\mu_1 > \mu_0), \tag{3}$$

which is

$$\begin{cases} \Phi\left(\frac{\lambda - \mu_0}{\sigma/\sqrt{n}}\right) = 1 - \alpha \\ \Phi\left(\frac{\lambda - \mu_1}{\sigma/\sqrt{n}}\right) = \beta \end{cases} (\mu_1 > \mu_0). \tag{4}$$

In the actual situation, sometimes we will consider the situation

$$\begin{cases} \Phi\left(\frac{\lambda - \mu_0}{\frac{\sigma}{\sqrt{n}}}\right) \ge 1 - \alpha \\ \Phi\left(\frac{\lambda - \mu_1}{\frac{\sigma}{\sqrt{n}}}\right) = \beta \end{cases} (\mu_1 > \mu_0) \tag{5}$$

4 Sampling Plan for Inspection by Variables

The Bayes method considers the product mean as a random variable and determines the prior distribution state of the product mean based on the product quality test data obtained earlier. In the new sampling test of product quality, the a priori distribution state and the new product sample information are used to obtain the posterior distribution state of the product mean, and based on the maximum risk of the product, the posterior risk meets the risk of the manufacturer and the user. Determine the number of sample inspections for new samples of product quality.

4.1 Conventional Sampling Inspection Program

In this paper, the conventional product quality sampling test scheme we use is the "k" type sampling plan (n, k) under the "σ" method in the international standard ISO 3951-4 [5].

In the ISO 3951-4, the "k" type sampling plan (n, k) satisfies

$$\begin{cases} \Phi\left(-\sqrt{n}\left(k+\Phi^{-1}(p_0)\right)\right) \geq 1-\alpha \\ \Phi\left(-\sqrt{n}\left(k+\Phi^{-1}(p_1)\right)\right) = \beta \end{cases} \qquad (6)$$

Here, n is the sample size under the "σ" method, k is the k acceptable constant under the "σ" method, p_0 is the claimed quality level (DQL), and p_1 is the limit quality, i.e. the product of the claimed quality level (DQL) and the limit quality ratio (LQR). α is the supplier's risk; β is the buyer's risk.

4.2   Sampling Inspection Program Based on Bayes Method

It is assumed that the prior distribution of the product mean μ approximates a normal distribution, i.e. $\mu \sim N\left(\mu_m, \sigma_m^2\right)$. Then the probability density function of μ is

$$P(\mu) = \frac{1}{\sqrt{2\pi}\sigma_m} e^{-\frac{(\mu-\mu_m)^2}{2\sigma_m^2}} \qquad (7)$$

The μ_m and σ_m in Eq. (7) can be determined by the moment method. Let $\hat{\mu}_m = \bar{\mu}$, $\hat{\sigma}_m^2 = s_\mu^2$, where $\bar{\mu}$ is the sample mean of the quality mean, s_μ^2 is the sample variance of the quality mean.

Assuming that the product quality X approximates a normal distribution, i.e. $X \sim N(\mu, \sigma^2)$, then $\bar{X} \sim N\left(\mu, \frac{\sigma^2}{n}\right)$. Then the probability density function of $\bar{X}$ is

$$P(\bar{x}|\mu) = \frac{1}{\sqrt{2\pi}\frac{\sigma}{\sqrt{n}}} e^{-\frac{(\bar{x}-\mu)^2}{2\frac{\sigma^2}{n}}} \qquad (8)$$

The Bayesian formula can be used to obtain the posterior distribution of product quality mean μ:

$$P(\mu|\bar{x}) = \frac{P(\bar{x}|\mu)P(\mu)}{P(\bar{x})} \qquad (9)$$

Substituting (7) and (8) into (9), we can obtain

$$P(\mu|\bar{x}) = \frac{1}{\sqrt{2\pi}B} e^{-\frac{(\mu-A)^2}{2B^2}} \qquad (10)$$

Where $A = \dfrac{\frac{n}{\sigma^2}\bar{x} + \frac{\mu_m}{\sigma_m^2}}{\frac{n}{\sigma^2} + \frac{1}{\sigma_m^2}}$, $B = \dfrac{1}{\sqrt{\frac{n}{\sigma^2} + \frac{1}{\sigma_m^2}}}$.

It means that the posterior distribution of the product quality means μ also obeys the normal distribution, and the posterior mean is A, and the posterior standard deviation is B.

According to Table 2, we can get the possibility of making the type I error is $P(\mu \leq \mu_0 | \bar{x} > \lambda)$; the probability of making the type II error is $P(\mu > \mu_1 | \bar{x} \leq \lambda)$. In the actual sampling test, we hope that the possibility of making mistakes is as small as possible.

$P(\mu \leq \mu_0 | \bar{x})$ is decreasing $\bar{x}$. By the monotonicity of $P(\mu \leq \mu_0 | \bar{x})$, we have

$$P(\mu \leq \mu_0 | \bar{x} > \lambda) \leq P(\mu \leq \mu_0 | \bar{x} > \lambda + 1) \leq P(\mu \leq \mu_0 | \bar{x} = \lambda)$$

At this time, we let $P(\mu \leq \mu_0 | \bar{x} = \lambda) \leq \alpha$, that is, there is a possibility of making a first type of error $P(\mu \leq \mu_0 | \bar{x} > \lambda) \leq \alpha$.

Similarly, $P(\mu > \mu_1 | \bar{x})$ is increasing $\bar{x}$. By the monotonicity of $P(\mu > \mu_1 | \bar{x})$, we have

$$P(\mu > \mu_1 | \bar{x} \leq \lambda) \leq P(\mu > \mu_1 | \bar{x} = \lambda)$$

At this time, we let $P(\mu > \mu_1 | \bar{x} = \lambda) \leq \beta$, that is, there is a possibility of making a second type of error $P(\mu > \mu_1 | \bar{x} \leq \lambda) \leq \beta$

So we hope

$$\begin{cases} P(\mu \leq \mu_0 | \bar{x} = \lambda) \leq \alpha \\ P(\mu > \mu_1 | \bar{x} = \lambda) \leq \beta \end{cases} \tag{11}$$

Combined with (10)

$$\begin{cases} \Phi\left(\frac{\mu_0 - A_0}{B}\right) \leq \alpha \\ \Phi\left(\frac{\mu_1 - A_0}{B}\right) \geq 1 - \beta \end{cases} \tag{12}$$

Where $A_0 = \dfrac{\frac{n}{\sigma^2}\lambda + \frac{\mu_m}{\sigma_m^2}}{\frac{n}{\sigma^2} + \frac{1}{\sigma_m^2}}$, $B = \dfrac{1}{\sqrt{\frac{n}{\sigma^2} + \frac{1}{\sigma_m^2}}}$.

4.3 Determination of Sampling Quantity and Receiving Constant λ

When determinate the value of sampling quantity n and receiving constant λ, we mainly use the nonlinear programming theory to solve

$$\begin{cases} \Phi\left(\frac{\mu_0 - A_0}{B}\right) \leq \alpha \\ \Phi\left(\frac{\mu_1 - A_0}{B}\right) \geq 1 - \beta \end{cases} \tag{13}$$

where $A = \dfrac{\frac{n}{\sigma^2}\lambda + \frac{\mu_m}{\sigma_m^2}}{\frac{n}{\sigma^2} + \frac{1}{\sigma_m^2}}$, $B = \dfrac{1}{\sqrt{\frac{n}{\sigma^2} + \frac{1}{\sigma_m^2}}}$.

We have a transformation of (13), then can obtain

$$\left\{ \begin{array}{l} \lambda \geq \frac{\sigma^2}{n} \left\{ \left(\frac{n}{\sigma^2} + \frac{1}{\sigma_m^2} \right) \left[\mu_0 - \frac{\Phi^{-1}(\alpha)}{\sqrt{\frac{n}{\sigma^2} + \frac{1}{\sigma_m^2}}} \right] - \frac{\mu_m}{\sigma_m^2} \right\} \\[4mm] \lambda \geq \frac{\sigma^2}{n} \left\{ \left(\frac{n}{\sigma^2} + \frac{1}{\sigma_m^2} \right) \left[\mu_1 - \frac{\Phi^{-1}(1-\beta)}{\sqrt{\frac{n}{\sigma^2} + \frac{1}{\sigma_m^2}}} \right] - \frac{\mu_m}{\sigma_m^2} \right\} \end{array} \right. \tag{14}$$

By plotting the function graph of accepting the constant λ the sample size n and, a range of values similar to the one below can be obtained. We can find a minimum of n and λ in the shaded part (Fig. 3).

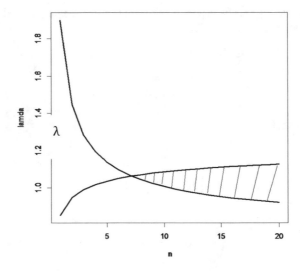

Fig. 3. The range of values of the sampling range n and the acceptance constant λ

5   Application of Bank Service Time Sampling Inspection Scheme Based on Bayes Method

Generally, the bank has declared that no more than 4 of service times should exceed 5 min. The manager of a branch of a bank is concerned that one of his tellers, who is currently on probation, provides a service that is too slow. The manager wishes to obtain objective evidence of the teller's incompetence. Therefore, the manager needs to conduct a sample test study on her service time to determine whether he believes that the cashier's service is incompetent. But the manager does not know how many sample sizes to choose to control the probability of committing one or two types of errors within a certain range.

5.1 Sampling Plan in ISO 3951-4

The ISO 3951-4 program is intended to be used when the quality characteristics are measurable variables that are independent and normally distributed, and where the quality of interest is the fraction of items that are nonconforming.

It is found from previous experience that the natural logarithm of service time approximately follows a normal distribution with a standard deviation of 0.5 (assuming known standard deviation).

The value 4% of the above cases corresponds to the claimed quality level (DQL) in ISO 3951-4. The sampling plan in ISO 3951-4 is jointly indexed for the claimed quality level (DQL) and the limit quality ratio (LQR). The sampling plan based on the "s" and "σ" methods are also provided.

This paper mainly studies the sampling scheme under the "σ" method. There are three LQR levels in the ISO 3951-4. The following sampling scheme was obtained from Tables 1, 2, 3 and 4 of ISO 3951-4, see Table 3.

Table 3. Sampling plan with a declared quality level (DQL) of 4%

LQR level	n	k	LQR	Probability of falsely contradicting a correct DQL (α)
LQR level I	6	0.786	9.9	0.91%
LQR level II	8	1.127	6.25	3.90%
LQR level III	17	1.442	5.86	0.90%

5.2 Sampling Plan Based on Bayes Method

Since we are studying the sampling inspection plan based on the mean value of the quality, while ISO 3951-4 is studying the sampling inspection plan based on the rate of nonconforming, we need to convert the "k" sampling scheme (n,k) of ISO 3951-4 into "excess" sampling scheme (n, λ).

5.2.1 Conversion of "k" Type Sampling Plan (n, k) and "λ" Type Sampling Plan (n, λ)

We need to use some techniques to convert the "k" type sampling plan (n, k) into a "λ" type sampling plan (n, λ).

Non-conforming product rate relative to the upper specification limit U is

$$p = P(X > U) = P\left(\frac{X - \mu}{\sigma} > \frac{U - \mu}{\sigma}\right) = 1 - \Phi\left(\frac{U - \mu}{\sigma}\right)$$

That is

$$\mu = U + \sigma \Phi^{-1}(p) \tag{15}$$

When the process standard deviation σ is known, since $\Phi^{-1}(x)$ is a strictly monotonic function, p and μ correspond one-to-one. Then we obtain

$$\begin{cases} \mu_0 = U + \sigma\Phi^{-1}(p_0) \\ \mu_1 = U + \sigma\Phi^{-1}(p_1) \end{cases} \tag{16}$$

Then let $\lambda = U - k\sigma$, the "k" type sampling scheme (n, k) in ISO 3951-4 can be converted into a "λ" type sampling scheme (n, λ). Correspondly, (5) can be translated to

$$\begin{cases} \Phi\left(\frac{\lambda-\mu_0}{\sigma/\sqrt{n}}\right) \geq 1 - \alpha \\ \Phi\left(\frac{\lambda-\mu_1}{\sigma/\sqrt{n}}\right) = \beta \end{cases} \tag{17}$$

Table 4. Correlation transformation between "k" type sampling plan and "λ" type sampling plan

LQR level	n	k	λ	p_0	μ_0	p_1	μ_1
Level I	6	0.786	1.216438	4%	0.7340949	0.396	1.477581
Level II	8	1.127	1.045938	4%	0.7340949	0.25	1.272193
Level III	17	1.442	0.8884379	4%	0.7340949	0.2344	1.247221

5.2.2 "λ" Type Sampling Plan (n, λ) Based on Bayes Method

After consulting relevant information, it was found that the mean μ of logarithmic service time approximate to normal distribution. Here, we use μ_m and σ_m^2 to represent the mean and the variance of μ, respectively. We use the moment method to determine the estimated value of μ_m and σ_m^2. Based on the prior experience we can obtain $\hat{\mu}_m = \mu = 0.7340949$, $\hat{\sigma}_m^2 = s_\mu^2 = 0.5$. Then, according to the nonlinear programming theory, the value range of the three levels of LQR and λ of LQR is drawn by the following formula:

$$\begin{cases} \lambda \geq \frac{\sigma^2}{n}\left\{\left(\frac{n}{\sigma^2} + \frac{1}{\sigma_m^2}\right)\left[\mu_0 - \frac{\Phi^{-1}(\alpha)}{\sqrt{\frac{n}{\sigma^2} + \frac{1}{\sigma_m^2}}}\right] - \frac{\mu_m}{\sigma_m^2}\right\} \\ \lambda \geq \frac{\sigma^2}{n}\left\{\left(\frac{n}{\sigma^2} + \frac{1}{\sigma_m^2}\right)\left[\mu_1 - \frac{\Phi^{-1}(1-\beta)}{\sqrt{\frac{n}{\sigma^2} + \frac{1}{\sigma_m^2}}}\right] - \frac{\mu_m}{\sigma_m^2}\right\} \end{cases} \tag{18}$$

According to Figs. 4, 5 and 6, we can get the sampling schemes (n, λ) at three LQR levels are (3, 1.292), (7, 1.078), (8, 1.056) respectively.

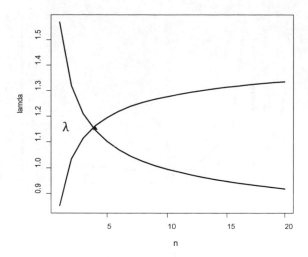

Fig. 4. Range of values for (n, λ) at the LQR level I. Note: The coordinates of the triangle is (3, 1.292)

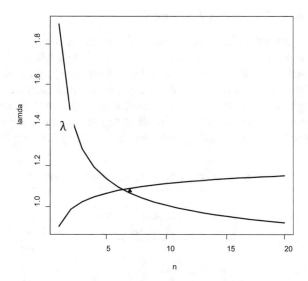

Fig. 5. Range of values for (n, λ) at the LQR level II. Note: The coordinates of the triangle is (7, 1.078)

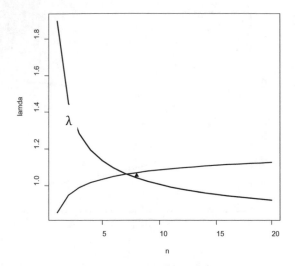

Fig. 6. Range of values for (n, λ) at the LQR level III. Note: The coordinates of the triangle is (8, 1.056)

5.3   Comparison of Two Sampling Plans

Comparing the conventional and Bayes methods, we found that the sample size n of different LQR levels decreased after Bayes method, but the LQR level I and the LQR level III decreased significantly. As shown in Table 5, the sample size reduction rate was 50% at the LQR level I and III, while the sample reduction rate is 12.5% at the LQR level II. Although the decrease is relatively small at the LQR level II, it is still reduced. Obviously, using the Bayes method has the effect of reducing the sample size, which is very helpful in reducing the sampling intensity and reducing the sampling cost.

Table 5. Comparison of two sampling plans

Method	n			λ		
	LQR Level I	LQR Level II	LQR Level III	LQR Level I	LQR Level II	LQR Level III
Conventional method	6	8	17	1.216438	1.045938	0.8884379
Bayes method	3	7	8	1.292	1.078	1.056

6   Conclusion

In this paper, the posterior distribution of the mean of bank tellers' logarithmic service time was obtained based on Bayes method, and then a new sampling of bank tellers' service time test was taken on the basis that the maximum posterior risk of the mean of bank tellers' log service time simultaneously met the risks of both the manufacturer and the user. We mainly look at the three levels of the corresponding LQR. Among them, the reduction rate of sample size reached more than 50% at the LQR level I and III, and the LQR level II was 12.5%. At the LQR level I and III, we can use the Bayes method to determine the sampling plan. In the future, the sampling plan of LQR Level II can be further studied to make it significantly reduce the sampling amount.

Acknowledgement. This research was supported by National Key Technology R&D Program (2017YFF0206503, 2017YFF0209004, 2016YFF0204205) and China National Institute of Standardization through the "special funds for the basic R&D undertakings by welfare research institutions" (522018Y-5941, 522018Y-5948).

References

1. Wang, H.: The relationship between bank service quality and customer satisfaction. J. Sun Yatsen Univ. (Soc. Sci. Ed.) **6**, 107–108 (2006)
2. Mao, S.: Bayesian Statistics, 2nd edn, pp. 3–4. China Statistics Press, Beijing (1999)
3. Hua, Z., Di, X.: Study on the quality sampling inspection scheme of water saving irrigation products based on Bayes method. Trans. Chin. Soc. Agric. Mach. **3**(3) (2010)
4. Xu, R.: The calculation of two kinds of errors in hypothesis testing and its relationship. J. Jilin Norm. Univ. (Nat. Sci. Ed.) **34**(2), 146–148 (2013)
5. International Organization for Standardization. ISO 3951-4: Sampling procedures for inspection by variables-Part 4: procedures for assessment of declared quality levels. International Organization for Standardization, 15 August 2011

Remaining Useful Life Prediction
for Components of Automated Guided Vehicles

Beata Mrugalska[1(✉)] and Ralf Stetter[2]

[1] Faculty of Engineering Management, Poznan University of Technology,
ul. Strzelecka 11, 60-965 Poznań, Poland
beata.mrugalska@put.poznan.pl
[2] Hochschule Ravensburg-Weingarten,
Doggenriedstraße, 88250 Weingarten, Germany
stetter@hs-weingarten.de

Abstract. This paper presents an approach to prediction of the Remaining Useful Life (RUL) for components of Automated Guided Vehicles (AGV). The focus is paid on the batteries which are a crucial element of these systems and influence the possible operation times considerably. For batteries, two aspects are taken into consideration, if the remaining useful life should be predicted: the State of Charge (SOC) and the State of Health (SOH). Both aspects include non-linearity and are influenced by many factors such as temperature and discharging velocity. To solve such problem a new estimator of SOC and SOH was developed. The proposed approach was applied for Health-Aware Model Predictive Control (H-A MPC) of two cooperate AGV.

Keywords: Remaining useful life · Prognosis · Automated guided vehicles

1 Introduction

The techniques of remaining useful life estimation of systems are mainly used to predict the moment in which a system failure occurs. This knowledge is usually used to take actions to counteract the occurrence of a failure or leading to a scheduled removal of its effects [1]. However, these techniques can also find other, not related to the detection of the moment of failure, areas of application, e.g. prediction of remaining useful life of technical objects using consuming resources. One such example is the depletion of energy in the batteries of the AGV [2–4], which are increasingly used to transport components between assembly stations in Flexible Manufacturing Systems (FMS) [5–7]. In this case, one cannot talk about a failure in the traditional sense, but rather about the inability to provide services for a given operated system. From a practical point of view, having knowledge about the current state of a given resource and predicting its behavior in the future is crucial for many industrial applications and can be used, for example, in H-A MPC.

W. Karwowski et al. (Eds.): AHFE 2019, AISC 971, pp. 420–429, 2020.
https://doi.org/10.1007/978-3-030-20494-5_39

2 Estimation Method of State of Charge of Batteries and Prediction of Their Remaining Useful Life for Control of AGV

The purpose of the conducted research was to develop a method that allows estimating the state of the AGV vehicle battery so that it is possible to predict its state of charge at any time. This task is considered in the context of diagnostics e.g. state of health being a measure of battery capacity for storing and supplying electricity [8]. It can be assumed that SOH denotes the number of work cycles an AGV vehicle can perform, starting with the maximum battery charge and ending with its full discharge. In such case, it is possible to intuitively determine the degree of degradation of the battery during its use what is related with its operating conditions. It can be assumed that a more used battery will allow for less work cycles for a given SOC than another unused battery. Such situation is illustrated in Fig. 1 where $k_{f,1}$ represents the feasible number of cycles of AGV_1 while $k_{f,2}$ stands for an analogous variable associated with AGV_2.

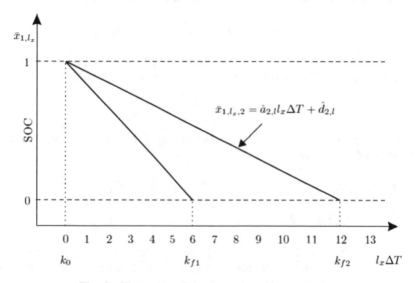

Fig. 1. Illustration of the state of health of batteries.

As it can be seen, the state of health of AGV_1 is definitely lower than the one of AGV_2.

The knowledge about the SOC and SOH of the battery allows to determine the remaining useful life of AGV. The developed methods can be applied in predictive control of cooperating AGVs taking into account their SOH and remaining useful life expressed in the form of the number of remaining possible transport cycles of the AGV.

During the research, a method of SOC estimation for non-linear model of the battery [9] described in the states space was developed. After few transformations and taking into consideration the system disturbances affecting of the system $\omega_l \in \mathbb{R}^{n_\omega}$ and its outputs $v_l \in \mathbb{R}$ takes the following form:

$$x_{l+1} = Ax_l + Bu_l + W_1 w_l, \tag{1}$$

$$y_l = Cx_l + Du_l + h(x_{1,l}) + W_2 v_l, \tag{2}$$

where:

$$A = \begin{bmatrix} 1 & 0 & 0 \\ 0 & e^{-\frac{\Delta t}{\tau Ci}} & 0 \\ 0 & 0 & e^{-\frac{\Delta t}{\tau Dif}} \end{bmatrix}, \quad A = \begin{bmatrix} 1 & 0 & 0 \\ 0 & e^{-\frac{\Delta t}{\tau Ci}} & 0 \\ 0 & 0 & e^{-\frac{\Delta t}{\tau Dif}} \end{bmatrix},$$

$C = [0, 1, 1]$, $D = R_0$, $h(\cdot) = V_{OCV}(\cdot)$, where $(\cdot)$ represents an average cell open circuit voltage. The input of the model is the current $u_l = I$ and its output is the voltage $y_l = V$. The remaining model parameters are defined in [9].

For such defined battery model, a new estimator structure is proposed:

$$\hat{x}_{l+1} = A\hat{x}_l + Bu_l + K\left(y_l - C\hat{x}_l - Du_l - h(\hat{x}_{1,l})\right), \tag{3}$$

where $\hat{x}_l$ is the state estimate, and K represents the gain matrix of the designed estimator. This matrix can be determine using the following relationship:

$$K = P^{-1}N \tag{4}$$

The estimator design problem boils down to solving (6) with respect to P and N for given values of $\alpha \in (0, 1)$:

$$\begin{bmatrix} -P - \alpha P & 0 & A^T P - (\bar{C})(\beta)^T N^T \\ 0 & -\alpha Q & \bar{W}_1^T P - \bar{W}_2^T N^T \\ PA - N(\bar{C})(\beta) & P\bar{W}_1 - N\bar{W}_2 & -P \end{bmatrix} \prec 0, \quad i = 1, 2, \tag{5}$$

where $\beta_1 > 0i$, $\beta_2 < 1$, $Q \succ 0$, $\bar{W}_1 = W_1[I_{n_w} 0_{n_w \times 1}]$, $\bar{W}_2 = W_2[0_{1 \times n_w} 1]$ and $\bar{C}(\beta) = [\beta\bar{\gamma}, 1, 1]$ whereas $\bar{\gamma} > 0$. Finally, the estimator design procedure can be realized as follows:

Algorithm 1
Off-line:
    Step 1: Select the upper bound $\bar{\gamma}$ and the overbounding matrix $Q \succ 0$,
    Step 2: Select $\alpha \in (0,1)$,
    Step 3: Solve LMIs (5) and obtain the gain matrix $K = P^{-1}N$.
On-line:
    Step 4: Set $\hat{x}_0$ and $l = 0$.
    Step 5: Obtain the state estimate $\hat{x}_{k+1}$ according to (3).
    Step 6: Set $l = l + 1$ and go to **Step 5**.

It is worth noting that the proposed solution allows to overcome the disadvantages of the traditional methods used in estimation tasks, i.e. EKF, the use of which may lead to the occurrence of a lack of convergence of the estimator which may result from the use of linearization. The new estimator was designed on the basis of Luenberger's observer and the approach of Quadratic Boundedness (QB) was developed [10]. While designing the estimator, it was assumed that all the uncertainty resulting from the occurrence of disturbances and the estimation error can be overbounded by the ellipsoid.

The developed state estimator allows to propose a new prediction method of remaining useful life of AGV battery. Due to the fact that the relation (1) describing the state of the battery is linear, the following form of the expression shaping the relation between time $l\Delta T$ and SOC $x_{1,l}$ for the battery of ith AGV was assumed:

$$\hat{x}_{1,l,i} = a_i l\Delta T + d_i, \quad i = 1, 2, \tag{6}$$

where a_i and d_i are unknown parameters which have to be estimated, and $\hat{x}_{1,l,i}$ is an estimate of SOC of ith AGV battery obtained with the estimator developed in the preceding section. To cope with the estimation problem of a_i and d_i, a celebrated Recursive Least Square (RLS) algorithm is employed, where the parameter estimation vector along with its regressor are defined by $\hat{p}_{l,i} = \left[\hat{a}_{l,i}, \hat{d}_{l,i}\right]^T$ and $r_{l,i} = [l\Delta T, 1]^T$ To summarize, to obtain a_i and d_i, the following algorithm is used:

Algorithm 2:
  Step 1: Set $\hat{p}_{0,i} = [0, 1]^T$, $P_{0,1} = \delta I_2$ and $l = 0$, with $\delta > 0$ being a sufficiently large positive constant
  Step 2: Obtain parameter estimates using:

$$\hat{p}_{l,i} = \hat{p}_{l-1,i} + K_{l,i}(\hat{x}_{1,l,i} - r_{l,i}^T \hat{p}_{l-1,1}), \quad (7)$$

$$K_{l,i} = P_{l-1,i} r_{l,i} \left(1 + r_{l,i}^T P_{l-1,i} r_{l,i}\right)^{-1}, \quad (8)$$

$$P_{l,i} = [I_2 - K_{l,i} r_{l,i}^T] P_{l-1,i}, \quad (9)$$

  Step 3: Set $l = l + 1$ and go to **Step 2**.

Note that by observing (6) with the initial estimate $\hat{p}_{0,i} = \left[\hat{a}_{0,i}, \hat{d}_{0,i}\right]^T = [0, 1]^T$ it is evident that it corresponds to SOC equal one. Indeed, this is a natural approach to assume that the battery is fully loaded at the beginning of the operation of AGV.

Finally, by bridging the algorithm of estimating SOC along with the one of estimating unknown parameters of (6), the following SOC predictor can be obtained:

$$\bar{x}_{1,l_x,i} = \hat{a}_{i,l} l_x \Delta T + \hat{d}_{i,l}, \quad i = 1, 2, \tag{10}$$

where l_x is the discrete time of prediction which is based on the parameter estimates obtained up to l. It implies that $l_x \geq l$.

On the basis of (10) the total remaining number of forward and backward cycles of ith AGV can be estimated as:

$$k_{f,i} = \left\lfloor -\frac{\hat{a}_{i,l} l \Delta t + \hat{d}_{i,l}}{\hat{a}_i \left(f_i(k) + \hat{b}_i(k) \right)} \right\rfloor \tag{11}$$

where $f_{i(k)}$ and $b_{i(k)}$ are the forward and backward transportation time between assembly stations for kth event counter and $\lfloor \cdot \rfloor$ rounds the resulting value to the smallest positive integer.

3   H-A MPC of Cooperating AGVs

The developed methods make it possible to predict the remaining useful life of AGVs for given battery sets. This result enabled to developed methods of H-A MPC of cooperating AGVs, taking into account the SOH of their battery and their remaining useful life.

The considered problem boils down to the transportation of the seat frame with all mechanical and electrical components such as airbag, belt retractor and belt lock from the seat frame assembly station located in the Building A, where these components were fitted, to the complete seat assembly station located in the Building B where the foam, trim and covers will be added. Between the buildings there is a pathway for AGVs. In the sample scenario two AGVs called AGV_1 and AGV_2 could perform the transportation tasks. Note that these two cooperating AGVs introduce concurrency into the overall production system. It should be underlined that the considered problem comes down to dispatching the transport tasks between two cooperating AGVs,

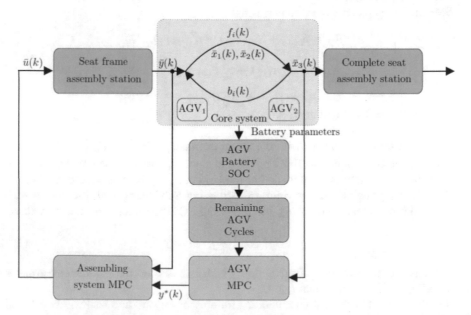

Fig. 2. The methods of battery SOC and RUL estimation in the assembling system MPC and AGV MPC.

taking into account the status of tasks performed in the previous cycle of the transportation process in such a way as to optimally use the AGVs remaining useful life resulting from their battery SOC and SOH.

The described transportation scenario, which is a sequential process, can be depicted in the form of a flowchart presented in Fig. 2.

The variables $\bar{x}_1(k)$ and $\bar{x}_2(k)$ are the start time of the first and second AGV for k-th event counter associated with the realization of k seat frame transportation. The variable $\bar{x}_3(k)$ denotes the seat frame arrival time k-th event counter, and $\bar{y}(k)$ represents the delivery time of seat frame, which has to be transported for k-th event counter. Moreover, $f_i(k)$ and $b_i(k)$ are the forward and backward transportation time between assembly stations.

In order to develop the H-A MPC for cooperating AGVs, the appropriate scheduling method of production system operating under fault conditions at the assembly station located in the building A should be used. To achieve it an appropriated predictive control method allowing for the optimal control of production process $\bar{u}(k)$ taking into account many limitations, i.e. work efficiency of individual assembly stations, uncertainty in the production process or possibilities of AGVs involved in the transportation process can be used. To solve such problem the scheduling method presented in [11] can be used.

Thus, having the appropriate MPC method, which allows for the optimal control of the assembly system, it is possible to develop an algorithm which enables to calculate $\bar{y}(k)$ in such a way as to guarantee that $\bar{x}_3(k)$ will follow a predefined schedule $t_{ref}(k)$ [9]. The main assumption behind further developments is that a single AGV is not able to follow $t_{ref}(k)$, and hence, its cooperative functioning is required. As it was already mentioned, such a system can be understood as one with concurrency.

The developed method of the H-A MPC for the cooperating AGVs allows to obtain the series of values $\bar{y}(k)$, which are used as output constraint $\bar{y}(k)$ of the seat frame assembly station in Building A. As a result, the optimal input $\bar{u}(k)$ is calculated according to the MPC approach for assembling system presented in [11]. It should be emphasized that the key elements enabling the implementation of the developed H-A MPC for cooperating AGVs are the developed battery SOC estimation method and prediction of the RUL method. The efficiency of the developed methods was verified on the example of the Li-NMC battery used in the AGV. As part of the research, an estimator was designed which allowed to determine the SOC of the battery. During the experiment, the AGV was moving and, as a consequence, the battery was gradually discharged. In Fig. 3 the current state of charge of the battery and its estimate obtained with the application of the developed method are presented.

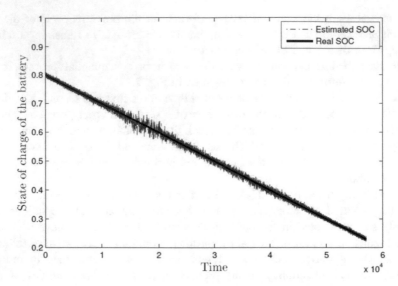

Fig. 3. Actual SOC and its estimate.

Moreover, the prediction of the SOC of the battery was made using the developed Algorithm 2, to enable the determination of the RUL of the AGV. The results of the experiment are shown in Fig. 4. The obtained results confirm the high effectiveness of the developed methods.

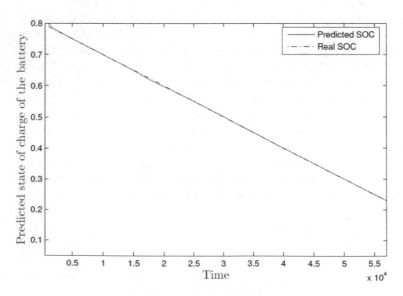

Fig. 4. SOC and its prediction.

The experiment confirming the relevance of the application of the developed methods in the control tasks of cooperating AGVs was carried out in an environment imitating the exemplary transport process. During the experiment, various H-A MPC scenarios in transportation task were tested taking into account the remaining useful life of battery of a real AGVs shown in Fig. 5.

Fig. 5. AGV applied to transportation of the seat frames.

In one of the scenarios, a different degree of battery SOH was assumed, which is expressed in the lower value of the parameter $k_{f,i}$ calculated with (11) for AGV_1. The results of the experiment are presented in Fig. 6, which shows the AGV vehicle battery consumption expressed in the form of the remaining cycles of their use, whereas Fig. 7 presents the activity of each AGV in the form of transport events.

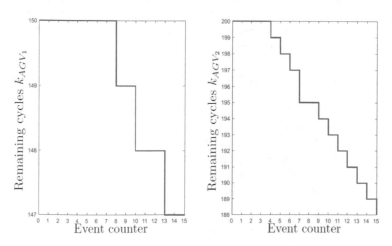

Fig. 6. SOC of AGVs expressed by the remaining cycles.

From this result it is an obvious fact that AGV_2 is more exploited while AGV_1 is used only while it is absolutely necessary. These results clearly prove that the proposed H-A MPC algorithm, which take into consideration the SOC estimate and the prediction of their RUL, allows balanced exploitation of the collaborative AGV system, which undoubtedly increases their operational abilities.

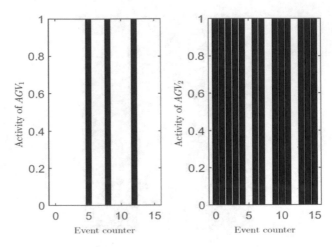

Fig. 7. Activities of AGVs.

4   Conclusions

In the paper a method of SOC estimation of AGV battery was presented. In particular, a robust estimator was shown on the basis of the Luenberger observer and the QB approach. The proposed solution allows to overcome the disadvantages of traditional methods used in estimation tasks, which often omit battery nonlinearity or use linearization, which may lead to the occurrence of high uncertainty or a lack of convergence of the developed estimator. Furthermore, the methods of the SOC and remaining useful life prediction based on the recursive least squares method were developed. The advantage of the developed methods is their low computational burden, which allows to apply them to predict the remaining useful life of batteries in AGVs. Moreover, the attention was paid to a measure allowing to express the current SOC of the battery in the form of the number of remaining AGV transport cycles. The proposed measure makes it possible to determine the AGV remaining useful life. This result allows to apply the developed methods in the H-A MPC of cooperating AGVs, taking into account the SOH and remaining useful life of the AGV batteries. The effectiveness of the developed methods of battery SOC and SOH estimation and prediction was verified on the example of cooperating AGV.

References

1. Mrugalska, B.: Remaining useful life as prognostic approach: a review. In: Ahram, T., Karwowski, W., Taiar, R. (eds.) Human Systems Engineering and Design, IHSED 2018. AISC, vol. 876, pp. 689–695. Springer (2019)
2. Kawakami, T., Takata, S.: Battery life cycle management for automatic guided vehicle systems. In: Design for Innovative Value Towards a Sustainable Society, pp. 403–408. Springer (2012)
3. Liu, X., Li, W., Zhou, A.: PNGY equivalent circuit model and SOC estimation algorithm for lithium battery pack adopted in AGV vehicle. IEEE Access 6(23), 639–647 (2018)
4. Mrugalska, B., Stetter, R.: Health-aware model-predictive control of a cooperative AGV-based production system. Sensors 19(3), 532–559 (2019)
5. Gnanavel Babu, A., Jerald, J., Noorul Haq, A., Muthu Luxmi, V., Vigneswaralu, T.: Scheduling of machines and automated guided vehicles in FMS using differential evolution. Int. J. Prod. Res. 48(16), 4683–4699 (2010)
6. Kumar, R., Haleem, A., Garg, S.K., Singh, R.K.: Automated guided vehicle configurations in flexible manufacturing systems: a comparative study. Int. J. Ind. Syst. Eng. 21(2), 207–226 (2015)
7. Ullrich, G., et al.: Automated Guided Vehicle Systems. Springer, Berlin (2015)
8. Le, D., Tang, X.: Lithium-ion battery state of health estimation using Ah-V characterization. In: Proceedings of the Annual Conference of Prognostics and Health Management (PHM) Society, Montreal, QC, Canada, vol. 2529, pp. 367–373 (2011)
9. Taborelli, C., Onori, S., Maes, S., Sveum, P., Al-Hallaj, S., Al-Khayat, N.: Advanced battery management system design for SOC/SOH estimation for e-bikes applications. Int. J. Powertrains 5(4), 325–357 (2016)
10. Alessandri, A., Baglietto, M., Battistelli, G.: Design of state estimators for uncertain linear systems using quadratic boundedness. Automatica 42(3), 497–502 (2006)
11. Majdzik, P., Akielaszek-Witczak, A., Seybold, L., Stetter, R., Mrugalska, B.: A fault-tolerant approach to the control of a battery assembly system. Control Eng. Pract. 55, 139–148 (2016)

Discussion on the Range of Cut-Off Values in Robust Algorithm A

Fan Zhang[1,2], Yue Zhang[3(✉)], Jing Zhao[1,2], Chao Zhao[1,2],
Gang Wu[1,2], Xinyu Cao[1,2], and Haitao Wang[1,2]

[1] Center for Quality and Statistics, China National Institute of Standardization,
Beijing, China
{zhangfan, zhaoj, zhaochao, wugang, caoxy,
wanght}@cnis.gov.cn
[2] AQSIQ Key Laboratory of Human Factors and Ergonomics (CNIS),
Beijing, China
[3] Capital Normal University, Beijing, China
15652668478@163.com

Abstract. ISO 13528-2005 "Statistical methods for use in proficiency testing by interlaboratory comparisons" introduced a robust method for calculating the mean and standard deviation—Algorithm A. In cases where it is not possible to use the classical statistical method to eliminate outliers, Algorithm A can be used without deleting the abnormal data and give a good estimate of population mean and population standard deviation. This method is widely used in proficiency testing to calculate assigned value and standard deviation for proficiency assessment. The iterative calculation of Algorithm A has a parameter c, called cut-off values, which usually takes between 1 and 2. In the actual application process, the cut-off values is often misused, and is taken to more than 2 which is not within the recommended range. By using simulation, this paper shows that the reasonable value of parameter c should be between 1 and 2. When the value exceeds 2, it does not satisfy the actual assumption.

Keywords: Robust statistics · Outliers · Algorithm A · Cut-off values

1 Introduction

In practical applications, when there are outliers in the data set, it is common practice to remove them directly from the data set. However, in most cases, we cannot accurately determine which are exactly outliers. Direct deletion may remove correct values. So it is not applicable to delete the outliers directly in practice. Robust statistics is a good way to get good estimates of population mean and population standard deviation from a data set that contains some outliers (wrong or abnormal data points).

The following is a brief introduction to a theory of robust estimation. Usually in the experiment, all the experimental data we hope to obtain is accurate, but often there will be some deviations in the operation due to measuring instruments or measurement routine. If we refer to the results of correct measurement as "good" results, the incorrect results in the experiment are called "bad" results. μ is the mean of "good"

W. Karwowski et al. (Eds.): AHFE 2019, AISC 971, pp. 430–434, 2020.
https://doi.org/10.1007/978-3-030-20494-5_40

measurement, σ is standard deviation of the measurement result of "good", and ε is the proportion of "bad" measurement result, the statistical distribution produced by the overall measured values ("good" and "bad" measurements) is a contaminated normal distribution $F(x) = (1 - \varepsilon)\Phi(\frac{x-\mu}{\sigma}) + \varepsilon H(x)$, where $\Phi(t)$ is standard normal distribution, $H(x)$ is an unknown pollution distribution and $0 \leq \varepsilon \leq 1$. Huber [1] proposed a new theory of robust estimation, discussed the asymptotic theory for estimating location parameters of contaminated normal distribution, and formulated three heuristic methods for the solution of the problem of the simultaneous determination of μ and σ for a given ε. Algorithm A [2] is "Method 2" from Huber, a process of iterative calculations. In the iteration, experimental data needs to be intercepted by the parameter c. The previous article describes that the cut-off value c usually takes between 1 and 2. However, in practical application, many people in the experiment think that taking c as more than 2 or larger numbers had no effect on the experimental results and the test effect was also excellent.

This paper first introduces basic idea and calculation process of Algorithm A, and then give some known distributions as pollution distributions $H(x)$, generate data by F (x), and using Algorithm A to simulate. Finally, through data simulation, the rationality of choosing c value between 1 and 2 is specified.

2   Algorithm A

Firstly, sort the data in increasing order, so that the outliers are mainly concentrated at both ends. Set a range, then the numbers outside the range are replaced by the end value of the range, and the original data inside the range do not change. Robust estimates of population mean, and population standard deviation are obtained through continuous iterative calculations. The calculation process of Algorithm A will be specifically described below.

Denote the n items of data, sorted into increasing order, by:

$$x_1, x_2, \ldots, x_n$$

Denote the robust average and robust standard deviation of these data by x^* and s^*. Calculate the initial values of x^* and s^* as:

$$x^* = medx_i \quad i = 1, 2, \ldots, n \tag{1}$$

$$s^* = 1.483 \times med|x_i - x^*| \quad i = 1, 2, \ldots, n \tag{2}$$

Update the values of x^* and s^* according to the following steps.
For each $x_i \, i = 1, 2, \ldots, n$, calculate:

$$x_i^* = \begin{cases} x^* - cs^* & x_i < x^* - cs^* \\ x_i & |x_i - x^*| \leq cs^* \\ x^* + cs^* & x_i > x^* + cs^* \end{cases} \tag{3}$$

Then calculate the new values of x^* and s^* by the following formula:

$$x^* = \sum_{i=1}^{n} x_i^*/n \tag{4}$$

$$s^* = \sqrt{\frac{1}{\beta(c)(n-1)} \sum_{i=1}^{n} (x_i^* - x^*)^2} \tag{5}$$

For the parameters ε, c, β have the following relationship:

$$\frac{1}{1-\varepsilon(c)} = \theta(c) + 2\frac{\varphi(c)}{c} \tag{6}$$

$$\beta(c) = \theta(c) + c^2(1 - \theta(c)) - 2\varphi(c) \tag{7}$$

where $\theta(c) = \int_{-c}^{c} \varphi(t)dt$, $\varphi(t)$ is the density function of standard normal distribution.

3 Simulation

In the following, the pollution distribution H(x) is taken as the chi-square distribution, t-distribution and uniform distribution, the actual data is generated by the contaminated distribution F(x). Then estimate the mean and standard deviation by robust Algorithm A, and compared with the mean and standard deviation of true value. Finally, explain the results of the simulation.

Let the contaminated normal distribution be

$$F(x) = (1 - \varepsilon)\Phi\left(\frac{x - \mu}{\sigma}\right) + \varepsilon H(x) \tag{8}$$

where $\varphi(t)$ is the density function of standard normal distribution and H(x) is an unknown pollution distribution and $0 \le \varepsilon \le 1$.

In the simulation, let pollution distribution H(x) take the chi-square distribution $(\chi^2(4))$, t-distribution (t(8)) and uniform distribution (U(1, 3)), respectively, and take c, 0.5, 1, 1.3, 1.4, 1.5, 1.7, 2, 2.5, 3, 3.5, 4. Then, the corresponding ε is calculated by the formula (6) and bring it into Eq. (8), randomly generate 1000 numbers as the original data.

The iteration is performed using Algorithm A. Stop the iteration until the number of four digits after decimal point of the robust estimate of population mean and population standard deviation does not change in two consecutive iterations and the process is considered to be convergent.

Tables 1, 2 and 3 shows the comparison between true mean and true standard deviation and the corresponding robust estimates after the iterative convergence of Algorithm A.

The smaller the c, the smaller the interval $(x^* - cs^*, x^* + cs^*)$ in Algorithm A, the bigger ε, the greater the proportion of outliers. By observing the above three tables,

Table 1. The comparison between true mean and standard deviation of chi-square distribution and its robust estimates

Chi-square distribution $\chi^2(4)$					
c	ε	Mean	Robust mean	Standard deviation	Robust standard deviation
0.5	0.4417	1.7212	1.5633	1.3052	1.1272
0.7	0.2899	1.1351	1.0827	1.0504	0.9799
1	0.1428	0.5675	0.5588	0.9350	0.9190
1.3	0.0655	0.2688	0.2641	0.9449	0.9391
1.4	0.0498	0.2083	0.2020	0.9530	0.9504
1.5	0.0376	0.1613	0.1546	0.9607	0.9602
1.7	0.0211	0.0974	0.0918	0.9730	0.9747
2	0.0084	0.0486	0.0451	0.9838	0.9891
2.5	0.0016	0.0223	0.0219	0.9902	0.9956
3	0.0003	0.0171	0.0169	0.9915	0.9932
3.5	$3.34 \times e^{-5}$	0.0163	0.0163	0.9917	0.9921
4	$3.57 \times e^{-6}$	0.0161	0.0161	0.9917	0.9918

Table 2. The comparison between true mean and standard deviation of t-distribution and its robust estimates

t-distribution $t(8)$					
c	ε	Mean	Robust mean	Standard deviation	Robust standard deviation
0.5	0.4417	−0.0121	−0.0155	0.7575	0.7665
0.7	0.2899	−0.0024	−0.0095	0.7846	0.7854
1	0.1428	0.0070	0.0001	0.8689	0.8609
1.3	0.0655	0.0119	0.0045	0.9313	0.9220
1.4	0.0498	0.0129	0.0058	0.9452	0.9384
1.5	0.0376	0.0137	0.0070	0.9562	0.9528
1.7	0.0211	0.0148	0.0091	0.9716	0.9727
2	0.0084	0.0156	0.0118	0.9836	0.9888
2.5	0.0016	0.0160	0.0156	0.9901	0.9957
3	0.0003	0.0161	0.0159	0.9914	0.9932
3.5	$3.34 \times e^{-5}$	0.0161	0.0161	0.9917	0.9921
4	$3.57 \times e^{-6}$	0.0161	0.0161	0.9917	0.9918

when c are 0.5 and 0.7, the proportion of outliers is 44.17 and 28.99% respectively. Although the mean value and standard deviation estimation results are better, but because the length of interval $(x^* - cs^*, x^* + cs^*)$ is short, more real data is removed. The correct data points may also be removed during the process, and a lot of information about the original data is lost. When c exceeds 2, the proportion of outliers is approximately 0. Especially when c takes more than 3.5, the estimate of mean is exactly equal to its true value. The estimation of standard deviation divided by $1/\sqrt{\beta(c)}$ is equal to the true value. No iterative update occurs, and no outliers exist. The calculation process uses all original data. When c is between 1 and 2, the proportion of outliers is 0.84–14.28%. The estimates

Table 3. The comparison between true mean and standard deviation of uniform distribution and its robust estimates

Uniform distribution U(1,3)					
c	ε	Mean	Robust mean	Standard deviation	Robust standard deviation
0.5	0.4417	0.8912	0.8875	0.6164	0.6342
0.7	0.2899	0.5904	0.5840	0.7282	0.7225
1	0.1428	0.2991	0.2951	0.8561	0.8372
1.3	0.0655	0.1458	0.1392	0.9285	0.9193
1.4	0.0498	0.1147	0.1075	0.9435	0.9358
1.5	0.0376	0.0906	0.0835	0.9552	0.9510
1.7	0.0211	0.0579	0.0519	0.9712	0.9718
2	0.0084	0.0328	0.0290	0.9835	0.9888
2.5	0.0016	0.0193	0.0189	0.9901	0.9957
3	0.0003	0.0166	0.0164	0.9914	0.9932
3.5	$3.34 \times e^{-5}$	0.0162	0.0162	0.9917	0.9921
4	$3.57 \times e^{-6}$	0.0161	0.0161	0.9917	0.9918

of mean and standard deviation are similar to the mean and standard deviation of real data, and the estimation effect is better.

4 Conclusion

Therefore, according to the above simulation results and corresponding analysis, in practical application, when c is between 1 and 2, the proportion of outliers is more in line with the actual situation, and the estimates of mean and standard deviation are similar to the mean and standard deviation of real data. As c value becomes larger, the length of interval $(x^* - cs^*, x^* + cs^*)$ increases, which makes more outliers participate in the calculation. In proficiency testing, this condition increases the standard deviation and decreases z value, makes the experimental results seem more convinced. But from the simulation we can see that it is unreasonable to take the value of c more than 2 or even larger. In the application, the c value can be selected according to the proportion ε of outliers in accordance with the values in the table, the commonly used value is 1.5.

Acknowledgement. This research was supported by National Key Technology R&D Program (2017YFF0206503, 2017YFF0209004, 2016YFF0204205) and China National Institute of Standardization through the "special funds for the basic R&D undertakings by welfare research institutions"(522018Y-5941, 522018Y-5948, 522019Y-6771).

References

1. Huber, P.J.: Robust estimation of a location parameter. Ann. Math. Stat. **35**(1), 73–101 (1964)
2. ISO 13528-2005 Statistical Methods for Use in Proficiency Testing by Interlaboratory Comparisons

Model for Dilution Control Applying Empirical Methods in Narrow Vein Mine Deposits in Peru

Luis Salgado-Medina[1](✉), Diego Núñez-Ramírez[1](✉),
Humberto Pehovaz-Alvarez[1](✉), Carlos Raymundo[2](✉),
and Javier M. Moguerza[3](✉)

[1] Escuela de Ingeniería de Gestión Minera, Universidad Peruana de Ciencias
Aplicadas (UPC), Lima, Peru
{u201320882, u201414907, pcgmhpeh}@upc.edu.pe
[2] Dirección de Investigaciones, Universidad Peruana de Ciencias Aplicadas
(UPC), Lima, Peru
Carlos.raymundo@upc.edu.pe
[3] Escuela Superior de Ingeniería Informática, Universidad Rey Juan Carlos,
Mostoles, Madrid, Spain
javier.moguerza@urjc.es

Abstract. Empirical methods play an important role in the field of geomechanics due to the recognized complexity of the nature of rock mass. This study
aims to analyze the applicability of empirical design methods in vein-shaped
hydrothermal mining deposits (narrow vein) using Bieniawski and Barton
classification systems, Mathews stability graphs, Potvin and Mawdesley
geomechanics classification systems, and mining pit dilution based on the
equivalent linear overbreak/slough (ELOS). In most cases, these methods are
applied without understanding the underlying assumptions and limits of the
database in relation to the inherent hidden risks. Herein, the dilutions obtained
using the empirical methods oscillate between 8% and 11% (according to the
frontal dimension), which are inferior to the operative dilution of the mine at
15%. The proposed model can be used as a practical tool to predict and reduce
dilution in narrow veins.

Keywords: ELOS · Dilution · Empirical methods · Narrow veins

1 Introduction

In 2014, the Peruvian Society of Geoengineering (SPEG) and the National Group of
the International Society of Rock Mechanics organized the International Congress of
Mining Design by Empirical Methods for the first time, wherein international keynote
speakers, such as Pakalnis [1], Potvin [2], and Villaescusa [3], marked their presence.
This event reached out to both the national and international mining sectors since the
lectures purely focused on the empirical methods of design. In the last 30 years,
empirical derivations have attracted considerable attention due to the predictive
capacity that they provide for the study of rock masses [1].

© Springer Nature Switzerland AG 2020
W. Karwowski et al. (Eds.): AHFE 2019, AISC 971, pp. 435–445, 2020.
https://doi.org/10.1007/978-3-030-20494-5_41

Currently, the most commonly used rock mass classification systems the rock mass rating (RMR) [4] and Q system [5]. These systems are used more than any other rock mass classification system [6]. On the contrary, empirical design methods based on the Mathews stability graphs [7], Potvin [8], Mawdesley [9], ELOS-based mining dilution [10], and the dilution percentage based on ELOS [11] are present. However, these systems have a series of limitations that should be overcomed to assess its applicability in any particular environment.

In Peru, the mining sector has emerged as one of the most relevant and growing industries with great geological potential. The Fraser Institute, Canada, has refereed Peru as a country with great mining geological potential owing to the large reserves of minerals, making it the second most attractive mining country in Latin America. In the southern coastal batholith of Peru, the Nazca–Ocoña gold belt features vein-shaped hydrothermal deposits (narrow vein) with a high-grade free gold content of 0.5–12 g/ton. To the best of our knowledge, the application of empirical methods to narrow veins in Peru has not been reported yet. This research is divided into the following sections: proposed model development, modeling, validation of results, and Conclusions. Herein, a model was implemented to assess the applicability of empirical underground mine design methods in these types of deposits.

2 State of the Art

Since the time when Pakalnis proposed the empirical methods, these methods have faced many detractors and supporters. The empiricists believe in the complexity of the rock mass and its interaction with the developed structures, which are implicitly worked upon using these methods to overcome the limitations and the uncertainties that are inherent to the study of the rock mass [12]. According to Melo [13], the number of case studies directly influences the quality of results according to the empirical method used.

Empirical methods such as the RMR system, which was proposed by Bieniawski [14], classifies the on-site rocks used in the construction of tunnels, excavations, slopes, and foundations. Q system, which was proposed by Barton [5], estimates the geotechnical mass parameters and designs tunnel and underground excavation supports. Godwin [15] claims that these systems are widely used as tools for rock mass characterization and for estimating excavation support requirements for the design and construction of tunnels, in particular in the case of tunnels excavated in hard rock masses and diaclasses.

Q, RMR, and rock quality designation (RQD) geomechanics classification systems are the most commonly used and accepted rock mass classification methods [6]. In addition, there are many cases where the RQD is the only rock mass classification index [16]. Therefore, a study was performed to describe the key aspects of RQD determination. The main objective of the rock mass classification systems is to divide a particular rock masses into groups of similar behavior and to provide a basis for understanding the characteristics of each group, thus obtaining quantitative data for engineering purposes [17].

On the contrary, empirical design methods were originally developed for the initial assessment of stability in pre-feasibility project stages. Currently, this method has become an established empirical design tool used in all mining pit sizing stages globally. However, the system has a series of limitations that must be overcome to assess its applicability in any particular environment. Over the years, the application and limitations of the method for the design of open pits have been studied by several authors [18–20].

The stability graphs methods are an empirical design tool where the size of the excavated geometry, competition of the rock mass, and stability of the excavation are related. Mathews [7] developed graphs based on on-site stresses and excavation geometry. The stability number (N') is one of the two parameters used by the graphical method proposed by Mathews to delimit the different stability zones determined in the stability graphs. The water reduction factor at the joints and the stress reduction factor are both assumed to be one. The second parameter is the shape factor or hydraulic radius defined as the ratio between the surface area and perimeter.

The Mathews stability graph method was modified and improved. The data were expanded, and a modified stability graph was proposed [8]. Additionally, the transition zone was also remarkably reduced. However, the adjustment factors used are different from those originally proposed by Mathews [7]. The term "caved" used in the stability graph refers to an unstable area instead of its common literal meaning of subsidence. Finally, we have the modification established by Mawdesley and Trueman [9], wherein three zones were proposed within the graph, which are stable, fault, and major fault. Further, 400 cases were considered for this graph with all the cases being developed according to the original factors defined by Mathews. The stability graph contributes to mining pit size determination and is used for wide mineralized bodies with the purpose of controlling dilution in large-scale mining. The graph is qualitative, and pits can only be described as stable, unstable, or subsiding. ELOS is quantitative but it is only applicable at average fault depths. Likewise, the graph can also be used for narrow vein deposits [11].

The ELOS, which was proposed by Clark and Pakalnis [10], is a measure of empirical dilution independent from mining width. ELOS approaches the quantification of dilution. Less competent and lower stress conditions tend to produce higher ELOS [2].

Recently, sublevel exploitation methods have been modified as an extraction strategy in competent underground mining. In Australia, 70% of the underground metal mines used sublevel and open stopes [21], and several other methods have been used for narrow veins and different types of sublevel and subsidence exploitation to extract ore [22].

The selected mining method excludes other options in terms of safety, productivity, recovery, and dilution control. The main objective of the design and planning of underground mines is to develop the project according to the company's business objectives and operate within the specific performance criteria to minimize financial and operational risks [3].

Since dilution in narrow veins is directly related to the pit width, establishing an optimum width is a very important parameter for predicting or analyzing dilution in narrow veins [23]. Ore dilution has an important influence on cost and on the viability of a mining operation. The costs associated with the dilution of sterile ore and ore below cut-off grade are deeply rooted in all mining and milling stages. Several research efforts have attempted to understand the ore dilution causes. Pit geometry was identified as one of the main factors that influence the ore dilution related to overbreak of pit walls [24].

Therefore, the following article is aimed at controlling the dilution of narrow veins of Caravelí's Capitana mine through the applicability of empirical methods.

3 Collaborations

The application of empirical methods in narrow veins is challenging and unique because no historical records have been reported in Peru.

The work methodology of this study is similar to the one suggested in the guidelines established by Pakalnis [1]. The proposed methodology consists of two stages:

(1) Stress analysis, rock mass classification, estimation of induced stresses, and subsequent assessment in Mathews [7], Mawdesley [9], and ELOS [10] stability charts.
(2) Impact of the design: For these purposes, three synthetic cases of 1.8 × 1.8 m, 2.1 × 2.4 m, and 2.4 × 2.4 forward drifts are made with different geometries and different stress conditions. Here, the stress factor (A), structure orientation factor (B), and the gravitational component factor (C) used both Mathews and Mawdesley graphs to observe the impact generated in the stability graph based on the N' and the maximum permissible hydraulic radius (RH). Likewise, the zones are located in the Clark and Pakalnis chart [10] in relation to ELOS. After this, ELOS is taken and expressed according to Suorineni (2015) [11] (Fig. 1).

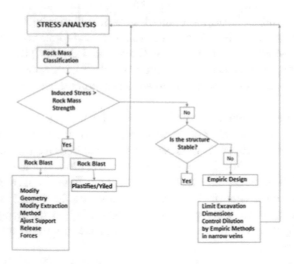

Fig. 1. Applied methodology

Given that the geomechanics information and geometrical characteristics of the mined sections of the Esperanza II vein are available, the parameters used by the stability graphs method and the extended stability graphical method of Mawdesley and Trueman may be assessed for obtaining hydraulic radii and stability numbers for each of the mined pits. This includes the geomechanics characterization and the empirical design of the pits, according to the current conditions of the Esperanza II vein, Capitana mining unit, Caravelí Mining Company, from the viewpoint of the on-site stress, dry conditions without groundwater, structural geology, and the rock mass structure itself. This information will be used to assess and determine the parameters that characterize the rock mass.

4 Validation

The development proposes using empirical methods for the assessment of the dilution in the Esperanza II vein, Esperanza mining unit, Caravelí Mining Company, located in the Department of Arequipa, Caravelí province. Previously, in the Capitana mining unit, a geomechanics study was conducted consisting of the assessment of the current state of the mine, the geomechanics characteristics, and the pit openings and sizing to ensure adequate stable conditions for the excavation's cavities associated with deep mining at the Esperanza II vein. Currently, there are no previously reported empirical method studies to control dilution at the Capitana mining unit, Caravelí Mining Company.

The exploitation method applied is cutting and filling, which presents dilution issues reaching values of up to 15%. To assess the applicability of these methods, a geomechanics study must be conducted for the narrow vein to classify and characterize the rock mass. Additionally, dilution values must be obtained by empirical ELOS-based methods [10]. Empirical methods may be applicable in narrow veins to control dilution since this directly affects metallurgical recovery, therefore increasing concentrate production costs.

4.1 Initial Values

The geomechanics zoning was executed by means of two methodologies. RMR was performed by assigning values to the five parameters involved, which are compression; strength; RQD; discontinuity spacing; and physical and geometric discontinuity conditions, such as opening, continuity, roughness, current wall, and filling states, in addition to the presence of water. While zoning the Esperanza vein, three different rock qualities were identified, namely good, fair, and bad or their respective GSI equivalents of moderately fractured/regular (F/R), very fractured/regular (MF/R), and extremely fractured/very poor (IF/P) (Fig. 2).

Competent quality rock (61 RMR) is determined by a compressive strength of 100 MPa, RQD in the range of 75–90%, spacing of up to 0.60 m, with slightly altered rough walls without the presence of water. This band of rock quality constitutes a large part of the mineralized structure, central part, and east side in the roof cavities of the vein. This rock quality must be supported with occasional safety struts.

Regular quality rock (45 RMR) has a strength of less than 100 MPa, RQD in the range of 25%–50%, fracture spacing up to 0.20 m, with slightly rough walls,

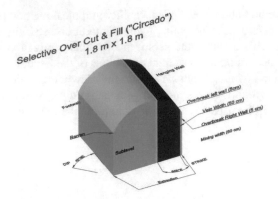

Fig. 2. Selective over cut & fill

moderately weathered and without water. This quality of rock extends as a band parallel to the floor of the mineralized structure on the western and central side of the vein. This rock must be supported by safety struts spaced at 1.50 m.

Poor quality rock (28 RMR) has a strength of less than 25 MPa, RQD less than 25%, fracture spacing less than 0.06 mm, and exhibits high-level alterations. It is presented as two parallel bands to the mineralized structure. One is on the mine floor, basically constituting the NW gable on the western and central side of the vein, and the other on the roof of the vein, on the west side presenting itself as a false cavity and constituting the area of greatest instability since it is prone to rock falls. The support should be with safety struts and wooden templates immediately spaced at 1.0 m.

The characterization of the rock mass indicates that the roof and floor rock cavities have an RMR that ranges from 30 to 62, typifying it as a poor to good grade rock. The RMR of the vein varies from 20 to 45, typifying it as a poor to regular grade rock. With these rock mass qualities, the applicability of the overhand cut and fill exploitation method is determined. Likewise, in pits where poor rock qualities are present, supports will be installed according to results from the geomechanics assessment.

4.2   Rock Mass Characterization and Classification

Empirical methods commonly used in the mining industry owing to their practical use in the design of mining structures [1]. They are a good reference for geomechanics design and should be used as a guide and always be subject to the local conditions of each mine or underground excavation. However, Pakalnis recommends that final designs should always be validated using analytical and/or numerical methods as verification methods. Prior to mine design by empirical methods, the rock mass should be classified through Bieniawski's RMR [25]. The data obtained from the mine such as the RMR were found using the current Bieniawski table [14]. Therefore, new RMR values were found with the first Bieniawski table.

With the data obtained from the geomechanics mapping of the Esperanza II vein (Level 1980) in Sect. 4.1, the RMR76 tables were prepared (Table 1).

Table 1. Comparison of results obtained

Area	RMR89	RMR76	Type of rock
1	61	66	Competent rock
2	45	42	Regular rock
3	28	23	Bad rock

The values show there are rocks with different quality, competent, regular, and poor. In this way, the values obtained from RMR76 will be used (Table 2).

Table 2. Field data according to assessed areas

RMR76	RQD	Spacing	Rugged wall
66	75–90%	0,6 m	Slightly changed
42	25–50%	0,2 m	Slightly rugged
23	<25%	<0,06 m	Extremely changed

5 Discussion

5.1 RMR Tables

In the three study areas, RMR studies were performed to know their classification, taking into account the three drifts with different dimensions. For the empirical mine design, the table used for the geomechanics rock mass classification is the one created by Bieniawski in 1976 and not the one improved in 1989. In these two tables, the main differences are the discontinuity spacing and underground water data. The current values of the Esperanza II, level 1980, geomechanics study were the values from the 1989 table:

RMR = 61, 45, and 28 for zones 1, 2, and 3, respectively.

Therefore, with the data obtained in the field, we proceed to find the 1976 RMR table, used by Pakalnis in their empirical methods, and thus verify the values obtained:

RMR = 66, 42, and 23 for zones 1, 2, and 3, respectively.

It was evidenced that the dilutions in a drift of section 1.80 × 1.80 m (vault) with an effective advance of 1.62 m. When increasing the area of the drift, the dilution obtained is lower than those found with a smaller width. Then, it is concluded that a higher dilution is reported for smaller areas by applying the selective method of overhand cut and fill.

For the different hydraulic radii taken according to the mining width of 60, 65 and 70 cm located at zones 1, 2 and 3, respectively, the locations of these points show us that they are not far away and that their average values may be used to improve correlation. However, the zones according to the quality of the rock do influence the location of the points [7, 9]. For this reason, a good geomechanics mapping is necessary to locate the points where the different zones between the stable and the transition zone are evident.

The Mawdesley and Trueman chart is much more accurate and provides a more accurate point location for the different zones. It also provides lines of fault isoprobability, which improves understanding of the rock mass and its possible collapses in certain areas previously assessed.

The assessment between the Diorite and Granodiorite rocks evidenced in the studied area did not present much influence since they have similar physical and mechanical characteristics, and there are no major differences in the data that may affect the location of the corresponding points in the Mathews or Mawdesley and Trueman stability graphs.

The hydraulic radii for narrow veins exhibit relatively small values and are not observed well when plotting points in the ELOS graph. The abacus lines may be extended in the Clark and Pakalnis [10] graph for a better understanding and approach of the Exact ELOS Values.

The dilution obtained by ELOS is lower than the values obtained by the 15% operative dilution. In terms of dilution percentage, they are lower with an error percentage of ±4%.

The induced stresses values recorded were 5.41 and 3.98 MPa in the roof and floor cavities, respectively.

6 Conclusions

The use of empirical methods as a dilution control method proved to be useful and applicable in the narrow vein scenario. However, there should be more geomechanics assessments of different mines with similar conditions to record historical data and should be able to create a specific correlated graph like the ELOS-based Clark and Pakalnis graph.

The operational dilution of the Esperanza II vein, Capitana mining unit, is 15%. At the time of applying the empirical mine design methods, dilution percentage was reduced to 8%. It was evidenced that the quantitative ELOS measures can be taken as a percentage, as required, through the conversion proposed by Suorineni [11].

In the proposed scenario, it was possible to plot the corresponding sections in the Clark and Pakalnis graph [10]. Based on this, it was evidenced that a smaller section is more stable and safer in terms of its dilution, according to the ELOS (Fig. 3).

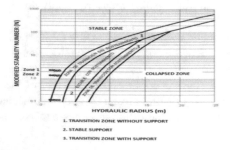

Fig. 3. Modified stability graph (Mawdesley & Trueman) applied to narrow veins

The empirical design approach will be used to calibrate existing databases. For the next analytical and empirical approaches, they have been modified according to the observed behavior of the mine. The tools have been used successfully to control the levels of dilution, predict pits stability, and limit their dimensions. The design methodology must identify the potential stress, the structural instability, and the rock mass. Therefore, the stability and ELOS graphs in the empirical design only represent a part of the overall design process. It is essential that empirical tools be used to predict the response of the rock through interpolation and not extrapolation when there is minimal data available. The approaches presented in this document should be used as a tool to increase the methodology used by the professional with his own database and decision-making process, through which a viable solution may be reached. The relations developed in this study will help operators/engineers to identify possible concerns, thus developing a safer work environment. The empirical methods have followed design guidelines, which have been implemented worldwide in partnership with researchers, mining engineers, operators, and legislators to arrive at design methodologies based on past practices, future implementations, and assessments in order to guarantee a safe and profitable mining operation. This article on empirical design is the first of many to come. Its greatest contribution may be that it has shed light on concerns from the Peruvian mining sector, specifically regarding vein-shaped deposits of hydrothermal origin (narrow veins) (Fig. 4).

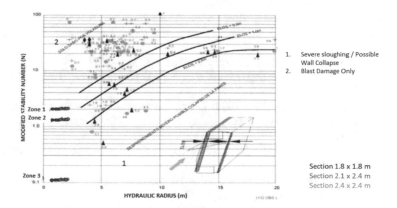

Fig. 4. Empirical ELOS estimation applied to narrow veins

References

1. Pakalnis, R.: Empirical design methods in practice. In: Proceedings of the International Seminar on Design Methods in Underground Mining, pp. 37–56. Australian Centre for Geomechanics (2015)
2. Potvin, Y., Grant, D., Mungur, G., Wesseloo, J., Kim Y.: Practical stope reconciliation in large-scale operations part 2, Olympic Dam, South Australia. In: Proceedings Seventh International Conference and Exhibition on Mass Mining (MassMin 2016), May, pp. 501–509 (2016)

3. Cepuritis, P.M., Villaescusa, E.: A reliability-based approach to open stope span design in underground mining. In: Proceedings of the MassMin (2012)
4. Pells, P.J., Bieniawski, Z.T., Hencher, S.R., Pells, S.E.: Rock quality designation (RQD): time to rest in peace. Can. Geotech. J. **54**, 825–834 (2017)
5. Barton, N., Lien, R., Lunde, J.: Engineering classification of rock masses for the design of tunnel support. Rock Mech. **6**(4), 189–236 (1974)
6. Azimian, A.: A new method for improving the RQD determination of rock core in borehole. Rock Mech. Rock Eng. **49**, 1559–1566 (2016)
7. Mathews, K.E., Hoek, E., Wyllie, D.C., Stewart, S.: Prediction of stable excavation spans for mining at depths below 1000 m in hard rock. CANMET DSS Serial No: 0sQ80-00081., Ottawa (1981)
8. Potvin, Y.: Empirical open stope design in Canada. Doctoral dissertation, University of British Columbia (1988)
9. Mawdesley, C., Trueman, R., Whiten, W.J.: Extending the Mathews stability graph for open–stope design. Min. Technol. **110**, 27–39 (2001)
10. Clark, L., Pakalnis, R.: An empirical design approach for estimating unplanned dilution from open stope hangingwalls and footwalls. In: Proceedings of the 99th Annual General Meeting. Canadian Institute of Mining, Metallurgy and Petroleum, Vancouver (1997)
11. Papaioanou, A., Suorineni, F.T.: Development of a generalised dilution-based stability graph for open stope design. Min. Technol. **125**, 121–128 (2016)
12. Suorineni, F.T.: Reflections on empirical methods in geomechanics–the unmentionables and hidden risks. In: Ausrock 2014 Third Ausrock 2014: Third Australasian Ground Control in Mining Conference, pp. 143–156 (2014)
13. Melo, M.: Mining, **67**(4), 413–419 (2014)
14. Bieniawski, Z.T.: Engineering Rock Mass Classifications: A Complete Manual for Engineers and Geologists in Mining, Civil, and Petroleum Engineering. Wiley, Hoboken (1989)
15. Godwin, W.H.: Encyclopedia of Engineering Geology, May 2017
16. Zhang, L.: Determination and applications of rock quality designation (RQD). J. Rock Mech. Geotech. Eng. **8**, 389–397 (2016)
17. Fereidooni, D., Khanlari, G.R., Heidari, M.: Assessment of a modified rock mass classification system for rock slope stability analysis in the Q-system. Earth Sci. Res. J. **19**(2), 147–152 (2015)
18. Stewart, S.B.V., Forsyth, W.W.: The Mathew's method for open stope design. CIM Bull. **88** (992), 45–53 (1995)
19. Suorineni, F.T., Henning, J.G., Kaiser, P.K.: Narrow-vein mining experiences at ashanti: case study. In: Proceedings of International National Symposium, Mining Techniques of Narrow-vein Deposits, Val'dor, Que., Canada. Canadian Institute of Mining, Metallurgy and Petroleum (2001)
20. Suorineni, F.T.: The stability graph after three decades in use: experiences and the way forward. Int. J. Min. Reclam. Environ. **24**, 307–339 (2010)
21. AUSTRADE: "Underground Mining," Sydney, Australia (2013)
22. Jang, H., Topal, E., Kawamura, Y.: Unplanned dilution and ore loss prediction in longhole stoping mines via multiple regression and artificial neural network analyses. J. South. Afr. Inst. Min. Metall. **115**, 449–456 (2015)
23. Stewart, P.C., Trueman, R.: Strategies for minimising and predicting dilution in narrow-vein mines–NVD method. In: Narrow Vein Mining Conference 2008, pp. 153–164. Australasian Institute of Mining and Metallurgy (2008)

24. El Mouhabbis, H.Z.: Effect of stope construction parameters on ore dilution in narrow vein mining. Doctoral dissertation, McGill University Libraries (2013)
25. Bieniawski, Z.T.: Rock mass classification in rock engineering applications. In: Proceedings of a Symposium on Exploration for Rock Engineering, vol. 12, pp. 97–106 (1976)

Management Approaches in Contemporary Enterprise

Management of Anthropopression Factors in Poland in the Context of the European Union Waste Economy

Jozef Fras[✉], Ilona Olsztynska, and Sebastian Scholz

Faculty of Management Engineering, Poznan University of Technology,
Strzelecka 11, 60-965 Poznan, Poland
Jozef.Fras@put.poznan.pl, ilona_olsztynska@o2.pl,
Scholz.S@outlook.com

Abstract. The modern man's living environment is shaped by many factors, including the factors of anthropopressure. Emissions, industrialization, urbanization, noise, waste water and waste, are some of the types of human pressure observed in the environment of our lives. With the development of societies, the amount of both industrial and municipal waste increases. Therefore, in recent years, waste management has become a subject of analysis and preventive measures. The activities carried out in the European Union to reduce the level of waste generation and storage are successful. The policy of the "recycling society" and the "closed circuit economy" is being implemented. Poland has also taken steps to improve waste management.

The aim of the article is to present the analysis and assessment of the current situation in the field of waste management in Poland, compared to other European Union countries, paying attention to the method of waste management, including their transboundary movement.

Keywords: Waste · Waste management · Cross-border shipment of waste · Recycling · Waste processing · Landfills · Anthropopressure · Management

1 Introduction

All forms of human activity, aiming at the use of natural resources for human needs, are determined by anthropogenic factors. These include agriculture, forestry, hunting and fishing, industry, mining, construction, communication, water regulation, tourism, as well as military activities [2]. Man, by factors anthropopression realizes its needs, including basic needs, such as get food, tools, the need for living, safety, development of personality, etc. These factors cause a number of specific impacts on the environment, this is the type of human pressure which may include: emissions of air pollutants, sewage and water pollution, littering, changing water relations, eroding the soil cover, noise emissions, destroying the plant cover, eliminating animals and other [2]. These factors should be controlled in order to limit the size and scope of their impact, which is implemented, among others through legal regulations. Today, an important role is played by the size and intensity of the induced emission anthropopressure air and noise pollution, water pollution, waste and littering [2].

© Springer Nature Switzerland AG 2020
W. Karwowski et al. (Eds.): AHFE 2019, AISC 971, pp. 449–458, 2020.
https://doi.org/10.1007/978-3-030-20494-5_42

With the increase in the number of city residents and changing consumption patterns, solid waste management has become a matter of growing global concern [11]. In industrialized countries, the determinants of the development of the importance of solid waste management are public health, environmental protection, scarcity of natural resources, climate change, and above all awareness and public participation. The situation is different in developing countries, where we have to deal with the uneven economic growth and urbanization, as well as diverse cultural and economic diversity of the society and the political system of the country, and international influences. All of these elements complicate the development of sustainable management systems in developing countries [11].

For many years, the European Union has been striving for sustainable development of the economy of individual regions of Europe. This is accomplished by setting targets in specific areas, including waste management, reducing greenhouse gas emissions, or increasing energy consumption from renewable energy sources in the total energy consumption of a given country and the entire European Union.

Building a society of recycling or a circular economy is one of the concepts for reducing the amount of waste by returning it to the economy. This is to suppress the existing relationship between economic growth and the production of waste affecting the environment. The challenge of the present times is to implement changes in the areas of product design, production systems as well as consumption in order to move to the low-carbon circular economy model as soon as possible.

By Decision 1600/2002/EC of 22 July 2002, the European Parliament and the Council established the sixth Community Environment Action Program which calls for the development or revision of waste legislation, including an explanation of the difference between waste and non-waste substances and objects, and to develop measures for the prevention and management of waste, including the definition of targets [7]. This is also the scope of such legal documents as: Directive 2008/98/EC of 19 November 2008 on waste, Directive 1999/31/EC of April 26, 1999 on landfill. In addition, a thematic strategy on the prevention and recycling of waste has been developed [8]. Legal regulations in the field of waste management are regulated by issues of naming and classification of waste, storage, handling, processing, utilization and recycling. The hierarchy of waste management has been specified in Directive 2008/98/EC of 19 November 2008 on waste and repealing certain directives and contains the following methods waste management:

(1) Prevention of waste formation;
(2) Preparation for reuse;
(3) Recycling;
(4) Other recovery processes;
(5) Disposal.

Waste means any substance or object that the holder discards or intends to discard or is required to discard. The amount of waste generated is influenced, among others, by demographic factors, the standard of living of residents, as well as ecological awareness of residents [13]. Total waste production per capita in 28 European Union countries (excluding mineral waste) in 2012 amounted to 1817 kg per person, which means a decrease by 7% to the level of 2004 [8]. The level of municipal waste generation per person has been reduced by 4% (until 2004) and in 2012 amounted to 481 kg per inhabitant [8].

The problem of municipal waste storage still exists in many European Union countries. This is associated with wastage of valuable resources, and can also lead to greater environmental impacts, compared with the benefits resulting even from energy recovery from biogas combustion from these landfills [10]. The impact of landfills on the health of people living in the neighborhood has also been proven [15]. That is why the European Union is aiming at the complete elimination of landfills in its policy. The amount of landfilled waste deposited in EU countries (including Iceland and Norway) decreased from 31% of waste generated (2004) to 22% (2010) (this applies to total waste excluding mineral waste, combustion products, animal and plant waste) [8].

Poland, after the accession to the European Union (1 May 2004), has taken measures to adapt its waste policy to set out the European Commission and the Council. For this purpose, the National Waste Management Plan 2014 (NWMP) was developed as well as the National Waste Prevention Program (KPZPO). Both of these documents have been developed for the entire territory of Poland. The NWMP contains provisions created based on the currently binding legal regulations, including requirements of Directive 2008/98/EC of November 19, 2008 on waste.

The starting point for determining Poland's objectives included in the NWMP was the state of waste management in 2000. At that time, in Poland, there were 139 340 thousand Mg of waste, including 13.860 Mg in the municipal sector and 125 480 thousand Mg in the economic sector (in including 1 578 thousand Mg of hazardous waste) [12].

Waste Act of 2001. Introduced the obligation to draw up plans for waste management both at the national, provincial, district and municipal level. It is required that these plans should be updated at least every 6 years.

The National Program for Waste Prevention in Poland, adopted by the Council of Ministers on 26 June 2014, also forms part of the European waste prevention and management program. Accordingly, waste is subject to recovery, treatment or disposal.

Waste is also the subject of trade, including international trade. The rules for cross-border shipments of waste are laid down in international agreements and treaties. In this regard, the Basel Convention on the Control of the Transboundary Shipments and Disposal of Hazardous Wastes is in force, Regulation (EC) No. 1013/2006 of the Parliament and of the Council of June 14, 2006 on shipments of waste (with later changes), as well as the Act of June 29, 2007 on international shipments of waste.

The aim of the article is to analyze issues related to waste management in Poland, against the background of the European Union, including cross-border shipments of waste. The information presented in this publication results from the analysis of data contained in the literature, including studies by the Central Statistical Office and Eurostat for the period 2000–2016.

The discussion was subject to information published in various available studies, monographs and conference publications, in the face of own observations regarding the management of waste by the European Union and Poland and the implementation of objectives in this area.

2 Waste Management in Poland Against the Background of the European Union Countries

The countries of the European Union are diverse in terms of both the size, the number of inhabitants, as well as the level of their wealth. This differentiation also applies to the level of production of solid waste, as well as to the way it is managed. Poland is the 9th country in Europe in terms of area (312 679 km^2) and 6 country in terms of population (38.5 million people) [5].

In the European Union in 2016, a total of 2,535 million Mg of waste was generated, of which recyclable waste accounted for 244 million Mg (approximately 9%) [9]. In the same period, Poland was in 4th place in terms of waste generation in the European Union, and the total amount of waste generated in this country amounted to 182 million Mg, including 13.5 million waste recycled (about 7%) [9]. The volume of waste generated in Poland increased by 32% compared to the level in 2004. This is due to the economic development of the country after joining the European Union and the growth of consumerism, characteristic of developing countries. The amount of waste generated in Poland compared to other European Union countries presented in Fig. 1.

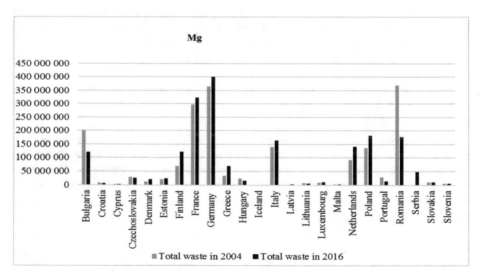

Fig. 1. Generation of waste in UE in 2004 and 2016 (quantity in Mg) [9]

In Poland, in 2016, the largest share in the total amount of waste was mineral and solid waste - about 73%, while for the European Union this share amounted to approx. 71%. Waste from combustion was also a high level - around 15.6% (to around 5% in the entire European Union).

Over the years, the share of individual categories of waste in the European Union and in Poland has changed. Some categories of waste, among others: textiles, used batteries and accumulators, or wood decreased their level compared to 2004. The categories of which the quantity increases include, among others, packaging waste, including plastics (an increase for the EU - approx. 52%, for Poland - about 660%), glass waste (growth for the EU - about 28%, for Poland - about 267%), used equipment (increase for the EU - 36%, for Poland - 306%) [9]. Such a significant increase for Poland results from the economic growth of the country and the wealth of the society.

Waste is generated in various sectors of the economy. In 2014, the amount of waste from mining and quarrying was the highest share in the European Union (around 50%) as well as in Poland (42%). The share of waste from individual sectors of the economy, such as mining and quarrying, energy, construction, industry, households for the European Union and Poland is presented in Fig. 2.

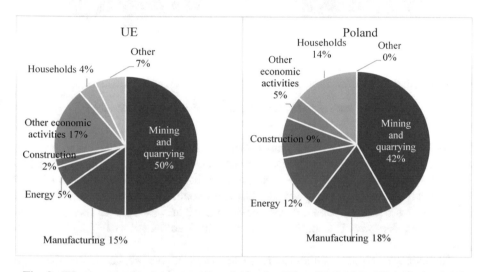

Fig. 2. Waste generation by economic activities for UE and Poland in 2014 (% share) [9]

The level of hazardous waste in the European Union remains stable (around 100 million Mg per year), which is about 4% of all waste. For Poland, this level is about 1% of all waste and it is about 1.9 million Mg per year [9]. In 2016, the largest share of hazardous waste in the European Union was mineral and solid waste (about 56%), as well as chemical and medical waste (about 27%), waste from combustion (about 12.6%) and waste equipment and equipment (about 11%). In the structure of hazardous waste for Poland, the share of individual categories was as follows: chemical and medical waste (about 47%), mineral and solidified waste (about 42%) [9].

As regards waste management, the following categories are distinguished: Recycling, Energy recovery, Backfilinig, Inciberation, Disposal. The method of waste management in Poland is close to the average for the European Union. Similarities and differences in this regard are shown in Fig. 3. It is important that the amount of waste deposited in landfills falls in the European Union. This is directly related to the improvement of the waste recycling rate, which increased from 28% (2004) to 36% in 2012 [8].

The most common method of waste management in the European Union is their disposal and recycling. Disposal of waste in Poland takes place through storage,

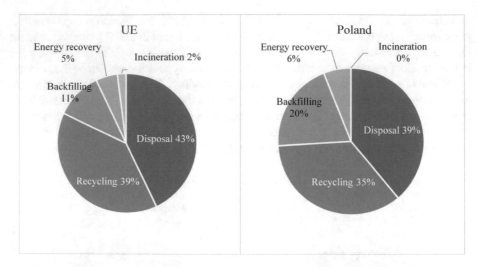

Fig. 3. Waste treatment for UE and Poland in 2014 (% share) [9]

thermal transformation, physical transformation, chemical or physic-chemical combined [13]. Unfortunately, the storage of waste (especially hazardous waste) is still a significant method of waste management, however, the amount of municipal waste sent to landfills is decreasing every year. In 2000, the amount of municipal waste deposited in Poland was 313 kg/inhabitant/year, while in 2013 it was only 157 kg/inhabitant/year. For the European Union, this indicator in 2013 amounted to 147 kg/inhabitant/year.

The average amount of municipal waste generated per capita in the European Union in 2016 amounted to 483 kg, while in Poland this amount was 303 kg/inhabitant/year. This amount increased in Poland in 2017 to the level of 311 kg/inhabitant/year [6].

3   Cross-Border Shipment of Waste in the Aspect of International Law Regulation

Waste is the subject of international trade. This is due to the fact that waste can be a reservoir of raw materials for the production of products, as well as the fact that many countries have staked on the development of recycling and technologies that process waste. Transportation of waste across national borders is regulated by a number of legal provisions. At the global level, it is the Basel Convention on the Control of the Transboundary Shipments and Disposal of Dangerous Waste, at European Union level - Regulation (EC) No. 1013/2006 of the Parliament and of the Council of June 14, 2006 on shipments of waste (with later changes), and at the national level (in Poland) - Act of June 29, 2007 on international shipments of waste.

Due to the type of hazard, the waste is classified into three groups, which have been included in the Annexes of Regulation (EC) 1013/2006. Non-hazardous waste is included in the Green List (Annex III to the Regulation). Hazardous or potentially hazardous waste, such as: batteries, asbestos-containing waste, used oils or activated glass waste, have been entered on the Amber List (this is Annex IV of the Regulation). Waste not entered on the Green and Amber List is included in Non-classified waste. These include waste from outside the lists, and waste being a mixture of different waste (the exception is mixtures of waste with code B1010 - waste of metals and B1050 - mixed non-ferrous metals).

The proceedings regarding the cross-border shipment of waste depends on the allocation of waste to groups of specified ranges. These ranges apply to the type of waste destination (for recovery or disposal), the type of List according to the requirements of Regulation (EC) No. 1013/2006, and the direction of waste shipments (within the EU, outside the EU, or outside the EU) and the country of destination (belonging to OECD, EFTA or ratifying the Basel Convention).

A cross-border shipment of waste is illegal if it is implemented in violation of any of the given elements and when the recovery or disposal method used is incompatible with European or international regulations (e.g.: the destination installation does not meet the emission standard requirements) or in violation of the procedures in force for shipments of waste from the EU to third countries and others.

In order to meet the legal requirements, two paths have been identified in the case of cross-border shipments of waste (Fig. 4).

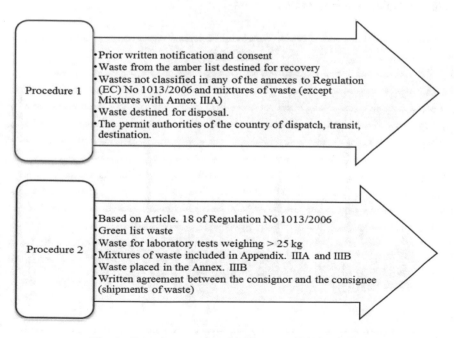

Procedure 1
- Prior written notification and consent
- Waste from the amber list destined for recovery
- Wastes not classified in any of the annexes to Regulation (EC) No 1013/2006 and mixtures of waste (except Mixtures with Annex IIIA)
- Waste destined for disposal.
- The permit authorities of the country of dispatch, transit, destination.

Procedure 2
- Based on Article. 18 of Regulation No 1013/2006
- Green list waste
- Waste for laboratory tests weighing > 25 kg
- Mixtures of waste included in Appendix. IIIA and IIIB
- Waste placed in the Annex. IIIB
- Written agreement between the consignor and the consignee (shipments of waste)

Fig. 4. Procedures to cross-border waste shipments [3]

4   Balance of Cross-Border Waste Transport in Poland

In 2016, 172 permits for the import of waste for a total mass of 720.3 thousand Mg, 50 permits for the export of waste (133.5 thousand Mg) and 15 permits for the transit of waste (142.8 thousand Mg) were issued in Poland [4]. Under the permits granted in 2016, 250.7 thousand Mg of waste were imported to Poland, 80.9 thousand Mg were exported, and transit was transported through Poland 13.3 thousand Mg [4].

The largest share of waste brought to Poland was waste from Germany (37%). Waste imports in 2016 included waste from thermal processes as well as waste from installations and devices used for waste management, from sewage treatment plants and treatment of drinking water and water for industrial purposes. The share of these categories was respectively 26% and 51% of the total amount of waste imported to Poland [5].

In 2016, the export of waste from Poland totaled 133 thousand Mg of waste, and the main target country, as in previous years, was Germany - 69% of exported waste. The real level of waste transit in 2016 amounted to 143 thousand Mg in Poland [5].

In summary, in Poland the dominance of waste imports over their exports has been maintained for years, and trade exchange in this area mainly takes place with Germany.

For comparison, the level of exports for the entire European Union has been increasing since 2004 and in 2015 it amounted to 19.26 million Mg of waste (for Poland - 93 thousand Mg) [9]. The level of exports in individual countries of the European Union is shown in Fig. 5.

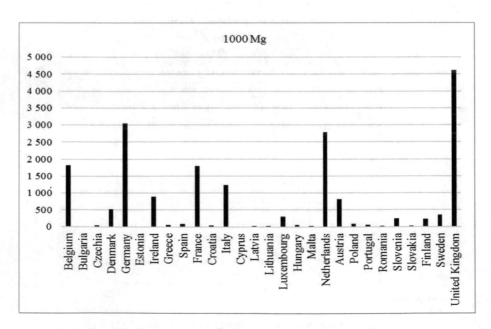

Fig. 5. Exports of waste from individual European Union countries in 2016 [9]

5   Conclusions

In order to limit the impact of the anthropogenic factor such as waste, a number of legal regulations have been established. They decide on how to handle waste, including transport, trade, processing or storage. The amount of waste is increasing in the European Union, although the share of waste is decreasing per capita.

Mining and quarrying play a significant share in the production of waste, followed by industry and energy. The level of waste generation in individual categories is changing. Currently, the increase in the amount of waste is recorded for plastics, waste glass and equipment (including electrical). While the increase in the average for the European Union amounts to several dozen percent by 2004, for Poland this increase amounts to several hundred percent. This is due to economic development that took place after the Polish accession to the European Union, as well as the growing wealth of society. The increase in the amount of waste is a characteristic of developing countries to which Poland belongs.

Thanks to the development of the industry that manages waste, the amount of waste, especially communal waste, has been significantly reduced in Poland. This was due to the development of recycling as well as the development of technologies enabling the recovery of raw materials from waste, as well as legal regulations imposing on producers the payment of environmental fees for the introduction of packaging, equipment and waste. The implementation of the idea of the Closed-Circuit Economy in various areas of the economy will be conducive to the development of national and local systems of conscious waste management in the coming years [8]. This will reduce the demand for raw materials, which will reduce the associated energy consumption and adverse environmental impacts [8]. This is also due to the development of new technologies in the reuse of waste, for example in the case of packaging and products made of synthetic polymers, such as films and other packaging (PSW) [1].

Data from recent years in the field of waste management in Poland indicate that the measures taken over the past 20 years are in line with the EU-wide trend of waste reduction. Certainly, better recycling technologies, improved infrastructure and more effective collection of secondary raw materials can improve this situation, while reducing pressure on the environment and dependence on extraction or import of raw materials [14]. For several years, more and more installations for fuel formation have been created in Poland [16], which is somewhat a response to the growing deficit of industrial fuels [17].

An area requiring further transformation is the method of dealing with waste in Poland. The level of recycling of waste is still below the average of the European Union countries, while the level of waste dumping is almost twice as high as the EU average. The same applies to the level of waste storage. In the management of waste already deposited on landfill sites and still emerging (especially municipal), incinerators in Poland will help. It is difficult to say how in the long run it will affect the level of recycling, because in some EU countries excessive processing capacities of incinerators are competition for recycling, that is, they make it difficult to pass to the higher levels of the waste hierarchy [8].

References

1. Al-Salem, S.M., Lettieri, P., Baeyens, J.: Recycling and recovery routes of plastic solid waste (PSW). Waste Manag. **29**(10), 2625–2643 (2009)
2. Balon, J., Maciejowski, W.: Geoecology for Landscape Architects, Institute of Landscape Architecture, Cracow University of Technology, Krakow (2012)
3. Chief Inspectorate of Environmental Protection: Transboundary Shipment of Waste - Collection of Regulations. Advertising and Publishing Agency A. Grzegorczyk, Warsaw (2015)
4. Chief Inspectorate of Environmental Protection. Waste Management (2014). http://www.gios.gov.pl
5. Central Statistical Office (CSO). Environment 2016. Regional and Environmental Surveys Department, Warsaw (2017)
6. Central Statistical Office (CSO). Environment 2017. Regional and Environmental Surveys Department, Warsaw (2018)
7. Decision 1600/2002/EC of the European Parliament and of the Council of 22 July 2002 establishing the Sixth Community Environment Action Program. Journal of EU Laws, No. 242 of 10 September 2002 (2002)
8. EEA.: Environment Europe 2015 - State and outlook. Synthesis. European Environment Agency, Copenhagen (2015)
9. Eurostat: Generation of waste. http://appsso.eurostat.ec.europa.eu/nui/show.do?dataset=env_wasgen&lang=en
10. Jeswani, H.K., Azapagic, A.: Assessing the environmental sustainability of energy recovery from municipal solid waste in the UK. Waste Manag. **50**(3), 346–363 (2016)
11. Marshall, R.E., Farahbakhsh, K.: Systems approaches to integrated solid waste management in developing countries. Waste Manag. **33**(4), 988–1003 (2013)
12. National Waste Management Plan 2014 (NWMP). Resolution No. 217 of the Council of Ministers of 24.12.2010 M.P. No. 101, pos. 1183 (2010)
13. Rosolak, M. Gworek, B.: State and assessment of waste management in Poland. Protection of the Environment and Natural Resources, 29, Warsaw (2006)
14. Rozalski, J.: Energetic use of waste as a remedy for fuel gaps and the solution of waste disposal. New Energy. Thematic Appendix **1**(2) (2009). https://nowa-energia.com.pl/2009/04/14/energetyczne-wyżytanie-odpadow/
15. Rushton, L.: Health hazards and waste management. Br. Med. Bull. **68**(1), 183–197 (2003). https://doi.org/10.1093/bmb/ldg034
16. Wandrasz, J.W.: Processes of burning fuels from waste as part of municipal waste management. New Energy. Thematic Appendix **1**(2) (2009). https://nowa-energia.com.pl/2009/07/04/procesy-spalania-paliw-z-odpadow-jako-element-gospodarki-odpadami-komunalnymi/
17. Wandrasz, J.W., Wandrasz, A.J.: Formed Fuels: Biofuels and Waste Fuels in Thermal Processes, p. 2006. Seidel-Przywecki, Warsaw (2006)

Human Factors/Ergonomics in eWorld: Methodology, Techniques and Applications

Oleksandr Burov[(⊠)]

Institute of Information Technologies and Learning Tools,
National Academy of Educational Sciences of Ukraine,
9 M. Berlyns'koho st., Kyiv 04060, Ukraine
ayb@iitlt.gov.ua

Abstract. The analysis of ergonomic properties has been made in relation to ergonomic evaluation of objects of digital world. The proposal is to extent three recognized domains of Ergonomic (physical, cognitive and organizational) by the new one: informational. Some appropriate related topics are proposed. Information/cyber security issues are formulated as ergonomic objects in the digital environment. It is propose to add to four recognized general ergonomic properties (learnability, serviceability, controllability, inhabitability) to add resilience of human-system integration as one more ergonomic property. The technique to measure "ergonomicity" are discussed with the appropriate 4-point scale that is based on use of 4 levels of ergonomic indices: integral complex, group and single.

Keywords: Human factors · Ergonomic properties · Ergonomic domains · Ergonomicity scale

1 Introduction

Ergonomics discipline promotes a holistic, human-centered approach to work systems design that considers the physical, cognitive, social, organizational, environmental, and other relevant factors independently on time and space of a human activity, and on a particular technology used [1]. It is world-recognized that changes in technologies led to that we live in digital world (eWorld), where not only compatibility of all components of the human-machine-environment system (HMES) can be critical, but the processes inside the HMES (information exchange, production of new information, etc.) play more and more significant role. Information obtained by the human-in-the-loop needs to be described not only by volume and flow rate ("external" characteristics), but its (cognitive) content ("internal"). This is accompanied by changes in the nature of HMES components where information became both tool ("machine") and environment at the same time. Besides, the network, where the human and the system activity are carried out, has new features [2]. Information content became the tool to impact on the human and (as a tool in HMES) could be an object for ergonomics intervention.

It is recognized that the System changes its feature over last decades [3], especially in digital space, and a human individual cognitive, creative and critical abilities became crucial for the humanity civilization [4]. Today's children were born, grow, study,

W. Karwowski et al. (Eds.): AHFE 2019, AISC 971, pp. 459–464, 2020.
https://doi.org/10.1007/978-3-030-20494-5_43

master their occupation, live and work in the world is increasingly losing the features of the material world and turning into the world of information and knowledge [5]. Taking into account the life-long learning trend in job market's needs, education and work become the mixed (to some extent) system of a new type – the system for production of knowledge and human talent as the intellectual capital. Because such a system has its own structure and functions, the general system performance can be described in terms of the systemic-structural activity theory [6].

Purpose. To analyze features and specifics of ergonomics challenges in the digital age as well as a human ergonomic needs in the eWorld.

2 Discussion of Results

The transformation of the role of information networks, their place in life led to a shift of attention of networks' designers towards human-centric nature of their creation and existence, the emergence of the need to use the concept of not only the "integration network", but the concept of "integrated person-centric network" with its corresponding features [2]. Integrated network is not a new type of passive element of innovative processes and active, because it is much more clearly manifests the changing nature of modern art, which is "currency" with its own laws of formation, development, traffic and the need to protect [7], especially when working in information environment [8].

It is recognized that the most often cited domains of specialization within HFE are physical, cognitive and organizational ergonomics (Fig. 1), where cognitive ergonomics focuses on mental activity. Above mentioned challenges can look like the cognitive ergonomics' domain, but the latter focuses on mental processes, and work in digital environment deals with content of the mental activity rather than with only the process of information flow. Ergonomics from the past to the present dealt with material world outside a human (even if mental processes reflected that world). Today's life and activity aims to a human brain and cognitive model of the world, because they produce new facts, new information and knowledge that can change the material world (mWorld).

At the same time, Ergonomics can be interpreted as the scientific discipline that study human, tools, environment and their interaction in activity (Fig. 2).

Environment is considered as not only natural (physical, chemical, biological) one, but item (human made), organizational, psychological, informational. Respectively, above mentioned Ergonomics domains can be associated with one or two types of environment. But in digital world we faced new challenges related to the specific role of information that represents human, tools/means and environment at the same time, and that cannot be separated from a human (as an external object of his/her activity).

"Information ergonomics" can be discussed as a new specialization within Human Factors/Ergonomics, and its relevant topics are: parameters of information stream perceived by the senses, emotional importance of the information for human activity, perceived and unperceivable by consciousness, density and pace, controlled and uncontrolled. The ergonomic problem is the possibility of assessing the hazard of

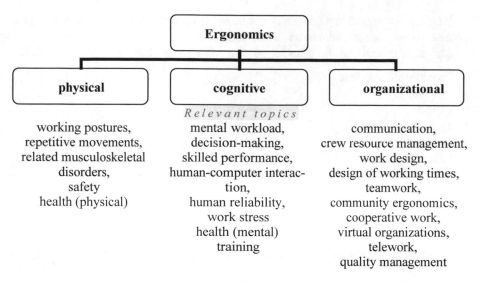

Fig. 1. Ergonomic domains and relevant topic (https://iea.cc) in digital world.

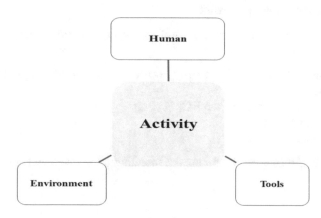

Fig. 2. Subject of ergonomics

information for a human life and activity, as well as the possibility of protecting against it or mitigating its negative impact, as well as to develop the information security culture [9].

Depending on the means of action, the problems (and appropriate means) of information/cyber security can be classified into five groups:

- Legal,
- Technical,
- Information,
- Organizational,
- Psychological.

The legal issues of cybersecurity are handled by specialized experts and organizations, so they are not addressed in this article.

Information tools can be categorized according to the tasks solved by the users:

- Protection/Remedies,
- Informing,
- Content,
- Learn how to use,
- Security,
- Life-span,
- Avoiding threats.

In the broadest sense possible targets for the impact of information/cyber security (in addition to critical infrastructure objects) can be:

- Databases
- Personal data, including financial
- Mass media
- Social networks
- Education and Training
- Textbooks, Historiographic editions.

Organizational tools for solving information/cyber security issues:

- informing,
- learning the culture of cybersecurity, professional staff of KB and the general population;
- creation of special means of the BC,
- distribution of KB facilities,
- control of use.

Psychological means can be grouped depending on the personal and interpersonal level:

- national,
- public,
- group,
- individual,
- cultural,
- cognitive,
- intellectual,
- habits.

One of the possible ergonomic ways to assist HMES design is discussed accounting above mentioned features of information with regards to a human safety, efficiency and comfort (wellbeing). Special attention in this regards should be paid to a human-integration system from point of view of the human.

How to measure the "ergonomicity" (ergonomic quality) of the HMES design for a human digital activity in eWorld?

It is proposed to discuss HMES' ergonomic properties (learnability, serviceability, controllability, inhabitability, resilience as well as cognitivity) and ergonomics indices to measure them (integral "ergonomicity", complex indices, group indices and single indices).

It is proposed the technique that uses measurable single indicators and those assessed by questionnaire (united in special group indices), group indices combined in complex indices ("ergonomic properties") accounting weights, and calculation of the integral "ergonomicity" (Table 1) that is normalized on a scale [0, 1]. This technique has been implemented in the form of ICT tools.

Table 1. Ergonomic indices

Level	Name				
Integral	Ergonomicity				
Complex	Ergonomic properties				
	Learnability	Serviceability	Controllability	Inhabitability	Resilience
Group (examples)	*Learning time, understanding of instruction, additional question need*	*Access to necessary tools, convenience*	*Visibility, opportunity to change parameters*	*Interface comfort, optimal parameters for human sensors*	*Ability to restore performance after damage*
Single (examples)	*Longevity of effective learning, scope of instruction*	*Number of objects in working field, access to adjustment*	*Optimal vision of necessary items, control panel*	*Comfort colours palette, size of items*	*Time to re-start, full recovery*

After calculation of number of critical single indices, the group, complex and integral indices are calculated in relation to the maximal numbers of corresponding indices. The integral ergonomicity is evaluated as the 4-point scale:

Value	Ergonomicity
0.75 … 1.0	Corresponds the best ergonomic designs
0.5 … 0.75	Good ergonomicity, but some parameters need to be improved
0.25 … 0.5	Relative ergonomicity. Need to re-design
<0.25	Unacceptably low level of ergonomics

Similar evaluation are made for every ergonomic property and could be used to compare the particular object of evaluation with competing ones, as well as for improving of ergonomic quality of the object.

3 Conclusion

The analysis of ergonomic properties has been made in relation to ergonomic evaluation of objects of digital world. The proposal is to extent three recognized domains of Ergonomic (physical, cognitive and organizational) by the new one: informational. Some appropriate related topics are proposed.

Information/cyber security issues are formulated as ergonomic objects in the digital environment.

It is propose to add to four recognized general ergonomic properties (learnability, serviceability, controllability, inhabitability) to add resilience of human-system integration as one more ergonomic property.

The technique to measure "ergonomicity" are discussed with the appropriate 4-point scale that is based on use of 4 levels of ergonomic indices: integral complex, group and single.

Acknowledgments. This research has been supported by the Institute of Information Technologies of the National Academy of Pedagogic Science.

References

1. Karwowski, W.: The discipline of human factors and ergonomics. Handb. Hum. Factors Ergon. **4**, 3–37 (2012)
2. Burov, O.: Virtual life and activity: new challenges for human factors/ergonomics. In: Symposium Beyond Time and Space STO-MP-HFM-231, STO NATO, pp. 8-1–8-8 (2014)
3. Wilson, J.R.: Carayon Pascale: systems ergonomics: looking into the future – editorial for special issue on systems ergonomics/human factors. Appl. Ergon. **45**(1), 3–4 (2014)
4. Strategies for the New Economy Skills as the Currency of the Labour Market. Report. World Economic Forum, 22 January 2019, Davos (2019). https://www.weforum.org/whitepapers/strategies-for-the-new-economy-skills-as-the-currency-of-the-labour-market
5. Pinchuk, O., Lyvynova, S., Burov, O.: Synthetic educational environment – a footpace to new education. Informacijni tekhnologhiji i zasoby navchannja: elektronne naukove fakhove vydannja **60**(4), 28–45 (2017). (in Ukrainian)
6. Bedny, G.Z., Karwowski, W.: A systemic-structural activity approach to the design of human-computer interaction tasks. Int. J. Hum.-Comput. Interact. **16**(2), 235–260 (2003)
7. Burov, O.Ju.: Educational networking: human view to cyber defense. Inst. Inf. Technol. Learn. Tools **52**, 144–156 (2016)
8. Da Veiga, A., Martins, N.: Information security culture and information protection culture: a validated assessment instrument. Comput. Law Secur. Rev. **31**(2), 243–256 (2015)
9. Glaspie, H.W., Karwowski, W.: Human factors in information security culture: a literature review. In: Nicholson, D. (ed.) Advances in Human Factors in Cybersecurity, AHFE 2017. Advances in Intelligent Systems and Computing, vol. 593. Springer, Cham (2018)

Production, Quality
and Maintenance Management

Production Management Model Based on Lean Manufacturing for Cost Reduction in the Timber Sector in Peru

Fiorella Lastra[1](✉), Nicolás Meneses[1](✉), Ernesto Altamirano[1](✉),
Carlos Raymundo[2](✉), and Javier M. Moguerza[3](✉)

[1] Escuela de Ingeniería Industrial, Universidad Peruana
de Ciencias Aplicadas (UPC), Lima, Perú
{u201417699, u201317454, pcinealt}@upc.edu.pe
[2] Dirección de Investigación, Universidad Peruana
de Ciencias Aplicadas (UPC), Lima, Perú
carlos.raymundo@upc.edu.pe
[3] Escuela Superior de Ingeniería Informática,
Universidad Rey Juan Carlos, Mostoles, Madrid, Spain
javier.moguerza@urjc.es

Abstract. At present, timber is the only commodity whose demand will increase at worldwide levels. Peru, despite being one of the countries with the highest forest potential, cannot compete with countries such as Brazil and China due to high production costs. Therefore, the aim of this article is to develop a production management model based on Lean Manufacturing techniques to increase production capacity by improving processes and reducing costs. For this, Knowledge Management, Change Management, and Production Management were implemented. The model was validated in a Peruvian timber company, where a 49% reduction in the cleaning and organization time was achieved. Calibration periods were reduced by 61%, and preventive maintenance periods by 72%.

Keywords: Lean Manufacturing · SMED · Preventive maintenance · 5S · Wood · Knowledge Management · Change Management

1 Introduction

In recent years, with respect to the main wood products, the timber industry has grown worldwide at a rate from three to six percent annually due to global economic development and the increased demand for renewable energy [1]. However, this trend is not applicable to countries such as Peru, with the ninth largest forest area on the planet where in the last 3 years, exports have dropped drastically. This has a great impact for the country, because this sector represents the third largest agro-export activity in terms of employment generation. This is mainly caused by the low competitiveness of national companies, due to high production and transport costs [2].

Faced with this need to enhance competitiveness, innovative proposals were presented, focused on improving processes and reducing costs, such as the use of Lean

© Springer Nature Switzerland AG 2020
W. Karwowski et al. (Eds.): AHFE 2019, AISC 971, pp. 467–476, 2020.
https://doi.org/10.1007/978-3-030-20494-5_44

Manufacturing techniques and of software to improve energy efficiencies along with the implementation of new tools to speed up procedures. We propose a production management model that applies SMED, 5S, and Preventive Maintenance techniques along with Knowledge Management and Change Management.

This document is divided into sections describing the research, analysis of the literature, contribution of the model, validation of the proposed model, and the conclusions of the model application.

2 State of the Art

Ample research was performed on aspects focused on the improvement of processes and their control, which are Production Management (GP), Change Management (CG), and Knowledge Management (GCT). (The acronyms are in Spanish.)

The purpose of Production Management is to increase productivity by optimizing the relationships between inputs and outputs using methods such as Lean Manufacturing, the Theory of Constraints, Total Quality Management, Six Sigma, and others. This methodology entails the elimination of *mudas*, or wastes, such as overproduction, waiting times, transportation, excess production and inventory, movements, and defects [3]. One of the most used techniques of this methodology is the Single Minute Exchange (SMED), which decreases the periods required to establish work teams [4]. An example of this is the case presented by Trojanowska et al. of a wood panel manufacturing company that applied this technique in order to reduce their high changeover times between assembly machines. This was achieved by making changes in production line design to reduce such periods by 50% [5].

The 5S technique will also be used, which includes conducting organizational and cleaning activities as well as detecting workplace defects [6]. A success case is that of a manufacturing company in which the 5S technique was applied to organize and reduce the search times of work tools and scientific instruments. By the end of the implementation, this led to an improvement in their performance indicators by 55 points in 20 weeks [7]. Finally, the TPM technique will be applied. It focuses mainly on improving team performance to avoid time wasted when failures occur [8]. An example of this is the case presented by Singh et al. of an absorbent manufacturer in India. In this case, a CNC machine was redesigned to perform various functions, thus reducing cycle times by 41% and increase productivity by 345% [9].

Moreover, it is necessary to implement GC in order to control the development of organizational change to improve system performance [10]. A success story is the development of a hierarchical change management system for the manufacturing sector, in which the execution is much faster and handles most temporary restrictions imposed by regulators. This produced an increase from 76% to 100% regarding decision quality [11].

According to Nonaka and Takeuchi, the only companies that will achieve success will be those that create new knowledge, disseminate it, and incorporate it quickly throughout the company [12]. That is why proper Knowledge Management is important, since it will help members of the organization to access all knowledge in an orderly, practical, and efficient manner such that it can be processed and used [13].

Articles such as the ones written by Sanjiv et al. explain the implementation of this management through a CRM program. It is a database that allows the acquisition and update of customer information, with which they determined that only 3.3% of the company's clients were highly profitable. This brought about changes in their marketing strategy [14].

3   Contribution

3.1   Proposed Model

The proposed management model combines GCT, GC, and GP. In the following figure, we can have a better idea of what was explained.

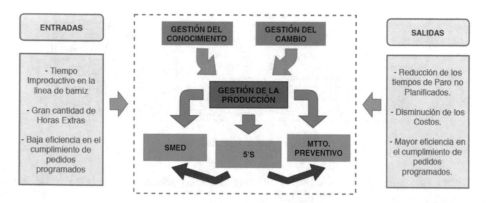

Fig. 1. General overview of the model

Figure 1 shows the application of GCT and GC in the entire GP implementation cycle. GCT will provide information such as work procedures, handbooks, and tool characteristics for the training on the techniques to be implemented. Additionally, CG will be used as a method to reduce the resistance to change of project stakeholders as much as possible.

Production management is based on 5S, SMED, and preventive maintenance techniques that are jointly focused on the same goal of reducing downtime. The contribution of the 5S technique is that when the red card is used, there will be a final arrangement to reduce production periods and create a more orderly workflow. The contribution of the SMED technique is to reduce downtimes by applying calibrations when this technique is combined with 5S. In addition, operators and coordinators will include improvement points for each procedure in the DOP format. Finally, another contribution is the use of preventive maintenance and 5S to reduce downtime caused by mechanical failures. This will be done with the classification and organization of resources in the maintenance area, which will increase tools' availability. On the other hand, mobile devices such as the iPad will be used for the optimal management and

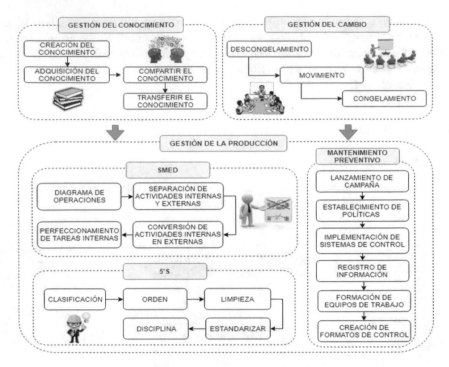

Fig. 2. Detailed process view

control of requirements and preventive maintenance schedules specifically for the implementation of preventive maintenance.

In Fig. 3, we can see the steps of each dimension of the model to be implemented. To start the implementation of techniques, the company GCT and GC will be needed. Subsequently, the 5S technique will be applied, which will support the stages of SMED and preventive maintenance.

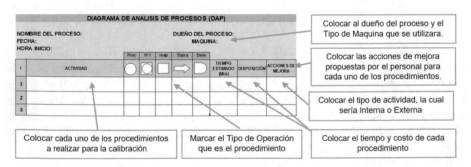

Fig. 3. Process analysis diagram form

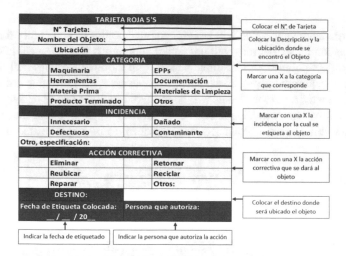

Fig. 4. 5S red card form

SMED

To implement SMED, a detailed record of each procedure will be made in a Process Analysis Diagram (DAP). In Fig. 2, we can see the form that will be used for DAP and how it will be filled out.

Subsequently, activities that need a machine to be stopped will be determined as well as those that can be done while the machine is in operation. Internal activities will be distinguished from external ones. The goal is to analyze each activity that can be performed with the machine in operation. In this manner, internal activities can be converted to external ones. To do this, improvement points proposed in the form were used and an evaluation of each procedure was added to the change acceptance form. It should be noted that at this stage, 5S was already implemented by the company; when staff started their daily tasks, tools in the closet were labeled for quick identification. Finally, the use of tool cards will facilitate the change process.

5S

To implement this stage, first, classification of the work area will be done. For this, necessary elements must be separated from unnecessary ones by using red cards for such elements. Subsequently, corrective actions will be taken to arrange them. Figure 4 shows a form that depicts this type of card-based process. Then, we will apply the usage frequency circle and an inventory form of the work area will be used to find a place for materials according to their use. Based on such information, shelves' and lockers' installation will be proposed. After this stage has been completed, a cleaning campaign for the area will be conducted and a cleaning manual will be made, which must be read by operators for its correct implementation. Subsequently, the first 3S will be standardized and work policies will be established. Finally, monthly audits will be executed to avoid any type of defect that might affect 5S operations.

TPM

A preventive maintenance plan will be implemented, and a responsibility form must be signed by personnel so that they will undertake continuous improvement in the project. Likewise, project policies will be implemented, and mobile resources will be allocated to streamline requirement processes. Then, the necessary information for the project will be documented and work teams will be trained. Finally, we will design a new form for line operators, who will have to record the various types of maintenance performed in their work shift. In the following figure, the form to be implemented is shown.

Fig. 5. Preventive maintenance form

Figure 5 shows the application of the project model, starting with process analysis and concluding with result analysis.

3.2   Validation

In recent years, any company that is part of the timber and/or its derivative industry has faced a lot of unproductive hours in the varnishing process, which could be reflected in low programming efficiencies and high production costs. Based on this, we analyzed why this occurred. These problems were caused by a large number of unplanned stoppages in the line, mainly due to long shutdown periods to conduct calibrations. The reason for this was incorrect work procedures, tool failures, and long common work stoppages. This was caused by tools that were unorganized: they lacked a defined space for each tool. Another problem was the long stoppage periods caused by machine malfunctions, which occur because there is no preventive maintenance in the company. In view of this situation, 5S, SMED, and preventive maintenance techniques were applied jointly and focused on each of the main unplanned stoppages (Fig. 6).

SMED

To implement the SMED technique, a process analysis diagram was made for calibration stoppages and a standard time of 45.60 min was obtained. The separation of the internal and external activities of the process was then performed; a standard time of 25.30 min was obtained. Subsequently, internal tasks were converted to external tasks using the 5S with tool cards to facilitate the change process. They will guide the calibrator to search for tools more quickly.

Fig. 6. Implementation flow

After the SMED tool was implemented with 5S by the company, we used the Arena software and obtained a standard time of 17.85 min. In the following figure, the SMED implementation flow is shown. As shown in the figure, during a one-year period of execution, 480 samples were obtained. The results of the analysis of the time spent showed that the value of the new indicator of change time was 89.2% of the standard established by the company [First row] (Fig. 7).

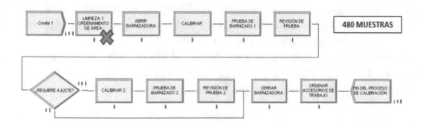

Fig. 7. SMED flow

5S

The application of the 5S technique was started with the classification of area elements. An inventory of the work area was then introduced to visualize different elements that are used based on their frequency and disposal time. After this stage, a cleaning plan was implemented, a cleaning manual was developed, and workers were assigned to different areas of the varnishing line. Subsequently, the first three Ss were standardized and work policies were established to stabilize the methodology. Finally, to complete this implementation, visual inspections and daily controls were performed to prevent any defect from affecting the implementation (Fig. 8).

Based on the above, the Arena simulator was used to calculate times for the organization and area cleaning. The goal was to reduce cleaning times and a standard time of 10.28 min was obtained: it was reduced by 49% compared to the initial standard time of 20.16 min. In the following figure, the cleaning flow is shown.

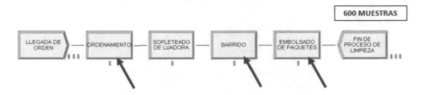

Fig. 8. Organization and cleaning flow

TPM

For the preventive maintenance proposal, the implementation began with the establishment of project policies and responsibilities. Then, we implemented mobile devices (iPads) that were distributed among bosses and coordinators of the area in order to streamline compliance with maintenance requirements. Subsequently, we collected all machine information and data on the types of technical personnel that are required for each machine. Likewise, work teams were formed. One team had three-line operators and the area coordinator. The other included the coordinator of varnishing and the maintenance coordinator. Both teams were trained for 2 weeks and their new functions as a team were explained, along with their responsibilities and their new work resources. Finally, after each worker and coordinator was trained on their work, they were held in charge of recording and writing down the compliance of the preventive maintenance of line operations. Different area maintenance tasks were planned to meet the organizational requirements in a timely manner.

After preventive maintenance was implemented along with 5S, we executed Arena software for the maintenance flow. A standard time of 13.82 min was achieved, which is a reduction by 72% compared to the initial standard time of 48.95 min (Fig. 9).

Fig. 9. Preventive maintenance requirement flow

Likewise, the preventive maintenance inspection flow was analyzed, and certain processes were eliminated. In the following figure, the preventive maintenance inspection flow is shown (Fig. 10).

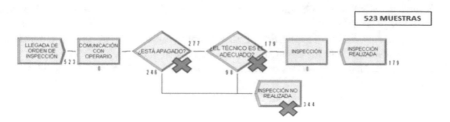

Fig. 10. Preventive maintenance inspection flow

Based on this flow, out of 523 annual samples, an 89% increase of the maintenance performance indicator was achieved.

Based on reduction times calculated for each stoppage, a total time available for floor production and processing of 26.197 min was achieved. This value corresponds to the year of implementation of the improvement proposal. The incremental value of the available time indicator for production was 77%.

4 Conclusions

A production management model was designed based on the Lean Manufacturing methodology, SMED, 5S, and preventive maintenance techniques. Its aim was to reduce high production costs of the varnishing process and to generate savings by implementing significant process improvement proposals for the company.

The contribution of this article is to present a production management model based on Change Management and Knowledge Management to improve production processes.

By applying Lean Manufacturing techniques and Arena software, it was possible to reduce calibration times by 61% and cleaning and organization times by 49%. In stoppages caused by mechanical failures, improvements in maintenance effectiveness were achieved by decreasing 72% of the management periods. An 89% fulfillment of machine maintenance and inspections was also attained.

References

1. FAO - Noticias: El crecimiento de la producción mundial de madera se acelera. http://www. fao.org/news/story/es/item/1073841/icode/. Accessed 18 Nov 2018
2. Programa «Contribución a las Metas Ambientales del Perú» (ProAmbiente)
3. Ōno, T.: Toyota Production System: Beyond Large-Scale Production. Productivity Press, Portland (1988)

4. Shingō, S.: A Revolution in Manufacturing: The SMED System. Productivity Press, New York (1985)
5. Trojanowska, J., Zywicki, K., Varela, M.L.R., Machado, J.M.: Shortening changeover time – an industrial study. In: 2015 10th Iberian Conference on Information Systems and Technologies (CISTI), pp. 1–6 (2015)
6. Sacristán, F.R.: Las 5S: orden y limpieza en el puesto de trabajo. Fundación CONFEMETAL (2005)
7. Gupta, S., Jain, S.K.: An application of 5S concept to organize the workplace at a scientific instruments manufacturing company. Int. J. Lean Six Sigma 6(1), 73–88 (2015)
8. Swanson, L.: Linking maintenance strategies to performance. Int. J. Prod. Econ. 70(3), 237–244 (2001)
9. Amin, S.S., Atre, R., Vardia, A., Sebastian, B.: Lean machine manufacturing at Munjal Showa limited. Int. J. Product. Perform. Manag. 63(5), 644–664 (2014)
10. Moran, J.W., Brightman, B. K.: Leading organizational change
11. Alrabiah, A., Drew, S.: Formulating optimal business process change decisions using a computational hierarchical change management structure framework. J. Syst. Inf. Technol. 20(2), 207–240 (2018)
12. Nonaka, I.: The Knowledge-Creating Company. Harvard Business Review Press, Brighton (2008)
13. de Deusto, U., Pazos, J., Rodríguez, E., Rodríguez-Patón, A., Suárez, S.: GESTIÓN DEL CONOCIMIENTO Anselmo del Moral
14. Srivastava, S.K., Chandra, B., Srivastava, P.: The impact of knowledge management and data mining on CRM in the service industry, pp. 37–52. Springer, Singapore (2019)

Production Management Model for Increasing Productivity in Bakery SMEs in Peru

Junior Huallpa[1(✉)], Tomas Vera[1(✉)], Ernesto Altamirano[1(✉)],
Carlos Raymundo[2(✉)], and Javier M. Moguerza[3(✉)]

[1] Escuela de Ingeniería Industrial, Universidad Peruana
de Ciencias Aplicadas (UPC), Lima, Perú
{u201413130, u201415618, pcinealt}@upc.edu.pe
[2] Dirección de Investigaciones, Universidad Peruana
de Ciencias Aplicadas (UPC), Lima, Perú
carlos.raymundo@upc.edu.pe
[3] Escuela Superior de Ingeniería Informática,
Universidad Rey Juan Carlos, Mostoles, Madrid, Spain
javier.moguerza@urjc.es

Abstract. There is a high margin of informality in small- and medium-sized companies in the bakery sector because of their lack of focus and poor standardization of their activities. Bakery SMEs, the activities of which have not yet been standardized, usually perform activities inefficiently, unnecessarily extending production times. The current average productivity of Lima-based SMEs is 1.7, a figure that, when compared with countries in the Pacific Alliance, is low. In addition, currently no methodologies seek continuous process improvement. Therefore, a Lean Process Management model was established to reduce activities and times. As validation, a production time simulation was performed in a warehouse, increasing the productivity to 2.08, with a percentage variation of 87.39% when compared with the initial productivity.

Keywords: Lean Manufacturing · Lean Thinking · Method Study ·
Process Management · Plant Layout

1 Introduction

Currently, the food industry is characterized by increasing demand and establishing procedures to guarantee food safety. This sector is immersed in the manufacturing sector, which represents 16.52% of the Gross Domestic Product (GDP) [1]. Within the food industry, the bakery industry represents 2.54% of all manufacturing industries [2]. In Lima, 65% of bakeries remain informal businesses mostly due to their lack of future focus and process organization [3]. This has triggered a reduction in the number of bakeries that open on a monthly basis. Within the Lima Greater Metropolitan Area, 56 bakeries start activities and 22 close down every month [4]. There are numerous bakeries in Lima. However, their processes are usually cost-intensive and generate losses during production. Therefore, greater control is required to reduce the number of reprocesses or downtimes during the processing and transportation of materials.

© Springer Nature Switzerland AG 2020
W. Karwowski et al. (Eds.): AHFE 2019, AISC 971, pp. 477–485, 2020.
https://doi.org/10.1007/978-3-030-20494-5_45

Peru, at the Latin American level, reports the lowest productivity level when compared with its Pacific Alliance counterparts. The average SME productivity in Peru divided by departments is as follows: Lima-Callao 1.7, Arequipa 1.7, Trujillo 1.8, Chiclayo 1.7, Iquitos 1.9, Huancayo 1.6, and Piura 1.6 [5]. At the national level, Lima reports average productivity, with 43% of all bakeries in Peru located within its borders. Currently, the Peruvian Association of Bakeries and Pastry Shops [6] regulates all working methodologies at bakeries and pastry shops and provides seminars and training sessions. In Latin America, previous studies have used Lean Manufacturing tools, such as the 5S and Kaizen, to improve productivity in companies.

The techniques used in previous literature have been assessed but have mostly failed because of the lack of support from authorities, the lack of personnel willing to implement changes, or economic limitations. We are aware that a vast majority of bakeries are informal. However, their processes could be streamlined to generate more income by increasing their competitiveness through the use of tools that could help them reorganize and adapt. Therefore, this article proposes a control alternative, which may foster thought changes in company managers and culture changes in SMEs in the food sector. The purpose of this study is to implement and evaluate a Process Management methodology in the production area of a bakery in order to optimize cost, time, and material transportation by reducing unnecessary activities. This methodology is expected to improve company profits, reduce unnecessary costs, and change workers' mindsets.

This study proposes the creation of a Lean Production Management Model to increase the productivity of SMEs in the nonprimary manufacturing sector. This model will use concepts defined within Lean Thinking and Plant Layout, starting with a current process analysis, identifying the activities currently being performed to assess their importance in bread production. For the analysis and extraction of information, visits were made to the company. Finally, unnecessary activities will be eliminated and processes will be standardized. Similarly, the production area layout will also be restructured, reducing distances between work areas and increasing tool proximity.

The following research proposal is distributed as follows: Introduction, Literature Review, Contribution, Model Validation, and Discussion of Results.

2   State of the Art

2.1   Lean Manufacturing

Lean Manufacturing research has been carried out in response to declining productivity because of inefficient management (SMED, 5S, VSM, PHVA, Lean Six Sigma, TQM, etc.) [7]. The literature presents studies in which these systems are assessed as solution tools that are based on a work philosophy or management system. These systems aim at reducing time between customer order and product shipment, thus increasing productivity while reducing costs [8, 9]. These studies show how to start implementation and obtain positive results in productivity, demonstrating that Lean Manufacturing tools may generate an improvement within processes. The first and second literatures describe SMED implementation coupled with the 5S in beverage and food companies with the objective of increasing batch size, minimizing production time, and implementing a culture of continuous improvement [8, 9].

2.2 Plant Layout

Previous Plant Layout-related studies are focused on improving production line productivity, reducing transportation time, eliminating waste, and removing unnecessarily occupied space. Some studies even focus on the proper use of work areas, arguing how planning tools may improve process flow, thereby taking into account greater production yield, improving productivity, and increasing the number of units produced [10, 11]. Through SLP and Lean tools, an increase was observed in overall material flow efficiencies. The impact is evidenced by reduced waiting time, increased production rate, and less cost because of material flow.

2.3 Method Study

The following studies propose a Reengineering Model that is based on the study of methods as a solution to inefficient process development. In the first case, the objective is to train staff members to obtain a quicker response to any atypical situation that may arise during workdays. To this end, work circles are created according to the competences of each worker. Then, the groups formed receive training that is based on their particular skills. The result indicates that competency-based training in industries improves reliability and increases competitiveness to the point of fostering workers to acquire new skills [12]. The second case attempts to predict the behavior of an uncontrollable variable, such as cooling. Cooling is deemed to be an uncontrollable variable as it is exposed to uncertain environmental conditions, which, in most cases, depend on location. To predict this behavior, software that can measure variables that are not perceptible to simple observation is used. This way, exact product cooling time is predicted, as well as specific final characteristics, thus reducing cooling time while increasing the production capacity of the organization [13].

2.4 Process Management

Finally, Process Management research is described. One of the methods used in previous studies is the integrated sustainable use of production resources or inputs. The main result obtained evidences how synergy in each of the processes included in the production line may minimize waste or material losses [14]. Another method that is used is Fuzzy Sensitivity Analysis, which is based on Dimensional Process Management. To implement this method, a simulation using the Extended Fourier Amplitude Sensitivity Test (eFAST) methodology is used. This methodology identifies weak points in production line stages, thus increasing product quality by increasing tolerances [15].

After performing an exhaustive analysis, we concluded that Lean models do not improve communications between the different areas of the company. Therefore, the model proposed in this article seeks to improve work methods and communication between workers.

3 Collaborations

3.1 Proposed Model

Per the research study and the comparison of methodologies performed during model development, where critical Production Management model dimensions were assessed, internal SME processes are the most important for product development. Therefore, we ask the following question: how can we improve SME processes and times to improve productivity?

This question can be answered by reviewing the previous Lean and Plant Layout models, which are focused on activity standardization. Consequently, a model for the improvement of processes and work area distribution was proposed. See Fig. 1.

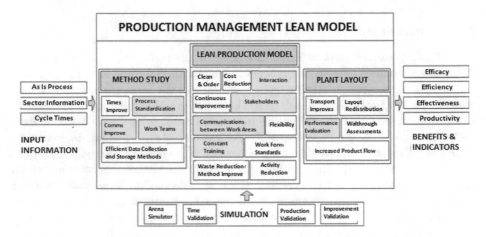

Fig. 1. Detailed view of the lean production management model.

3.2 Model Components

Process Standardization. The proposed model seeks to achieve complete process standardization to generate improvements in product production. Likewise, times are also expected to improve. Therefore, using Method Study, new forms are proposed and stored in the database.

Communication Improvement. One of the important aspects of the new model is providing new forms to facilitate communication between areas. As an artifact, a Suggestions Form was proposed for workers to provide their opinions to their area within the company.

Work Group Members. To improve current data collection for further analysis, we formed a Work Group consisting of five members (based on quality circles).

Interaction Between Areas. Regarding the Lean model, interactions were proposed to improve the flow of information between work areas, thus expanding the Lean Management philosophy of fostering interaction between areas.

Stakeholder Planning and Identification. Stakeholders, that is, people involved in the realization of this project, were identified. In addition, we identified how each stakeholder influences project development positively and negatively. The purpose of this activity was to improve interpersonal relationships.

Development of Communications Between Work Areas. To support the communication improvements proposed in the Method Study, meetings were proposed with participation from key members from the different company areas to improve operations and synergy, focusing on customer satisfaction.

3.3 Proposed Model Process

Here, a flowchart establishes the procedures performed during project development. Figure 2 denotes how project processes will be carried out. As already discussed in item Sect. 3.2 above, artifacts should be used to improve interaction and facilitate the implementation of the new Lean Production Management Model among workers.

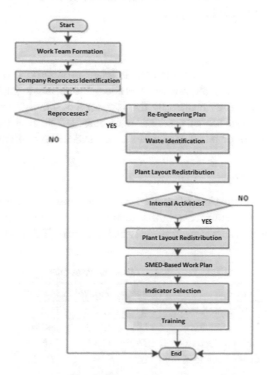

Fig. 2. Process flowchart.

3.4   Indicators

Productivity. This indicator will help us to determine how much productivity increases within the sector. Therefore, the productivity indicator reviews its influence on capital, raw materials, and labor. This indicator will be valued on the basis of production line activity.

$$\text{Productivity} = \text{Product (S/.)}/\text{Production Investment (S/.)}$$

4   Validation

To validate the proposed model, a simulation applied to an SME will be performed.

4.1   Panadería Américo E.I.R.L.

Panadería Américo E.I.R.L. is a small company that is dedicated to the mass production and sale of bakery products. It currently employs eight people for its four main products and reports an annual income of approximately S/. 353,541.34. It is located in the district of Chorrillos, Lima, Peru.

4.2   As-Is Process

When assessing the company under study, deficient productivity was observed in the production lines of its four main products: French, Yema, Karamanduka, and Ciabatta breads. This problem was evidenced by the low service quality, the nonfulfillment of orders, and the increased use of production resources. The root causes were unnecessary transportation, poor bread preparation, and excessive line changeover time.

4.3   Arena Simulation

The current company scenario will be used as the starting point for the simulation and to obtain quantitative results. Later, the results from the current scenario will be compared against the results from the enhanced scenario (Table 1).

Table 1. Current situation assessment.

Indicator	August	September	October
Global productivity	1.2	1.13	1.11
MP productivity	43.4	42.13	40.63
MO productivity	162.24	160.1	160.2
Capital productivity	7.02	6.98	6.85
Efficacy	73%	70%	68%
Efficiency	79%	75%	74%
Effectiveness	58%	53%	51%

4.4   Results

To demonstrate the positive results of the proposed solution, we proceeded to simulate both the current situation and the proposed scenario. The following is the TO-BE process or scenario proposed as a result of the "SLP" described above, with the objective of validating the results provided by the "SMED" and "Method Study" methodologies.

Once all the information was synthesized and analyzed, a new scenario was proposed using Plant Layout Redistribution, SMED, and Method Study. This new scenario will optimize the three main factors measured in this project: labor, resources used, and investment capital.

Finally, production area indicators were assessed for the bakery under study (Table 2).

Table 2. Productivity indicators.

Indicator	Type	Calorimetry			Frequency	Situation after improvement
		Red	Amber	Green		
Global productivity	Growing	<1.00	[1.00–1.70]	>1.70	Monthly	2.8
MP productivity	Growing	<45	[45–50]	>50	Monthly	55.74
MO productivity	Growing	<165	[165–170]	>170	Monthly	180.27
Capital productivity	Growing	<7.50	[7.50–8.00]	>8.00	Monthly	9.03
Efficacy	Growing	<70%	[70%–75%]	>75%	Monthly	78%
Efficiency	Growing	<70%	[70%–75%]	>75%	Monthly	83%
Effectiveness	Growing	<55%	[55%–60%]	>60%	Monthly	65%

On the basis of the Plant Layout Redistribution simulation, the SMED philosophy, and Method Study, or Methods Reengineering, direct productivity results were obtained, both globally and for each subindicator (capital productivity, raw material productivity, and labor productivity), as well as for efficiency, efficacy, and effectiveness (Table 3).

The improvement in each of the indicators is remarkable: productivity increased by 87%, MP productivity increased by 37%, MO productivity increased by 13%, and Capital productivity increased by 32%. In addition, efficacy, efficiency, and effectiveness increased by 10%, 9%, and 14%, respectively. This evidences how the simulated implementation of the research project has improved productivity.

Table 3. Current scenario vs proposed scenario.

Indicator	October	Validation	Variation (%)
Global productivity	1.11	2.08	87%
MP productivity	40.63	55.74	37%
MO productivity	160.2	180.27	13%
Capital productivity	6.85	9.03	32%
Efficacy	68%	78%	10%
Efficiency	74%	83%	9%
Effectiveness	51%	65%	14%

5   Conclusions

The techniques used were SMED, SPL, and Method Study, increasing productivity to 2.08 UNIT against the standard productivity of 1.70 for greater Lima.

The obtained productivity value of 2.08 represents an increase by 87.39% points with respect to the initial productivity. Likewise, efficacy increased by 10%, efficiency by 9%, and effectiveness by 15%, which translates into higher income and fewer losses for the company under study.

References

1. National Peruvian Institute of Statistics (2018)
2. National Peruvian Institute of Statistics (2016)
3. Daily Economy and Business Management (2016)
4. National Peruvian Institute of Statistics (2014)
5. National Peruvian Institute of Statistics (2017)
6. Peruvian Association of Bakeries and Pastry Shops (2018)
7. Katayama, H.: Legend and future horizon of lean concept and technology. Procedia Manuf. **11**, 1093–1101 (2017). FI:23519789
8. Ferreira, W., Maniçoba, A., Zampini, F., Pires, C.: Applicability of the lean thinking in bakeries. Espacios (2017). ISSN 07981015
9. Lozano, J., Saenz-Díez, J., Martínez, E., Jiménez, E., Blanco, J.: Methodology to improve machine changeover performance on food industry based on SMED. Int. J. Adv. Manuf. Technol. **90**, 3607–3618 (2017)
10. Gómez, J., Tascón, A., Ayuga, F.: Systematic layout planning of wineries: the case of Rioja region (Spain). J. Agric. Eng. **49**, 34–41 (2018)
11. Syed, A., Fahad, M., Atir, M., Zubair, M., Musharaf, M.: Productivity improvement of a manufacturing facility using systematic layout planning. Cogent Eng. **3**, Article no. 1207296 (2016)
12. Smith, J., Veitch, B.: A better way to train personnel to be safe in emergencies. ASCE-ASME J. Risk Uncertainty Eng. Syst. Part B: Mech. Eng. **5**, Article no. 011003 (2019)
13. Fu, J., Ma, Y.: A method to predict early-ejected plastic part air-cooling behavior towards quality mold design and less molding cycle time. Robot Comput. Integr. Manuf. **56**, 66–74 (2019)

14. Singh, S.K., Kulkarni, S., Kurmar, V., Vashistha, P.: Sustainable utilization of deinking paper mill sludge for the manufacture of building bricks. J. Clean. Prod. **204**, 321–333 (2018)
15. Oberleiter, T., Heling, B., Schleich, B., Willner, K., Wartzack, S.: Fuzzy sensitivity analysis in the context of dimensional management. ASCE-ASME J. Risk and Uncertainty Eng. Syst. Part B Mech. Eng. **5**, Article no. 011008 (2019)

Conditioning of Computerized Maintenance Management Systems Implementation

Zbigniew Wisniewski$^{(\boxtimes)}$ and Artur Blaszczyk

Faculty of Management and Production Engineering,
Lodz University of Technology, ul. Wolczanska 215, 90-924 Lodz, Poland
zbigniew.wisniewski@p.lodz.pl,
artur.blaszczyk89@gmail.com

Abstract. Flow maintenance management in manufacturing companies is a complex process that has a significant impact on the profitability of operations. Effective supervision is nowadays almost impossible without IT support, but on the other hand, successful implementation of information systems requires a multi-stage project and entails overcoming a number of barriers. Therefore, it is not surprising that even large organizations from around the world achieve low efficiency in the implementation of computerized maintenance management systems (CMMS) in order to increase reliability and efficiency. This article presents the conditions for implementing computerized solutions and outlines the factors that should be considered during the implementation of CMMS. They have been determined by a research carried out as part of the project, the purpose of which was to investigate what features of the organization and external factors facilitate and which hinder the implementation of CMMS. The project was carried out in the SYDYN research group.

Keywords: CMMS · Computerized maintenance management system ·
Flow maintenance

1 Introduction

The modern world is increasingly dependent on technology and automation. In many industries, the equipment necessary for the production of goods and services is expensive to acquire as well as use and maintain. Damage to even one device can lead to the hindering of all production. Apart from direct production losses, downtime caused by machine failures can damage the company's reputation and, in some cases, even lead to the imposition of fines and penalties [1, 2]. The above aspects make maintenance of equipment a key activity for many organizations and in some industrial companies spend between 15% and 40% of production costs as maintenance costs [3]. In today's competitive environment, understanding the total cost of ownership and optimizing these costs has a major impact on the company's profitability [4]. Considering that technical service is an important function for an organization, flow maintenance management requires a multidisciplinary approach combined with a business perspective [5–7].

© Springer Nature Switzerland AG 2020
W. Karwowski et al. (Eds.): AHFE 2019, AISC 971, pp. 486–494, 2020.
https://doi.org/10.1007/978-3-030-20494-5_46

Along with the industrial evolution, information has been one of the most important resources that determines the development of the maintenance functions which can be defined as a set of data transferred to the recipient, in order to make decisions [8]. In organizations, they appear as strategic resources, important for better dynamics and coordination between employees.

Computerized systems originate from information technology and cover many activities ranging from information technologies to organizational activities, such as the use of appropriate techniques to determine user requirements and create appropriate solutions [8]. Information systems supporting the maintenance business function are referred to as computerized maintenance management systems (CMMS).

The implementation of CMMS can enable fast and effective communication and bring many benefits, such as improved planning and scheduling, easier access to historical data, and generating reports enabling reduction of costs related to spare parts and maintenance [9]. Using the system in the organization will probably facilitate the effective implementation of TPM (Total Productive Maintenance) [10, 11].

The research was set in typically Polish business reality. It was taken up with the benefits in mind that can be provided by a correctly implemented CMMS, which would serve the reliability of machinery parks in organizations and cost optimization, and thus increasing the competitiveness of domestic companies. The main goal of the study was to diagnose the conditions of the implementation of CMMS in manufacturing enterprises. On the basis of literature studies and preliminary research based on individual interviews and conversations with managers of maintenance departments and managers of technical departments as well as experts in the field of practical implementation of IT maintenance management systems, a questionnaire was built, which was used to conduct actual research.

2 Computerized Maintenance Management Systems

CMMS is a tool supporting a maintenance strategy based on the combination of an information system with a set of functions that process data to develop indicators supporting maintenance activities. Computerized Maintenance Management Systems usually have the following range of functionalities and applications [12–14]:

- Management of preventive maintenance, supporting planning, scheduling and control of activities;
- Asset management, consisting of the creation of a register of all available resources and collecting data on inspections and failures;
- Management of spare parts ensuring their availability;
- Report management, allowing for processing large amounts of data and generation of relevant indicators and guidelines;
- Work order management, enabling the creation and assigning tasks to be performed by technicians.

The above functionalities allow the increase of the efficiency and effectiveness of maintenance through the use of information technology [15], however, taking into account the currently available systems, it is possible to identify several areas that need improvement [16, 17]:

- Analysis of data on device health monitoring;
- Diagnosis of resource failures;
- Support for decision-making.

Despite the importance of CMMS as a key tool for maintenance management, the success following skillful implementation of this solution, even in large, well-equipped organizations, is surprisingly low. According to research [9], the number of successful implementations is only around 25–40%, and the number of users using the system to the full extent is only 6–15%.

3   The Reasons for the Lack of CMMS Implementation

The conducted research indicates the existence of various conditions and barriers to CMMS implementation. Some of them lay in the organization itself, others in its surroundings. Some of them are connected with the human factor, some with the organizational element, and other yet with the economics [18, 19].

In a survey conducted among enterprises that do not have CMMS at the moment, respondents were asked about the reasons for not implementing a computerized maintenance management system. The results obtained are summarized in Fig. 1.

The most frequently chosen reasons are:

- maturity level of the Maintenance Department (66.1%),
- high implementation costs (59.6%),
- the specificity of technical infrastructure owned (49.5%),
- ignorance of the system (48.6%),
- unskilled workforce (47.7%),
- too few employees (44.0%),
- failure to determine the length of the computerized maintenance management system implementation process (43.1%),
- difficulties in changing organizational culture (41.3%),
- too much uncertainty (41.3%).

Other, less common reasons are:

- obsolete planning and management methods (28.4%),
- resistance of trade unions (22.9%).

In analyzed companies, comparing the frequency of choosing particular reasons for the lack of CMMS implementation showed two statistically significant differences. Enterprises that plan to implement computerized maintenance management systems in the future, significantly more often, indicated high costs of system implementation than lack of plans of implementation (p < 0.05). The fractions were respectively 0.77 and 0.33. Additionally, wherever companies had difficulties in changing old methods and habits, this particular problem was mentioned by the organizations planning the

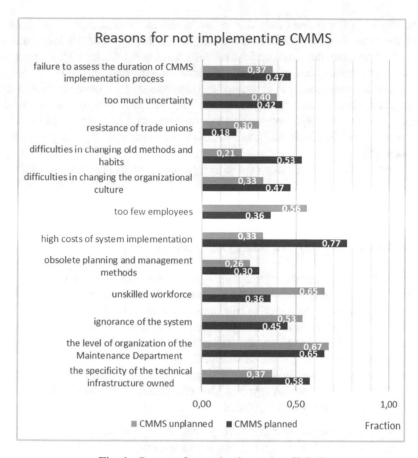

Fig. 1. Reasons for not implementing CMMS

implementation of CMMS significantly more often than by companies that had no implementation plans (p < 0.05). The fractions were 0.53 and 0.21, respectively.

On the other hand, enterprises that do not intend to implement CMMS in the future point more frequently to such factors as:

- resistance of trade unions - fraction 0.30,
- too few employees - fraction 0.56,
- unskilled workforce - fraction 0.65,
- ignorance of the system - fraction 0.53,
- maturity level of the Maintenance Department - fraction 0.67.

The frequency of other choices in the compared types of companies was similar (p > 0.05). The following reasons are indicated on an almost identical level:

- too much uncertainty - 0.40 and 0.42 fractions,
- obsolete planning and management methods - fractions 0.26 and 0.30,
- maturity level of the Maintenance Department - fractions 0.67 and 0.65.

Summarizing the above-mentioned research results, it should be emphasized that one of the most frequently indicated reasons for the lack of CMMS implementation were high implementation costs and the maturity level of the Maintenance Department.

It is worth noting that out of 193 surveyed enterprises, only 84 (43.5%) have implemented a computerized maintenance management system. Organizations working with standard, conservative management methods still have a dominant share in any economy. On the other hand, companies that are interested in applying technical asset management standards, including the ISO 55001 standard, are beginning to appear on the market [20, 21].

Anyhow, in the case of the surveyed companies, it turns out that one of the main reasons for the lack of CMMS implementation are high costs of solution implementation, which is an area for possible improvement.

4 Factors Adversely Affecting the Implementation of CMMS

When implementing computerized maintenance management systems, one may encounter a number of difficulties. This is associated with a number of factors that can significantly jeopardize the effective implementation of CMMS.

In the opinion of the surveyed organizations, conditions that adversely affect the implementation of the system can vary extensively, due to its complexity, among other things. The test results are presented in Fig. 2.

The factors mentioned in the questionnaire were the result of preliminary tests carried out in manufacturing enterprises. The assessment of the determinants has been described on a scale from 1 to 5, where:

- 1 - determines that the factor is completely unimportant,
- 2 - determines that the factor is not relevant,
- 3 - determines that the factor is mildly important,
- 4 - determines that the factor is important,
- 5 - determines that the factor is very important.

The degree of importance of the factors is grouped in the following way:

- 1–2 - determinant invalid,
- 3 - determinant moderately important,
- 4–5 - determinant very important.

The study basically made it possible to classify the most important factors into two groups:

I. Factors that significantly hinder the implementation are:

- general reluctance of employees towards change (85% of respondents considered the factor as important or very important),
- employees unprepared to use the system (85% of respondents considered the factor as important or very important),
- lack of awareness of the benefits associated with system implementation (85% of respondents considered the factor as important or very important),

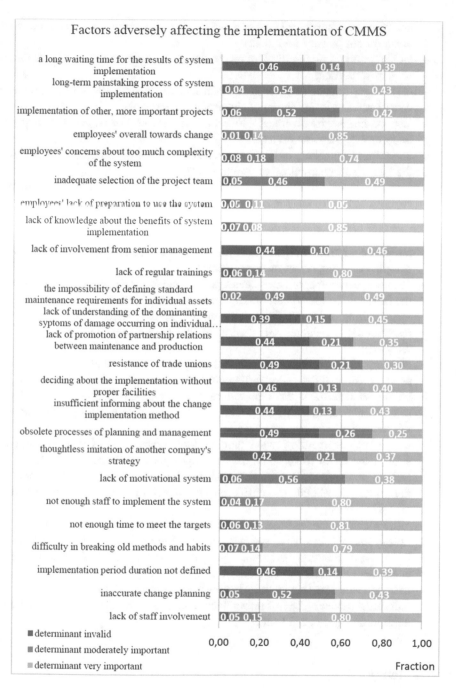

Fig. 2. Factors adversely affecting the implementation of CMMS

- deadlines too short to achieve goals (81% of respondents considered the factor as important or very important),
- lack of regular training (80% of respondents considered the factor as important or very important),
- insufficient human resources to implement the system (80% of respondents considered the factor as important or very important),
- lack of human involvement (80% of respondents considered the factor as important or very important),
- difficulties in breaking old methods and habits (79% of respondents considered the factor as important or very important),
- employees' fears of too much complexity of the system (74% of respondents considered the factor as important or very important).

II. Factors that have a minor impact on the implementation are:

- resistance of trade unions (49% of respondents considered the factor to be completely unimportant or not relevant),
- obsolete planning and management processes (49% of respondents considered the factor to be completely unimportant or not relevant),
- long waiting time for the results of the system implementation (46% of respondents considered the factor to be completely unimportant or not relevant),
- deciding on implementation without prepared background (46% of respondents considered the factor to be completely unimportant or not relevant),
- failure to determine the duration of the implementation process (46% of respondents considered the factor to be completely unimportant or not relevant),
- lack of management's involvement in the project (44% of respondents considered the factor to be completely unimportant or not relevant),
- lack of promotion of partnership relations between maintenance and production (44% of respondents considered the factor to be completely unimportant or not relevant),
- insufficiently wide information on the manner of implementing changes (44% of respondents considered the factor to be totally unimportant or not relevant),
- thoughtless imitation of another company's strategies (42% of respondents considered the factor to be totally unimportant or not relevant),
- lack of understanding of the dominant symptoms of damage occurring on individual devices (39% of respondents considered the factor to be completely unimportant or not relevant).

5 Summary

The conducted research and wider observation of enterprises prove that companies make a lot of organizational and implementation errors. This may likely be due to insufficient knowledge of the entire implementation process. According to the research, managers who decide to implement CMMS, do not pay enough attention to the appropriate preparation of the enterprise. The mistakes are made at the very beginning, because, as research results indicate, there is a definite lack of employee education in the field of knowledge about the system itself.

The lack of thorough knowledge results in difficulties with changing awareness among the employees when it comes to the use of computerized systems in flow maintenance. It also makes employees reluctant to changes [22]. Naturally, the implementation of any new solution, such as CMMS, meets resistance, and that is why factors such as general reluctance of employees towards the change, employees unprepared to use the system and lack of knowledge about the benefits of system implementation were the most frequently mentioned barriers to the implementation of CMMS.

Modern organizations are increasingly more aware of their problems with the maintenance department and try to eliminate them. Among other things, the trend of increasing demand for CMMS is noticeable, which may mean that enterprises also see real benefits that are achievable through these solutions and try to use them to improve their financial condition and competitive position.

References

1. Stamboliska, Z., Rusinski, E., Moczko, P.: Proactive Condition Monitoring of Low-Speed Machines. Springer, Cham (2015)
2. Muchiri, P., Pintelon, L., Gelders, L., Martin, H.: Development of maintenance function performance measurement framework and indicators. Int. J. Prod. Econ. **131**, 295–302 (2011)
3. Dunn, R.L.: Advanced maintenance technologies. Plant Eng. **40**, 80–82 (1987)
4. Campbell, J.D.: Uptime. CRC Press, Boca Raton (2016)
5. Murthy, D.N.P., Atrens, A., Eccleston, J.A.: Strategic maintenance management. J. Qual. Maint. Eng. **8**, 287–305 (2002)
6. Polak-Sopinska, A., Wrobel-Lachowska, M., Wisniewski, Z., Jalmuzna, I.: Physical work intensity of in-plant milk run operator. Part I - guidelines for assessment. Adv. Intell. Syst. Comput. **793**, 66–76 (2019)
7. Polak-Sopinska, A.: Physical work intensity of in-plant milk run operator. Part II – case study. Adv. Intell. Syst. Comput. **793**, 77–89 (2019)
8. Laudon, K.C., Laudon, J.P.: Management Information Systems. Prentice Hall, Upper Saddle River (2000)
9. Wienker, M., Henderson, K., Volkerts, J.: The computerized maintenance management system - an essential tool for world class maintenance. Procedia Eng. **138**, 413–420 (2016)
10. Ying, Z.C.L.: CMMS integrated system and its application based on TPM and RCM synergic relationship. World Sci-Tech R&D **6**, 735–738 (2008)
11. Yee, V.: Advancing CMMS to support performance based asset management. In: Proceedings of the Water Environment Federation, pp. 894–902 (2007)
12. Cato, W.W., Mobley, R.K.: Computer-Managed Maintenance Systems: A Step-by-Step Guide to Effective Management of Maintenance, Labor, and Inventory. Butterworth-Heinemann, Oxford (2001)

13. O'Donoghue, C.D., Prendergast, J.G.: Implementation and benefits of introducing a computerised maintenance management system into a textile manufacturing company. J. Mater. Process. Technol. **153–154**, 226–232 (2004)
14. Zhang, Z., Li, Z., Huo, Z.: CMMS and its application in power system. Int. J. Power Energy Syst. **26**, 75–82 (2006)
15. Carnero, M.C., Noves, J.L.: Selection of computerised maintenance management system by means of multicriteria methods. J. Prod. Planning Control **17**, 335–354 (2006)
16. Labib, A.W.: A decision analysis model for maintenance policy selection using a CMMS. J. Qual. Maint. Eng. **10**, 191–202 (2004)
17. Rastegari, A., Mobin, M.: Maintenance decision making, supported by computerized maintenance management system. In: Annual Reliability and Maintainability Symposium (2016)
18. Krason, P., Maczewska, A., Polak-Sopinska, A.: Human factor in maintenance management. Adv. Intell. Syst. Comput. **793**, 49–56 (2019)
19. Bielecki, M., Hanczak, M.: Evaluation of aspects of logistics efficiency of a product in the production enterprise. In: International Conference on Industrial Logistics, ICIL - Conference Proceedings, pp. 24–31 (2016)
20. Blaszczyk, A., Wisniewski, Z.: Changes in maintenance management practices - standards and human factor. Adv. Intell. Syst. Comput. **348–354**, 606 (2017)
21. Blaszczyk, A., Wisniewski, Z.: Requirements for IT systems of maintenance management. Adv. Intell. Syst. Comput. **793**, 531–539 (2019)
22. Wisniewski, Z., Polak-Sopinska, A., Rajkiewicz, M., Wisniewska, M., Sopinski, P.: Modelling formation of attitudes to change. In: Goossens, R. (ed.) Advances in Social & Occupational Ergonomics, pp. 117–127. Springer, Cham (2017). CH-6330

Analysis of Operational Efficiency in Picking Activity on a Pipes and Fittings Company

Vanina Macowski Durski Silva[1(✉)] and Gustavo Henrique Moresco[2]

[1] Department of Mobility Engineering, Federal University of Santa Catarina, Rua Dona Francisca, Joinville 8300, Brazil
vanina.durski@ufsc.br
[2] Tigre Tubos e Conexões, Joinville, Brazil
morescogh@gmail.com

Abstract. Faced with the need to reduce costs to remain competitive in the market, several companies are seeking internal process improvements, focusing on process optimization coupled with the importance of speed in logistics activities. The present work, characterized as a case study, aims to evaluate the operational efficiency of the picking activity in the Distribution Center of a multinational company located in Joinville, Brazil, presenting proposals for improvements in the processes and layout of the warehouse. Using Operational Research tools, it was possible to simulate different approaches for the picking sector in order to compare them identifying the most productive one. The results showed that it was possible to reduce the total picking time of the company up to 25%, reducing the company's costs in this activity by approximately R$ 120.000,00 per year, in the best scenario found in the development of the work.

Keywords: Inventory management · Logistics · Warehouse · Picking · Operations research

1 Introduction

Logistics has become a prominent sector for companies facing an increasingly competitive scenario. As stated by [1], traditional performance measurements systems are commonly based on cost and management accounting developed in the late nineteenth and early twentieth centuries. In addition, due to the need to reduce their costs, several companies seek to improve their logistics performance in order to gain competitive advantage over their competitors. Thus, the management of a Distribution Center (DC) has become an increasingly recurring theme in the accompaniment and planning of a company. The performance analysis of their workers and the picking process as a whole allow corrective measures to be taken to improve the process. Given this scenario, this study aims to analyze the different methodologies of picking (separating products) in a DC in order to minimize the spent time, the traveled distance and the average idle time per picker during the activity.

© Springer Nature Switzerland AG 2020
W. Karwowski et al. (Eds.): AHFE 2019, AISC 971, pp. 495–506, 2020.
https://doi.org/10.1007/978-3-030-20494-5_47

2 Theoretical Review

According to [2], the strategies adopted by the companies have a strong influence on the internal factors of itself, and the importance of logistics in the strategic planning of a company is highlighted, aiming at reducing costs and improving processes before the product is shipped. Among a variety of activities carried out within a distribution center, according to [3], picking is the activity in which products are withdrawn at specific locations in the warehouse, and [4] states that the product separation activity is usually the most laborious stage of the warehouse and it has a high impact on the costs of the DC.

2.1 Picking

2.1.1 Single Order Picking

This is the most used picking methodology, mainly due to its simplicity and effectiveness, where each picker (worker responsible for separating the products) is responsible for a collection order, traversing, if necessary, the whole distribution center collecting each item of the order. As advantage, it is possible to reduce the chance of errors, since each picker is responsible for all the products requested by a client in a collection order. On the other hand, this methodology can become unproductive because the pickers can travel great distances within the Distribution Center, resulting in an unproductive time.

2.1.2 Zone Picking

This methodology has the objective of dividing the Distribution Center into zones of products and assigning to the pickers a specific zone of work. So, each picker will be responsible for part of an order of a client, and after the products are separated, they are grouped in an area to continue the checkout process. As advantage, it allows the reduction of the traveled distance of the pickers in the DC. On the other hand, it is greater the risk of errors in the fulfillment of the client's request.

2.1.3 Batch Picking

This methodology focus on productivity, since each picker must accumulate several collection orders and then proceed to separate a larger quantity of certain products, thus reducing the amount of times a storage position in the DC should be visited. According to [5], the methodology of picking by batch usually results in less displacement time per item.

2.1.4 Wave Picking

It has a great similarity to batch picking, since each picker is responsible for collecting certain products, which may be in different collection orders. In this case, the orders are grouped for a period of time, and the quantity is summed so that the picker collects the total quantity of the materials.

2.2   Performance of a Distribution Center

According to [6], the DC operation represents a major part of the material handling operation in manufacturing and may vary between 15% and 70% of the manufacturing cost of a product. Consequently, arises the importance of adjusting logistics processes in order to obtain greater operational efficiency.

The improvement of the performance of a distribution center goes through the revision of its processes and activities aiming to increase the productivity of the picking and checkout teams, for example. These two processes have great importance in the shipping activity, but the other activities developed in a warehouse also deserve monitoring and constant evaluation of performance. All activities, upon receipt, can directly influence the productive performance of a distribution center. In this way, maintaining a high level of service is the duty of managers in order to maintain control over the company's productive performance.

2.3   Quantitative Concepts of Logistical Support

2.3.1   Manhattan Distance

Manhattan Distance is a tool that assists in calculating the distance between two points based on the coordinates of these two points and, in this study, it will be used to calculate the distance between all storage positions of the DC. The Manhattan Distance is calculated by the Eq. (1):

$$D_{i,j} = |x_i - x_j| + |y_i - y_j|. \tag{1}$$

The distance between the points i and j is calculated by the module of the difference of the coordinates x, summing the module of the difference of the coordinates y.

2.3.2   Knapsack Problem, Traveling Salesman Problem, the Nearest Neighborhood Method and 2-Opt Exchanges

In order to determine the storage positions of each product minimizing the total traveled distance, it is possible to use mathematical formulations such as the *knapsack problem methodology*. According to [7] item j has a profit p_j and a weight w_j and the capacity of the problem of the knapsack is represented by c. The model proposed by the authors is presented as follows:

$$\max \sum_{j=1}^{n} p_j.x_j. \tag{2}$$

subject to:

$$\sum_{j=1}^{n} w_j.x_j \le c. \tag{3}$$

$$x_j \in \{0, 1\} \quad j = 1, 2, \ldots, n. \tag{4}$$

where:

$$x_j = \left\{ \begin{array}{l} 1, \text{ if } j \text{ is loaded} \\ 0, \text{ otherwise} \end{array} \right\}$$

The first restriction (2) represents the boundary of the knapsack, that is, its capacity. The variable x_j can assume only the values of 1, if the item is included in the knapsack, and 0, otherwise, as the last restriction shows. For the present study, as an academic contribution, the mathematical model will be adjusted to a minimization problem where $p_{(i,\,j)}$ represents the weight that will be considered if product i is allocated in position j. This weight consists of the multiplication of the demand of product i by the distance from the storage position j to the checkout position. The mathematical model is presented below:

$$\min \sum_{i=1}^{n} \sum_{j=1}^{n} p_{i,j} \cdot x_{i,j}. \tag{5}$$

subject to:

$$\sum_{i=1}^{n} x_{i,j} = 1 \quad \forall j = 1, \ldots, n. \tag{6}$$

$$\sum_{j=1}^{n} x_{i,j} = 1 \quad \forall i = 1, \ldots, n. \tag{7}$$

$$x_{i,j} \in \{0, 1\} \quad \forall i, j = 1, \ldots, n. \tag{8}$$

where:

$$x_{i,j} = \left\{ \begin{array}{l} 1, \text{ if } i \text{ is allocated at position } j \\ 0, \text{ otherwise} \end{array} \right\}$$

The problem has a binary variable that can assume values of 1 if product i is allocated in position j and 0, otherwise. The first constraint (6) refers to the constraint of the problem which ensures that every storage position j can only receive one product. Similarly, the second constraint (7) ensures that every product i can only be allocated at a position j. The third constraint (8) identifies the binary variable, which will only assume values of 1 if product i is allocated in position j and 0, otherwise.

The Traveling Salesman Problem (TSP) is a programming problem that aims to determine a path with minimal cost in a network, covering all the vertices that are in it and returning to the initial vertex. Given a set of cities and the distance between each pair of them, the TSP aims to find the shortest route to visit all cities returning to the origin point. This is one of the combinatorial problems most known and researched due to its application in several areas such as circuit manufacture, production scheduling, telecommunications, DNA sequencing, among others [8]. Due to the difficulty of solving the problem [9] affirm that TSP is an NP-hard problem, thus, the present study will use the Nearest Neighborhood and 2-opt exchange heuristics to solve the problem.

The Nearest Neighborhood method is a heuristic method that aims to present a route considering the smallest possible path. The method can be defined by a set S of n points to visit in a space X. The objective of the problem is, from a starting node, to select the closest node until all nodes in the set S are visited. The heuristic method of 2-opt exchange aims, from a previously defined path, to make changes in the nodes positions in order to improve the route.

3 Case Study

The company where this study was carried out is a multinational company located in Joinville, Brazil, that produces pipes and fittings and it has a Distribution Center (DC) with 807 pallet positions, which represent the 807 storage positions available to be considered in the problem under study, where in each position only one product can be stored. Thus, in order to improve the efficiency of the picking activity performed by the company, the present study suggests a new allocation for the items, and it also simulates different picking methodologies as previously described.

3.1 Products Allocation Problem

The first step of this study refers to the definition of a new allocation for the products in the storage positions in the DC, aiming to allocate the products that have the greatest demand near to the checkout position. This step was necessary in light of the perception that not all products with higher demand were allocated in the vicinity of the checkout position. Thus, the Manhattan Distance calculation was used to calculate the distances between all the DC storage positions as well as the distances between the checkout position and each position. These data were calculated using the MATLAB software and form an array of distances of dimensions 808×808.

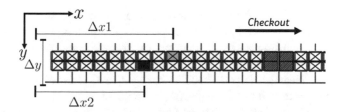

Fig. 1. Variations in Manhattan distances calculation. Source: the authors.

Figure 1 shows one of the possible variations of the Manhattan distance, where the total displacement is given by the sum of the components $\Delta x1$, $\Delta x2$ and Δy. The next step is to calculate the weight of the combinations between product and storage position, where the demand for each product and the distance from each position to the checkout position are used. For the calculation of the weight of the products, it is used the multiplication of the distance of the position j of the DC by the demand of the product i. The demand used for the products was analyzed in the period of 10 months from the beginning of 2017. Moreover, the weight p_{ij} is obtained for each position and item. These data were used in the "Knapsack Problem" in order to minimize the total weight, that is, the products that have the highest demand are allocated in the positions with the shortest distance to the checkout position.

To solve the model, we used the online Neos Server tool and the Gurobi solver for linear programming models. Thus, the resolution of the problem presented 1.614 constraints, 651.249 variables and it performed 385.238 iterations of the simplex method. As a result from the simulation model, the products with higher demand were allocated near to the checkout position, as shown in Fig. 2. The different scenarios obtained in this step are used in the next section to compare the products allocations and their results in the operational performance of the DC.

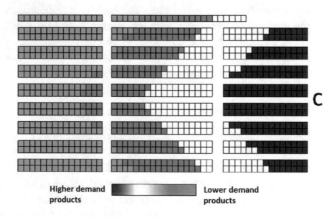

Higher demand products   Lower demand products

Fig. 2. Thermal map of the proposed situation in the DC. Source: the authors.

3.2   Model of Performance Evaluation

In order to evaluate the different scenarios proposed by this study the "Traveling Salesman Problem" was used to define the route for the picking activity. Considering its difficulty for proposing an exact solution, some heuristics were implemented in MATLAB for obtaining an approximate solution.

Initially, the "Nearest Neighborhood" was used, and it considered the checkout position as the starting point of the collection route. The products in a collection order are listed, and from that list, it is defined which storage locations on the DC must be visited on the picking route. From the starting point, the next point of the route is defined, selecting the position of the DC closest to the starting point. The next collection point of the route to be selected refers to the collection position that has not yet been visited closest to the current point. This procedure is repeated until all previously selected storage points are present in the route, and finally the last point of the collection route is connected to the checkout position, ending the route. The method is applied to each picking order of the selected period, generating an initial route for each order.

In order to determine an optimized picking route, the local search heuristic "2-opt exchanges" is applied. This method was chosen due to the fast convergence to a response, even if it does not guarantee the optimum solution, only to compare the different picking methodologies.

3.3   Modeling Scenarios

The scenarios used for modeling differ in the methodology used for the picking activity and also in the allocation of the products in the storage positions, be the current one used by the company or the one proposed by this work. In order to simulate a service period, 10 orders of collections received by the company were randomly selected. These orders were used to evaluate each of the scenarios proposed by this work and to measure the total time to perform the picking of the products belonging to the orders. The collection orders are presented in Table 1. Considering that the company has four pickers per shift to carry out the picking activity, the scenarios will be evaluated by the indicators of picking time and also by the picking productivity.

Table 1. Quantity of products per picking order

Picking order	1	2	3	4	5	6	7	8	9	10
# of products	21	19	22	20	30	21	20	20	24	15

3.3.1   Scenario A – Single Order Picking

Scenario A refers to the current configuration of the company, which uses the single order picking strategy. Using this methodology, the company processes the collection orders, with each picker being responsible for a complete order. The distances of each route, per picking order, are presented in Graph 1.

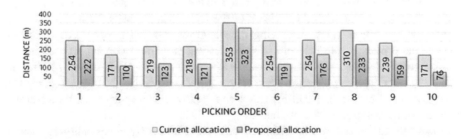

Graph 1. Distances from the picking routes of Scenario A. Source: the authors.

It is possible to notice that the distances of the picking routes generated by the model vary around 76 to 323 m and that the values of the distances are directly related to the quantity of products in the orders, presented in Table 1. It was possible to reduce the distance covered in until 55,7% in the collection order 10. This reduction in distance generates an increase in the productivity of the picking since the pickers must travel smaller distances to carry out the same activity. In this scenario, each picker is responsible for a picking order and only after finalizing the separation of the products in this order and delivering them to the checkout operators, it is given a new order to start the next collection. The separation time is directly related to the quantity of products present in the collection order, that is, for a collection order with 19 items (order 2), the separation time is 537 s in the current model and 435 s in the proposed model, and the separation time of an order containing 24 items (order 9) is 734 s in the current model and 600 s in the proposed model. The total picking time for each picker (separator) is shown in Graph 2. Thus, the total time for picking all the collection orders is approximately 1.966 s for the current scenario and 1.462 s for the proposed scenario.

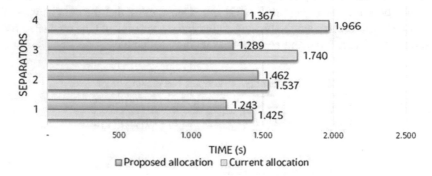

Graph 2. Work times per picker in Scenario. Source: the authors.

3.3.2   Scenario B – Zone Picking

The scenario B refers to the picking by zone, where each picker is responsible for a section of the Distribution Center and must separate only the items belonging to its separation zones. In this way, the division of the DC zones has been defined in order to balance the separation times. Analyzing the distances traveled to carry out the separation of the products from the defined orders it was possible to obtain a reduction in the average traveled distance between 12,9% in zone 2 and 31,1% in zone 3. From the distance traveled in each collection order it was possible to define the times to carry out the separation of each order, considering a constant rate of displacement between the positions and separation of the products.

According to the selected and divided collection orders by zones, only one order was placed under the responsibility of picker (separator) 4 in the current scenario. Thus, in order to optimize the total separation time, this separator was responsible for collection order 8, which belonged to the orange zone. With this change it was possible to reduce the inactive time of picker 4, and the total work time to carry out the separation of all orders is approximately 1.966 s for the current scenario and 1.508 s for the proposed scenario.

3.3.3   Scenario C – Batch Picking

In this scenario the orders of collection from several clients are accumulated and reprocessed forming new orders with batches of the same kind of product. In this way, in this scenario it is possible to group the products contained in different orders, avoiding that the storage position in which this product is located be visited again. The results show that the distance reduction ranged from 9,3% in order of collection 9 to 43,1% in collection order 7. The total distance was reduced by approximately 575 m, that is, 23,5% of the collection distance performed by the current scenario C.

In all the collection orders used in this work there was a reduction in the collection time between the current scenario C and proposed scenario C. This reduction ranged from 6% in order 9 to 25,8% in order 5, and it was possible to reduce approximately 14,8% of the order collection time. In order to determine the total separation time for all the collection orders, a service order was defined where each picker is responsible for one collection order at a time, resulting in a total separation time of 2.163 s in the current scenario and 1.806 in the proposed scenario.

Regarding the economic result of these reductions and considering the salary of a picker as R$ 6,00/h, with a reduction of 14,4% of the working time, it can be concluded that the change in the position of the products in the warehouse of the company results in a monthly savings of approximately R$ 6.856,00. This reduction allows firing up to 2 pickers, considering the labor costs that the company has.

3.3.4   Scenario D – Wave Picking

To simulate these scenarios, the chronological order of arrival of the requests was considered. As a service interval, two waves of 30 min each were defined, where each wave has 5 clients requests. The configuration of the new orders aimed at grouping, in the same order, the total quantity requested of the same item by several clients. The majority of collection orders were reduced, ranging from 7,8% in order 10 to 45,6% in order 2. There was an increase of the distance in three orders, ranging from 0,5% in order 9 to 8,4% in order 6. Regardless of the total distance traveled by the pickers, a reduction of approximately 12% was obtained.

As in the other scenarios, a working scale was performed for each picker according to the collection times of each order. Considering that all orders must be collected, it is necessary 2.079 s to perform the current scenario and 1.917 s for the proposed scenario.

4   Final Considerations

Due to the increased competitiveness between companies, it is necessary to improve the optimization in their main processes to obtain productivity gains and reduce their costs. Graph 3 illustrates the total traveled distance in each scenario, showing that considering the same picking methodology, better results could be obtained in the proposed scenarios.

As can be seen in Graph 3, the proposed scenario B is the one with the lowest total traveled distance per picker in view of the use of the "picking by zone" methodology that aims to divide the DC into exclusive collection zones for certain pickers. In such

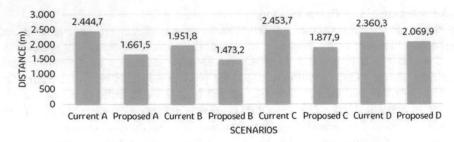

Graph 3. Total traveled distance per scenario. Source: the authors.

case, a single picker needs to collect products that are located in his zone, reducing the distance traveled. The scenario with the second lowest total traveled distance refers to the proposed scenario A, which uses the "single order picking" methodology, and has a total distance approximately 11% higher than the proposed scenario B. The good performance of these two scenarios is also visible when analyzing the total collection time, shown in Graph 4.

The two scenarios that have the best results, as well as in the distance, are the scenarios proposed A and proposed B. However, in this case the proposed scenario A is the one with the shortest total collection time, that is, the picking of all products contained in the collection orders is finalized more quickly when using the configuration addressed in the proposed scenario A. The proposed scenario B, which has the shortest total distance traveled, presents the second lowest total collection time, being approximately 3% larger than the scenario proposed A. This difference in time is a result of the accumulation of products in zones 2 and 3, close to the checkout position, and the time spent by the workers for handling the products of these zones becomes greater.

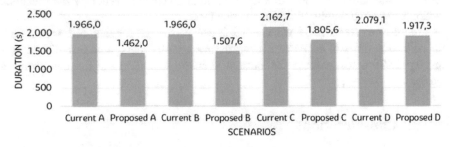

Graph 4. Comparison of the total time of collection for the different scenarios. Source: the authors.

Thus, there is a need for a third indicator that is able to identify the best work scenario for the proposed situation. Therefore, on account of the divergence of working time presented for each picker in the various scenarios presented in this paper, the importance of evaluating the average inactive time per picker is highlighted. The proposed scenario

A was the one that presented the least average inactive time per picker, where the "single order picking" methodology manages to equalize the amount of work among all workers of the separation area. It is also worth noting that the proposed scenario B, which has good results in the other indicators, also presented a good result related to the mean inactive time, since it represents the third lowest value among all the scenarios.

Scenarios C and D did not show significant improvements when compared to the current work model performed in the company. Thus, as presented in the previous section, the methodology of "batch picking", used in the current scenarios C and proposed C, did not present satisfactory results in this work due to the large variety of products present in the portfolio of the company under study. In addition, the collection orders used by the company have a high volume of products and this also damages the operational efficiency of this methodology.

Likewise, the methodology used in the current scenarios D and proposed D, "wave picking", also did not present good results in this study. This is due to the characteristics of the methodology, since the company has a short lead time for picking, generating collection orders a few hours before the vehicle is loaded. In addition, by grouping quantities of products from different collection orders, this methodology tends to increase the checkout time (not considered in this work) of the products, since it is necessary to differentiate products in the same order of collection that will not be dispatched in the same vehicle.

As a result of the three indicators presented by this work, we highlight the proposed scenario A as the best work scenario for the company under study. This scenario was the one that presented the best results in the indicators of total collection time and inactive mean time of the pickers. Furthermore, the choice of this scenario allows a reduction of approximately 23% of the total collection time in relation to the current scenario A, which represents the current scenario of the process carried out by the company under study. This imminent collection time results in a reduction of approximately R$ 9.677,00 per month, which would make it possible to reduce up to four pickers. However, this reduction of pickers can affect the productivity of the company in the final periods of each month, where there is usually a peak in sales volume, generating more demand for work by the pickers. In this case, it is suggested that these pickers can be used in other activities in the DC and during peak periods they must assist the picking activity.

Therefore, the presented results were satisfactory, since all the proposed scenarios generated improvements in the productivity results if compared to the current scenarios (considering the same methodology of picking between the compared scenarios). This is due to the new thermal map proposed by this work, which defined new storage positions for the company products, allocating the products with higher demand near to the checkout position, starting and ending positions of the collection routes. Future work should extend the modeling scenarios with a higher number of collection orders and different layouts of the DC in order to verify the relationships of the picking methodologies regarding the different configurations of the storage positions. Finally, it would be useful to simulate the segmentation of collection orders so after fulfilling all attributed orders, one picker can help another picker in finalizing his orders. For this change, it would be necessary a control system for the picking activity that understands and allows the segmentation of collection orders dividing it to more than one picker.

References

1. Khadem, M., Ali, S.A., Seifoddini, H.: Efficacy of lean metrics in evaluating the performance of manufacturing systems. Int. J. Ind. Eng. **15**(2), 176–184 (2008)
2. Oliveira, D.D.P.R.D.: Planejamento estratégico: conceitos, metodologias e práticas. In: Planejamento estratégico: conceitos, metodologias e práticas. Atlas, São Paulo (2010)
3. da Silva, G.Q., et al.: Análise de estratégias de picking aplicada a armazém de empresas de autopeças por meio de simulação discreta. In: Simpósio em excelência em gestão da tecnologia. Resende (2015)
4. Giustina, A.D.: O processo de expedição de um centro de distribuição de produtos acabados. Monografia (Bacharel em Engenharia de Produção) - Universidade Tecnológica Federal do Paraná. Campus Medianeira (2013)
5. Petersen, C.G.: An evaluation of order picking policies for mail order companies. Prod. Oper. Manag. **9**(4), 319–335 (2000)
6. Ballou, R.H.: Gerenciamento da cadeia de suprimentos: logística empresarial, 5th edn. Bookman Editora, Porto Alegre (2006)
7. Kolhe, P., Christensen, H.: Planning in logistics: a survey. In: 10th Performance Metrics for Intelligent Systems Workshop, pp. 48–53. ACM (2010)
8. Arenales, M., et al.: Pesquisa Operacional para cursos de engenharias, 6th edn. Elsevier, Rio de Janeiro (2007)
9. das Merces Calado, F., Ladeira, A.P.: Problema do caixeiro viajante: Um estudo comparativo de técnicas de inteligência artificial. E-xacta **4**(1), 1–12 (2011)

Author Index

© Springer Nature Switzerland AG 2020
W. Karwowski et al. (Eds.): AHFE 2019, AISC 971, pp. 507–509, 2020.
https://doi.org/10.1007/978-3-030-20494-5